Versatile Video Coding
Latest Advances in Video Coding Standards

RIVER PUBLISHERS SERIES IN SIGNAL, IMAGE AND SPEECH PROCESSING

Series Editors:

MONCEF GABBOUJ
Tampere University of Technology
Finland

THANOS STOURAITIS
University of Patras, Greece
and
Khalifa University, UAE

Indexing: All books published in this series are submitted to Thomson Reuters Book Citation Index (BkCI), CrossRef and to Google Scholar

The "River Publishers Series in Signal, Image and Speech Processing" is a series of comprehensive academic and professional books which focus on all aspects of the theory and practice of signal processing. Books published in the series include research monographs, edited volumes, handbooks and textbooks. The books provide professionals, researchers, educators, and advanced students in the field with an invaluable insight into the latest research and developments.

Topics covered in the series include, but are by no means restricted to the following:

- Signal Processing Systems
- Digital Signal Processing
- Image Processing
- Signal Theory
- Stochastic Processes
- Detection and Estimation
- Pattern Recognition
- Optical Signal Processing
- Multi-dimensional Signal Processing
- Communication Signal Processing
- Biomedical Signal Processing
- Acoustic and Vibration Signal Processing
- Data Processing
- Remote Sensing
- Signal Processing Technology
- Speech Processing
- Radar Signal Processing

For a list of other books in this series, visit www.riverpublishers.com

Versatile Video Coding
Latest Advances in Video Coding Standards

Humberto Ochoa Domínguez
IIT-UACJ
Mexico

K. R. Rao
University of Texas at Arlington
USA

LONDON AND NEW YORK

Published 2018 by River Publishers
River Publishers
Alsbjergvej 10, 9260 Gistrup, Denmark
www.riverpublishers.com

Distributed exclusively by Routledge
4 Park Square, Milton Park, Abingdon, Oxon OX14 4RN
605 Third Avenue, New York, NY 10158

First published in paperback 2024

Versatile Video Coding Latest Advances in Video Coding Standards / by Humberto Ochoa Domınguez, K. R. Rao.

© 2018 River Publishers. All rights reserved. No part of this publication may be reproduced, stored in a retrieval systems, or transmitted in any form or by any means, mechanical, photocopying, recording or otherwise, without prior written permission of the publishers.

Routledge is an imprint of the Taylor & Francis Group, an informa business

Publisher's Note
The publisher has gone to great lengths to ensure the quality of this reprint butpoints out that some imperfections in the original copies may be apparent.

While every effort is made to provide dependable information, the publisher, authors, and editors cannot be held responsible for any errors or omissions.

ISBN: 978-87-7022-047-7 (hbk)
ISBN: 978-87-7004-367-0 (pbk)
ISBN: 978-1-003-33999-1 (ebk)

DOI: 10.1201/9781003339991

Contents

Preface	xi
Acknowledgements	xiii
List of Figures	xv
List of Tables	xvii
List of Abbreviations	xix
Number of References and Projects	xxix
Introduction	1

1 Beyond High Efficiency Video Coding (HEVC)   3
     1.1   Introduction 3
     1.2   Standard Dynamic Range 4
         1.2.1   Video Test Sequence Formats and Frame Rates . . . 4
         1.2.2   Coding Conditions of Submissions 4
         1.2.3   Anchors 6
     1.3   High Dynamic Range 7
         1.3.1   Video Test Sequence Formats and Frame Rates . . . 7
         1.3.2   Coding Conditions of Submissions 7
         1.3.3   Anchors 7
     1.4   360° Video 9
         1.4.1   Video Test Sequence Formats and Frame Rates . . . 9
         1.4.2   Coding Conditions of Submissions 9
     1.5   Test Sequences 10
     1.6   Distribution Formats for Test Sequences and Decoded Results, Delivery of Bitstreams and Binary Decoders, Utilities and Cross-check Meeting 11
     1.7   Testing Environment and Methodology 13
         1.7.1   Selection of the Test Method 13

	1.7.2	DSIS	13
	1.7.3	DSCQS	13
	1.7.4	How to Express the Visual Quality Opinion with DSIS	15
	1.7.5	How to Express the Visual Quality Opinion with DSCQS	15
	1.7.6	Training and Stabilization Phase	16
	1.7.7	The Laboratory Setup	17
	1.7.8	Viewing Distance, Seats and Monitor Size	17
	1.7.9	Viewing Environment	17
	References	18

2 HEVC Encoder — 19
- 2.1 Joint Collaborative Team on Video Coding (JCT-VC) 19
- 2.2 Analysis of Coding Tools in HEVC Test Model, HM 1.0 – Intra Prediction 30
- 2.3 HEVC Encoder 32
  - 2.3.1 Intra Prediction 35
  - 2.3.2 Transform Coefficient Scanning 39
  - 2.3.3 Luma and Chroma Fractional Pixel Interpolation .. 40
  - 2.3.4 Comparison of Coding Tools of HM1 and HEVC Draft 9 41
- 2.4 Extensions to HEVC 42
- 2.5 Profiles and Levels 43
- 2.6 Performance and Computational Complexity of HEVC Encoders 43
- 2.7 System Layer Integration of HEVC 45
- 2.8 HEVC Lossless Coding and Improvements [E86] 45
- 2.9 Summary 47
- 2.10 Projects 59

3 Screen Content Coding for HEVC — 139
- 3.1 Introduction to SCC 139
- 3.2 Screen Content Coding Tools 142
  - 3.2.1 Intra Block Copy 143
  - 3.2.2 Palette Mode 145
    - 3.2.2.1 Palette Derivation 145
    - 3.2.2.2 Coding the Palette Entries 147
    - 3.2.2.3 Coding the Palette Indices 148

	3.2.3	Adaptive Colour Transform (ACT)	148
		3.2.3.1 Colour Space Conversion	149
		3.2.3.2 Encoder Optimization	150
	3.2.4	Adaptive Motion Vector Resolution	150

4 Lossless and Visually Lossless Coding Algorithms   153
- 4.1 Residual DPCM 153
  - 4.1.1 Sample-Based Weighted Prediction with Directional Template Matching 154
  - 4.1.2 Sample-Based Angular Intra-Prediction 156
  - 4.1.3 Sample-Based Angular Intra-Prediction with Edge Prediction 157
- 4.2 Fast Coding Algorithms 158
  - 4.2.1 Adaptive Motion Compensation Precision 158
  - 4.2.2 Fast Intra Coding 160
- 4.3 Visual Quality Assessment 161
  - 4.3.1 Screen Image Quality Assessment 162
  - 4.3.2 Objective Quality Assessment 163
  - 4.3.3 Subjective Quality Assessment 165
- 4.4 Other SCC Algorithms 165
  - 4.4.1 Segmentation 165
  - 4.4.2 Rate Control 166
- 4.5 Summary 168
- 4.6 Projects 169
- References 172

Appendix   181
H.264 Advance Video Coding (AVC)/MPEG-4 Part 10
    References 181
Books on H.264 189
H.264 Standard, JM SOFTWARE 190
DCT References 190
HEVC (High Efficiency Video Coding) References 208
Software Repository; Scalable Extensions of HEVC 224
    Build System 225
    Software Structure 225
ATSC Advanced Television Systems Committee 225
Video Coding: Recent Developments for HEVC
    and Future Trends 242
MPEG Internet Video Coding 249

Conclusion	250
MPEG-DASH	273
HEVC Overview Online	287
Test Sequences/Data Bases	288
IEEE Journal on Emerging and Selected Topics in Circuits and Systems (JETCAS)	292
How to Access JCT-VC Documents	294
Performance Comparison of Video Standards (PCVS)	295
Subjective Evaluaton of Compression Algorithms and Standards	296
Books on HEVC and Related Areas	304
Overview Papers	305
Tutorials	308
Special Sessions	314
Transcoders	316
Encryption of HEVC Bit Streams	320
File Format	320
Online Courses (OLC)	320
Open Source Software	321
X265 Source Code	324
vTune Amplifier by Intel	325
General	325
JVT REFLECTOR	327
HEVC Quality Evaluation	327
JCT-VC Documents	327
References on SSIM	328
SSIMPLUS Index for Video Quality-of-Experience Assessment	331
Bjontegaard Metric	332
VP8, VP9	333
JPEG 2000	337
Digital Cinema	341
JPEG 2000 Related Work	341
JPSearch	342
Sweden Records Missile Tests Using JPEG2000 Technology April 2009	343
JPEG XR	345
JPEG-LS	347
JPEG	348
JPEG XT	359
JPEG XT Projects	362
JPEG PLENO	364

JPEG XS	368
JPEG AIC	372
JBIG	372
LAR-LLC	372
PNG	373
WebP	373
WebM	373
DIRAC (BBC)	374
DAALA	375
Bjontegaard Metric	378
AVS China	378
Thor Video Codec	381
References on Screen Content Coding	382
Beyond HEVC	388
Projects on Beyond HEVC	389
Post – HEVC Activity	393
AV1 Codec (Alliance for Open Media – AOM)	393
Alliance for Open Media	395
Real Media HD (RMHD)	402
SMPTE	402
WMV-9 (VC-1)	404
VC-2 Standard	407
Legacy Codec	410
DSC by VESA	410
Joint Video Experts Team (JVET)	411
360 Degree Video Coding	416
RGB <=======> YCbCr Conversion	416
PSNRAVG	416
Weighted BDBR	417
Colour-Sensitivity-Based Combined PSNR (CSPSNR)	417
Cross-Media Retrieval	417
Multiple Description Coding	418

Index   **419**

About the Authors   **425**

Preface

The primary focus is on the developments beyond High-Efficiency Video Coding (HEVC). The new standard called versatile video coding (VVC) is projected to be finalized by 2020. A detailed description of the tools and techniques that govern the encoder indirectly and the decoder directly is intentionally avoided as a number of books (specially by the specialists who are directly involved in proposing/contributing/evaluating/finalizing the detailed processes that constitute the standards) are already available in this field, besides the overview papers, standards documents, reference software, software manuals, test sequences, source codes, tutorials, keynote speakers, panel discussions, reflector, and ftp/web sites – all in the public domain. Access to these categories is also provided. Since 2014, in the standards arena in HEVC, range extensions and new profiles (3D, multi-view, scalability, screen content coding – SCC) have been finalized. Also others such as MPEG-4 Internet Video Coding (ISO/IEC 14496-33) and AVS2 (IEEE 1857-4) have been standardized. Similarly, the industry has also finalized codecs such as, DAALA, THOR, VP9 (Google), VC1 (SMPTE), real media HD (Real Networks), and DSC (display stream compression) by VESA (Video Electronics Standards Association). AV1 (Alliance for Open Media, AOM) has been finalized. A brief introduction to the joint exploration model (JEM) by joint video experts team (JVET), versatile video coding (VVC), and future video coding (FVC) is provided. References related to 360-degree video coding are also listed. Other interesting and relevant topics such as cross-media retrieval and multiple description coding are also described.

This book provides access to all these developments. International, regional, and national standards organizations, industry groups (alliances), professional societies, etc. are continuously exploring more efficient video coding techniques.

A brief description of FVC beyond HEVC/H.265 is provided. Nearly 500 references related to HEVC and other emerging video coding standards are added along with 405 projects/problems. The latter are self-explanatory and govern the spectrum from a 3-hour graduate credit to research at the masters and doctoral levels. Some require a dedicated group of researchers with extensive computational (software) and testing facilities. Additional projects/problems based on image coding standards (both developed and some in final stages) such as JPEG2000, JPEG LS, JPEG XR, JPEG-XT, JPEG XS (call for proposals for a low-latency lightweight image coding system issued in March 2016 by JPEG and almost finalized in April 2018), JPEG-HDR, and JPEG-PLENO are added. References, overview papers, panel discussions, tutorials, open source software, test sequences, conformance bit streams, etc. emphasizing these topics are also listed. A brief description related to JVET, also known as next-generation video coding (NGVC), formed by both MPEG and VCEG, is targeted for a potential new standard by the end of 2020. This is also called VVC. Also, VESA issued a "call for technology" with the objective to standardize a significantly more complex codec called ADSC (advanced DSC) that is visually lossless at a bit rate lower than DSC. The AVS workgroup of China is on a fast forward track in adding SCC capability to AVS2. All these developments can immensely help the researchers, academia, and the graduate students and provide food for thought to delve deeply into the fascinating world of multimedia compression.

The reader is now well aware that this book is mainly at the research/reference level rather than as a textbook. It challenges the academic/research/industrial community regarding not only the present state-of-the-art but also, more specifically, the future trends and projections. Hence, it is an invaluable resource to this community.

Acknowledgements

This book is the result of a long-term association of Dr. Humberto Ochoa Domínguez with Dr. K. R. Rao. Special thanks to their respective families for their support, perseverance and understanding. Dr. Rao likes to acknowledge the support provided in various forms by Dr. Jean-Pierre Bardet, Former Dean, College of Engineering (COE), Dr. J. W. Bredow, Chair, Department of Electrical Engineering, and colleagues all in UTA. Dr. G. J Sullivan, Microsoft, Dr. Nam Ling, University of Santa Clara, Dr. Ankur Saxena and Dr. Zhan Ma (both from Samsung Research Labs), Dr. Wen Gao, Peking University, Dr. M. Budagavi, Samsung Research America (SRA), Dr. M. T. Sun, University of Washington, Dr. H. Lakshman, Fr. D. Grois of Fraunhofer HHI, Dr. T. Borer, BBC, Dr. Deshpande, Sharp Labs, Dr. Bankoski and Dr. D. Mukherjee (both from Google) Dr. Y. Reznik, Brightcove, Dr. H. Kalva, Florida Atlantic University, Dr. E. Izquierdo, Queen Mary University of London, Dr. E. Magli, Dept. of Electronics and Telecommunications, Politecnico di Torino, Italy, Dr. W.-K. Cham, Chinese University of Hong Kong, Hong Kong and P. Topiwala, FastVDO for providing various resources in this regard. Constructive review by Dr. Ashraf Kassim, National University of Singapore, is highly valuable. Shiba Kuanar, Harsha Nagathihalli Jagadish, Anusha Vasireddy and Swaroop Krishna Rao at UTA contributed countless hours in tying up all loose ends (references/copy right releases, proof reading and million other details). The graduate students and alumnae in multimedia processing lab (MPL) at UTA in various ways have made constructive comments.

List of Figures

Figure 1.1	DSIS BTC.	14
Figure 1.2	DSCQS BTC.	14
Figure 1.3	Example of DSIS test method scoring sheet.	15
Figure 1.4	Example of DSCQS test method scoring sheet.	16
Figure 2.1	Evolution of video coding standards (Alliance for open media).	20
Figure 2.2	Overall average MOS results over all Classes for Random Access coding conditions.	24
Figure 2.3	Overall average MOS results over all Classes for Low Delay coding conditions.	24
Figure 2.4	Available prediction directions in the unified intra prediction in HM 1.0.	27
Figure 2.5	HEVC encoder block diagram.	27
Figure 2.6	HEVC decoder block diagram.	32
Figure 2.7a	Recursive block structure for HEVC, where k indicates the depth for CU_k and TU_k.	34
Figure 2.7b	Quadtree partitioning of a 64×64 CTU. (a) 64×64 CU at depth 0. (b) Four CUs of 32×32 at depth 1. (c) 16 CUs of 16×16 at depth 2. (d) 64 CUs of 8×8 at depth 3.	34
Figure 2.7c	Partition mode for the PU of a CU.	34
Figure 2.8	Intra prediction angle definition.	37
Figure 2.9	Intra prediction mode directions.	38
Figure 2.10	(a) Diagonal scan pattern in an 8×8 TB. The diagonal scan of a 4×4 TB is used within each 4×4 sub-block of larger blocks. (b) Coefficient groups for 8×8 TB.	39
Figure 2.11	Integer and fractional positions for luma interpolation.	41
Figure 2.12	Future visual applications and demands. (NHK and Mitsubishi have developed a SHV-HEVC hardware encoder).	44
Figure 2.13	HEVC encoder block diagram with lossless coding mode.	46
Figure 2.14	Block diagram for DCT/ROT applied to intra prediction residuals only.	63

List of Figures

Figure 3.1	Image example from camera captured video content.	140
Figure 3.2	Images of screen content: (a) slide editing, (b) animation, (c) video with text overlay, (d) mobile display.	141
Figure 3.3	Encoding block diagram for screen content coding.	142
Figure 3.4	Intra block copy prediction in the current picture.	143
Figure 3.5	Candidates of block vector predictor.	145
Figure 3.6	Example of indexed representation in the palette mode.	146
Figure 3.7	Use of palette predictor to signal palette entries.	147
Figure 3.8	Coding the palette indices.	148
Figure 3.9	Location of the ACT in the encoder.	149
Figure 3.10	Location of the ACT in the decoder.	149
Figure 4.1	Secondary prediction (green line) after vertical RDPCM (red lines).	155
Figure 4.2	Causally neighbouring pixels for prediction and patches. Pixel X is to be predicted from the patch with pixels a, b, c, and d (left, centre); Patch around pixel b (right).	155
Figure 4.3	Classification of IQA depending on type of images, measurement method, and reference images.	163
Figure 4.4	Histogram of the MOS values for (left) SCC and (right) HEVC. Higher MOS values are achieved by SCC.	165

List of Tables

Table 1.1	Combinations of classes and constraint sets for the Standard Dynamic Range category	5
Table 1.2	Target bit rate points not to be exceeded for Class SDR-UHD1	5
Table 1.3	Target bit rate points not to be exceeded for Class SDR-HD1	5
Table 1.4	Target distortion points for Class SDR-UHD2 using Constraint Set 1	6
Table 1.5	Combinations of classes and constraint sets for the High Dynamic Range category	8
Table 1.6	Target bit rate points not to be exceeded for the High Dynamic Range category	8
Table 1.7	Combinations of classes and constraint sets for the 360° video category	9
Table 1.8	SDR test sequences	10
Table 1.9	SDR test sequence md5sums	10
Table 1.10	HDR test sequences	10
Table 1.11	HDR test sequence md5sums	11
Table 1.12	A 360° video test sequences	11
Table 1.13	360° video test sequence md5sums	11
Table 2.1	Test Classes and Bit Rates (constraints) used in the CfP	23
Table 2.2	Structure of tools in HM9 configuration	36
Table 2.3	Mapping between intra prediction mode direction and intra prediction angles	39
Table 2.4	Filter coefficients for luma fractional sample interpolation	40
Table 2.5	Filter coefficients for chroma fractional sample interpolation	41
Table 2.6	Summary of coding tools of high efficiency configuration in HM1 and HEVC	42
Table 2.7	Level limits for the main profile in HEVC	44
Table 2.8	Subjective viewing comparison results for sequences encoded using the HM and JM encoders	47
Table 2.9	Subjective viewing comparison results for sequences encoded using the eBrisk and ×264 encoders	47

List of Abbreviations

2D	Two Dimension
3D	Three Dimension
AAC	Advanced Audio Coding
ACM MoVid	Association for Computer Machinery Mobile Video
ACQP	Adaptive Chroma Quantization Parameter
ACT	Adaptive Colour Transform
ADMM	Alternating Direction Method of Multipliers
ADSC	Advanced DSC
AI	All Intra
AIC	Advanced Image Coding
AIP	Angular Intra Prediction
AHG	Ad Hoc Groups
AIF	Adaptive Interpolation Filter
ALF	Adaptive Loop Filter
AMT	Adaptive Multiple Core Transform
AMVP	Advanced Motion Vector Prediction
AOM	Alliance for Open Media
APIF	Adaptive Pre-Interpolation Filter
APSIPA	Asia Pacific Signal and Information Processing Association
AMVR	Locally Adaptive Motion Vector Resolution
AR	Augmented Reality
ARM	Advanced RISC Machines
ASIC	Application-Specific Integrated Circuit
ASIP	Application Specific Instruction Set Processor
ASM	Asymmetric Motion Partition(ing)
ASO	Arbitrary Slice Order
ATMVP	Alternative Temporal MV Prediction
ATR	Average Time Reduction
ATSC	Advanced Television Systems Committee
AVC	Advanced Video Coding—the official MPEG name is ISO/IEC 14496-10–MPEG-4 Part 10, and ITU-T name is ITU-T H.264
AVS	Audio Video Standard
AU	Access Unit

BBC	British Broadcasting Corporation
BD	Bjontegaard Delta
BDS	Bit Depth Scalability
BH	Beyond HEVC
IO	Bi-directional Optical Flow
BIQA	Blind Image Quality Assessment
BL	Base Layer
BLA	Broken Link Access
BMFF	Base Media File Format
BMSB	Broadband Multimedia Systems and Broadcasting
BMVP	Bottom-up Motion Vector Prediction
bpp	Bits per pixel
BRISQUE	Blind/Reference-Less Image Spatial Quality Evaluator
BS	Boundary Strength
BSTM	Butterfly Style Transform Matrices
BTC	Block Truncation Coding
BV	Block Vector
BVP	Block Vector Prediction
CABAC	Context Adaptive Binary Arithmetic Coding
CAVLC	Context-Adaptive Variable-Length Coding
CBF	Coded Block Flag
CBMV	Child Node's Best MV
CCLM	Cross Component Linear Model
CCP	Cross Component Prediction
CE	Consumer Electronics, Core Experiment
CfE	Call for Evidence
CfP	Call for Proposal
Cgs	Colour Gamut Scalability
CI	Confidence Interval
CIC	Compound Image Compression
CIF	Common Intermediate Format
CNN	Convolution Neural Networks
COFDM	Co-orthogonal Frequency Division Multiplexing
CPMV	Control Point MV
CPU	Central Processing Unit
CRA	Clean Random Access
CRI	Colour Remapping Information
CSPS	Collaborative Scheduling-Based Parallel Solution
CSPSNR	Colour-Sensitivity-Based Combined PSNR
CSVT	Circuits and Systems for Video Technology
CTC	Common Test Conditions
CU	Coding Unit

List of Abbreviations xxi

CUDA	Compute Unified Device Architecture
CWWSSIM	Complex-Wavelet Structural Similarity Index
DASH	Dynamic Adaptive Streaming over HTTP
DATE	Design, Automation and Test in Europe
DCC	Data Compression Conference
DCT	Discrete Cosine Transform
DCTIF	Discrete Cosine Transform Interpolation Filters
DDCT	Directional Discrete Cosine Transform
DF	Deblocking Filter
DIBR	Depth-Based-Image Rendering
DICOM	Digital Image and Communication on Medicine
DIP	Digital Image Processing
DiOR	Digital Operating Room
DIQA	Document Image Quality Assessment
DMB	Digital Multimedia Broadcasting
DMVD	Decoder Side MV Derivation
DMVR	Decoder Side MV Refinement
DMOS	Difference Mean Opinion Score
DPCM	Differential Pulse Code Modulation
DR	DIRAC (BBC)
DSC	Display Stream Compression
DSCQS	Double Stimulus Continuous Quality Scale
DSIS	Double Stimulus Impairment Scale
DSP	Digital Signal Processing
DST	Digital Sine Transform
DTM	Directional Template Matching
DTT	Discrete Tchebichef Transform
DTV	Digital Television
DVB-H	Digital Video Broadcasting – Handheld
EA	Early Access
EBU	European Broadcasting Union
EC	Error Concealment
EE	Electrical Engineering
EEG	Electroencephalography
EGK	Exp. Golomb Kth Order
EI	Electronic Imaging
EL	Enhancement Layer
EMT	Enhanced Multiple Transform
EPFL	Ecole Polytechnique Fédérale de Lausanne
ESIM	Edge similarity
ETRI	Electronics and Telecommunications Research Institute
EURASIP	European Association for Signal Processing

FCC	False Contour Candidate
FDAM	Final Draft Amendment
FDIS	Final Draft International Standard
FF	File Format
FIR	Finite Impulse Response
FME	Fractional Motion Estimation
FMO	Flexible Macroblock Ordering
FPGA	Field Programmable Gate Array
fps	Frames per second
FSIM	Feature Similarity
FVC	Future Video Coding
GALF	Geometry Transform Based ALF
GPU	Graphics Processing Unit
HD	High Definition
HDR	High Dynamic Range
HDTV	High Definition Television
HE-AAC	High Efficiency Advanced Audio Coder
HEIF	High Efficiency Image File Format
HEVC	High efficiency video coding—the official MPEG name is ISO/IEC 23008-2 MPEG-H Part 2 and ITU-T name is ITU-T H.265
HEVStream	High Efficiency Video Stream
HHI	Heinrich Hertz Institute
HLS	High Level Syntax
HM	HEVC Test Model
HOR	Horizontal
HP	High Profile
HTTP	Hyper Text Transfer Protocol
IASTED	International Association of Science and Technology for Development
IBC	Intra Block Copy
ICASSP	International Conference on Acoustics, Speech, and Signal Processing
ICCE	International Conference on Consumer Electronics
ICIEA	IEEE Conference on Industrial Electronics and Applications
ICIP	International Conference on Image Processing
ICME	International Conference on Multimedia and Expo
ICPC	International Conference on Pervasive Computing
ICPR	International Conference on Pattern Recognition
ICT	Integer Cosine Transform
IDR	Intra Decoding Refresh
IEC	International Electrotechnical Commission

IEEE	Institute of Electrical and Electronics Engineers
ILR	Inter Layer Interference
IME	Integer Motion Estimation
INTDCT	Integer Discrete Cosine Transform
intra HE	Intra High Efficiency
IP	Image Processing
IPTV	Internet Protocol Television
IQA	Image Quality Assessment
IS & T	Information Systems and Technology
ISCAS	International Symposium on Circuits and Systems
ISCCSP	International Symposium on Communications, Control and Signal Processing
ISDB-T	Integrated Services Digital Broadcasting - Terrestrial
ISO	International Organization for Standardization
ISOBMFF	ISO Based Media File Format
ITS	International Telecommunication Symposium
ITU-T	Telecommunication Standardization Sector of the International Telecommunications Union
IVC	Internet Video Coding
ITM	Internet Video Coding Test Model
IVMSP	Image, Video, and Multidimensional Signal Processing
J2K	JPEG 2000
JBIG	Joint Binary Image Group
JCI	JND-Based Compressed Image
JCTVC	Joint Collaborative Team on Video Coding
JEM	Joint Exploration Test Model
JETCAS	Journal on Emerging and Selected Topics in Circuits and Systems
JVET	Joint Video Experts Team
JLS	JPEG-LS
JM	Joint Model
JMKTA	JM Key Technology Areas
JND	Just Noticeable Distortion
JNQD	Just Noticeable Quantization Distortion
JPEG	Joint Photographic Experts Group
JPEG-XR	JPEG Extended Range
JSVM	Joint Scalable Video Model
JTC	Joint Technical Committee
JVCIR	Journal of Visual Communication and Image Representation
JVET	Joint Video Experts Team
JVT	Joint Video Team
JXR	JPEG-XR
JXT	JPEG-XT

KROCC	Kendall Rank-Order Correlation Coefficient
KTA	Key Technology Areas
LAR-LLC	Locally-adaptive Resolution Lossless low-complexity
LCICT	Low Complexity Integer Cosine Transform
LD	Low Delay
LDB	Low Delay with B pictures
LDM	Layered Division Multiplexing
LDP	Low Delay with P Pictures
LDPC	Low Density Parity Check
LHM	Light Weight HEVC Test Model
LIC	Local Illumination Compensation
LOCO	Low Complexity
LR	Low Resolution
LS	Lossless (near lossless)
LSC	Layer Segmentation Based Coding
L-SEABI	Low Complexity Back Projected Interpolation
M.S.	Masters
MANE	Media – Aware Network Element
Mbit/s	Megabit per second
MC	Motion Compensation
MCC	Mixed Content Coding
MCL	Media Communication Lab
MD	MPEG DASH
MDDCT	Modified Directional Discrete Cosine Transform
MDDT	Mode-Dependent Directional Transform
MDNSST	Mode-Dependent Non-Separable Secondary Transforms
ME	Motion Estimation
MJPEG	Motion JPEG
MMSP	Multimedia Signal Processing
MOMS	Maximal-Order interpolation with Minimal Support
MOS	Mean Opinion Score
MPEG	Moving Picture Experts Group—the official ISO name is ISO/IEC JTC1/SC29/Working Group 11.
Mpixel	Megapixel
Mpm	Most Probable Modes
MRSE	Mean Root Square Error
MSP	Main Still Profile
MV	Motion Vector
MVC	Multi View Coding
MVF	MV Field
NAB	National Association of Broadcasters
NAL	Network Abstraction Layer

NGBT	Next Generation Broadcast Television
NGVC	Next Generation Video Coding
NIQA	Natural Image Quality Assessment
NUH	NAL Unit Header
NTT	Nippon Telegraph and Telephone Corporation
OBMC	Overlapped Block MC
OP	Overview Papers
OR	Outlier Ratio
OSS	Open Source Software
P2SM	Pseudo 2D String Matching
PBMV	Parent Node's Best MV
PCM	Pulse Code Modulation
PCS	Picture Coding Symposium
PCoIP	PC over IP
PDPC	Position Dependent Intra Prediction Combination
PLCC	Pearsonlinear Correlation Coefficient
PPS	Picture Parameter Set
PSNR	Peak-to-peak Signal to Noise Ratio
PU	Prediction Unit
PVC	Perceptual Video Coding
PVQ	Perceptual Vector Quantization
QOE	Quality of Experience
QP	Quantizer Parameter
QTBT	Quadtree plus Binary Tree
RA	Random Access
RADL	Random Access Decodable
RASL	Random Access Skipped
RD	Rate Distortion
R&D	Research and Development
RDOQ	Rate-Distortion Optimized Quantization
RDPCM	Residual Differential Pulse Code Modulation
RDH	Reversible Data Hiding
RHCNN	Residual Highway CNN
RICT	Recursive Integer Cosine Transform
RL	Reference Layer
RMHD	Real Media HD
RMVB	Real Media Variable Bitrate
ROI	Region of Interest
ROT	Rotational Transform
RTC	Real Time Communications
RTP	Real-time Transport Protocol
RQT	Residual Quad Tree

SAO	Sample Adaptive Offset
SC	Sub Committee
SC	Stimulus Comparison
SCADA	Supervisory Control and Data Acquisition
SCC	Screen Content Coding
SCI	Screen Content Images
SCM	Screen Content Coding Test Module
SDCT	Steerable Discrete Cosine Transform
SDCT-AM	Steerable Discrete Cosine Transform-Alternated Minimization
SDCT-BT	Steerable Discrete Cosine Transform-Binary Tree
SDM	Structural Degradation Model
SDR	Standard Dynamic Range
SDSCE	Simultaneous Double Stimulus Continuous Evaluation
SDT	Signal Dependent Transform
SE	Subjective Evaluation
SEI	Supplemental Enhancement Information
SELC	Sample based weighted prediction for Enhancement Layer Coding
SG	Study Group
SHV	Super Hi-Vision
SHVC	Scalable High Efficiency Video Coding
SI	Switching I
SID	Society for Information Display
SIQA	Screen Image Quality Assessment
SIQM	Structure Induced Quality Metric
SIMD	Single Instruction Multiple Data
SIP	Signal and Image Processing
SMF	Sparse Matrix Factors
SMPTE	Society of Motion Picture and Television Engineers
SoC	System on Chip
SHV	Super Hi- Vision
SHVC	Scalable HEVC
SP	Switching P
SPA	Signal Processing: Algorithms, Architectures, Arrangements, and Applications
SPEC	Shape Primitive Extraction and Coding
SPL	Special
SPS	Sequence Parameter Set
Spl H	Special Issues on HEVC
SPIE	Society of Photo-Optical and Instrumentation Engineers
SR	Super Resolution
SROCC	Spearman Rank-Order Correlation Coefficient
SR-SIM	Spectral Residual-Based Similarity

SS	Single Stimulus
SSCQE	Single Stimulus Continuous Quality Evaluation
SSD	Spatio-Spectral Decomposition
SSIM	Structural Similarity
SSST	Southeastern Symposium on System Theory
SSVC	Spatially SVC
SSVIP	Steady-State Visual Evoked Potentials
STMVP	Spatial-Temporal MV Prediction
SVC	Scalable Video Coding
SVD	Singular Value Decomposition
SVM	Support Vector Machine
SVQA	Subjective VQA
SWP	Sample-Based Weighted Prediction
TB	Transform Block
TE	Tool Experiment
TS	Test Sequence, Transport Stream
TSM	Transform Skip Mode
TENTM	Tandberg, Ericsson and Nokia Test Model
TMP	Temporal Matching Prediction
TMuC	Test Model under Consideration
TMVP	Temporal Motion Vector Prediction
TMO	Tone Mapping Operator
TSA	Temporal Sub-layer Access
TSF	Transform Skip Flag
TU	Transform Unit
Tut	Tutorial
TX	Texas
TZSearch	Test Zone Search
UGC	User-Generated Content
UHD	Ultra High Definition
UHDTV	Ultra High Definition Television
UTA	University of Texas at Arlington
VBS	Variable Block Size
VC	Video Coding
VCB	Video Coding for Browsers
VCEG	Visual Coding Experts Group-the official ITU name is ITU-T/SG 16/Q.6- t
VCIP	Visual Communications and Image Processing
VCIR	Visual Communication and Image Representation
VDI	Virtual Desktop Infrastructure
VDP	Visible Difference Predictor
VER	Vertical

VESA	Video Electronics Standards Association
ViMSSIM	Video-modified Structural Similarity
VLSI	Very Large-scale Integrated circuit
VPS	Video Parameter Set
VQ	Vector Quantization
VQA	Video Quality Assessment
VQEG	Video Quality Experts Group
VQM	Video Quality Measure
VR	Virtual Reality
VSB	Vestigial Sideband
VTM	VVM Test Model
VUI	Video Usability Information
VVC	Versatile Video Coding
WCG	Wide Colour Gamut
WPP	Wavefront Parallel Processing
WD	Working Draft
WG	Working Group
WHP	Wavefront-Based High Parallel
WQVGA	Wide Quarter Video Graphics Array
WSIs	Whole Slide Images
WVC	Web Video Coding
WVGA	Wide Video Graphics Array
YC_bC_r	Y is the Brightness (luma), C_b is blue minus luma (B-Y) and C_r is red minus luma (R-Y)

Number of References and Projects

The following list summarizes the number of references cited in the book and the number of projects proposed in each category,

No.	Category	Number of References	Number of Projects
1.	High efficiency video coding (HEVC)	483	404
2.	Screen content coding (SCC)	97	15
3.	H.264/AVC	80	8
4.	DCT References	67	61
5.	Subjective Evaluation (SE)	45	7
6.	Encryption of HEVC (EH)	1	1
7.	VP 8, 9 (Google)	16	7
8.	JPEG2000	51	7
9.	JPEG-XR	10	1
10.	JPEG-LS	9	1
11.	JPEG	36	18
12.	JPEG-XT	18	13
13.	JPEG PLENO	8	5
14.	JPEG XS	8	6
15.	Locally-adaptive Resolution Lossless low-complexity (LAR-LLC)	1	6
16.	DIRAC (BBC)	12	3
17.	DAALA (Mozilla & other contributors)	5	3
18.	MPEG-DASH	2	1
19.	Advanced Video Standard (AVS) China	17	9
20.	Thor Video Codec (Cisco)	1	5
21.	Beyond HEVC	19	21
22.	AV1 Codec (Alliance for Open Media – AOM)	7	14
23.	Real Media High Definition (Real Networks)	–	1
24.	VC1	8	1
25.	Video Coding 2 (VC2) (SMPTE)	–	2
26.	Legacy Codec	1	1
27.	DSC by VESA	6	3
28.	Screen content Coding	50	2
29.	JVET (JEM)	13	4

Introduction

Whereas many international standards related to video/audio/image such as H.26X, MPEG and JPEG series have already been finalized, development of new standards reflecting the trends in interactive communication services, handheld/mobile devices, internet, video on demand, HDR encoder/decoder units, display/storage/transmission capabilities is an ongoing process. Our book versatile video coding addresses these challenges. It not only covers the various profiles/additions in high efficiency video coding (HEVC), but also the trends covering the newer applications, research areas strengthened by citing projects and problems. The projects cover the spectrum of 3 credit hours at the graduate level, theses and dissertations appropriate at the masters and doctoral requirements. Extensive and elaborate literature in terms of review papers, open source software, standards documents, special issues, web/ftp sites, test sequences etc is provided. Hence this book is an invaluable resource for researchers, faculty and graduate students.

1
Beyond High Efficiency Video Coding (HEVC)

1.1 Introduction

Video is the main driver of bandwidth use, accounting for over 80 percent of consumer Internet traffic. The majority of this video is coded using the international standards developed in collaboration with ITU-T Study Group 16 and MPEG.

In July 2017, the Joint Video Experts Team (JVET) released a Call for Proposals on video coding technology with compression capabilities that significantly exceed those of the HEVC standard (Rec. ITU-T H.265 |ISO/IEC 23008-2) and its current extensions. The new standard, called Versatile Video Coding (VVC), is a new generation of video compression technology that has substantially higher compression capability than the existing HEVC standard is planned to be finalized by 2020 [1, 2].

VVC will aid the deployment of higher-quality video services and emerging applications such as 360° omnidirectional immersive multimedia and high-dynamic-range (HDR) video. The San Diego meeting evaluated responses to the Call for Proposals. Responses to the call were received from 32 organizations, with some demonstrating compression efficiency gains over HEVC of 40 percent or more.

The gain was measured in extensive formal subjective tests conducted by independent test laboratories. Both 360° omnidirectional video and HDR video were tested as well as conventional dynamic range video. Particular effectiveness was shown on ultra-high definition (UHD) video test material.

The results of this very successful call led to the creation of a first draft of the VVC standard, a test model for simulation experiments, and a technology benchmark set for the VVC project.

The new standard is expected to enable the delivery of UHD services at bit rates that today are used to carry HDTV. Alternatively, using VVC would enable twice as much of video content to be stored on a server or sent through a streaming service [1, 2, 3]. Proposals submitted must comply with the following:

1.2 Standard Dynamic Range

In SDR all test material is progressively scanned and uses 4:2:0 color sampling with either 8 or 10 bits per sample,

1.2.1 Video Test Sequence Formats and Frame Rates

The initial proposed classes of test video sequences are:

Class SDR-UHD1: 3840 × 2160p 60 fps 10 bits: "FoodMarket4", "CatRobot1", "DaylightRoad2"
3840 × 2160p 50 fps 10 bits: "ParkRunning3"
3840 × 2160p 30 fps 10 bits: "CampfireParty2"
Class SDR-HD1: 1920 × 1080p 50 fps 8 bits: "BasketballDrive", "Cactus"
1920 × 1080p 60 fps 8 bits: "BQTerrace"
1920 × 1080p 60 fps 10 bits: "RitualDance", "MarketPlace"
Class SDR-UHD2: 3840 × 2160p 30 fps 10 bits: "MountainBay2"
(*Additional video sequences will be included*)

1.2.2 Coding Conditions of Submissions

Constraint cases are defined as follows:

- Constraint Set 1: no more than 16 frames of structural delay, e.g. 16 pictures "group of pictures (GOP)", and random access intervals of 1.1 seconds or less. A random access interval of 1.1 seconds or less shall be defined as 32 pictures or less for a video sequence with a frame rate of 24, 25 or 30 frames per second, 48 pictures or less for a video sequence with a frame rate of 50 frames per second, 64 pictures or less for a video sequence with a frame rate of 60 frames per second, and 96 pictures or less for a video sequence with a frame rate of 100 frames per second.
- Constraint Set 2: no picture reordering between decoder processing and output, with bit rate fluctuation characteristics and any frame-level multi-pass encoding techniques to be described with the proposal. (A metric to measure bit rate fluctuation is implemented in the Excel file to be submitted for each proposal.)

Submissions shall be made for the test cases (combinations of classes and constraint sets) as listed in Table 1.1.

Table 1.1 Combinations of classes and constraint sets for the Standard Dynamic Range category

	Class SDR-UHD1	Class SDR-HD1	Class SDR-UHD2
Constraint set 1	X	X	X
Constraint set 2	–	X	–

Submissions shall include encodings for all video sequences in all classes, and each decoding shall produce the full specified number of pictures for the video sequence (no missing pictures).

Submissions for Class SDR-UHD1 using Constraint Set 1, and submissions for class SDR-HD1 using Constraint Set 1 and Constraint Set 2 will be evaluated by means of a formal subjective assessment and BD PSNR and rate [3, 4] criteria.

Submissions for Class SDR-UHD2 using Constraint Set 1 will be evaluated by BD PSNR and rate criteria.

For Class SDR-UHD1 using Constraint Set 1 defined above, submissions to this Call shall submit results for the target bit rate as listed in Table 1.2. The submitted results shall not exceed the target bit rate points.

For Class SDR-HD1 using Constraint Set 1 and Constraint Set 2 defined above, submissions to this Call shall submit results for the target rate points as listed in Table 1.3. The submitted results shall not exceed the target bit rate points.

For Class SDR-UHD2 using Constraint Set 1, submissions to this Call shall submit results for the target distortion points. The average picture PSNR

Table 1.2 Target bit rate points not to be exceeded for Class SDR-UHD1[1]

Sequences	Target Bit Rates [kbit/s]			
	Rate 1	Rate 2	Rate 3	Rate 4
FoodMarket4	950	1400	2200	3800
CatRobot1	1500	2400	3800	6000
DaylightRoad2	1400	2200	3500	5500
ParkRunning3	2000	3300	6000	10000
CampfireParty2	2000	3300	6000	10000

Table 1.3 Target bit rate points not to be exceeded for Class SDR-HD1

Sequences	Target Bit Rates [kbit/s]			
	Rate 1	Rate 2	Rate 3	Rate 4
BQTerrace	400	600	1000	1500
RitualDance	900	1500	2300	3800
MarketPlace	500	800	1200	2000
BasketballDrive	800	1200	2000	3500
Cactus	500	800	1200	2000

---

[1] 1 kbit/s means 10^3 bits per second, and 1 Mbit/s means 10^6 bits per second.

6 Beyond High Efficiency Video Coding (HEVC)

Table 1.4 Target distortion points for Class SDR-UHD2 using Constraint Set 1

Sequences	Target PSNR [dB]			
	Distortion 1	Distortion 2	Distortion 3	Distortion 4
MountainBay2	TBD	TBD	TBD	TBD

Note: The values in Table 1.4 will be updated in the Final Call for Proposals. Additionally, it is anticipated that the target distortion range may also be modified.

of the submitted results, calculated over all pictures of the test sequence, shall be within a $+/-0.5$ dB range of the target distortion points. The target distortion points for Class SDR-UHD2 using Constraint Set 1 are listed in Table 1.4.

Submissions to this Call shall obey the following additional constraints:

1. No use of pre-processing.
2. Only use post-processing if it is part of the decoding process, i.e. any processing that is applied to a picture prior to its use as a reference for inter prediction of other pictures. Such processing can also be applied to non-reference pictures.
3. Quantization settings should be kept static except that a one-time change of the quantization settings to meet the target bit rate is allowed. When any change of quantization is used, it shall be described. This includes a description of the one-time change.
4. Proponents are discouraged from optimizing encoding parameters using non-automatic means as well as optimizing encoding parameters on a per sequence basis.
5. Proponents shall not use any part of the video coding test set as a training set for training large entropy coding tables, VQ codebooks, etc.

1.2.3 Anchors

Two anchors have been generated by encoding the above video sequences. The first anchor uses the HM 16.16 software package. Static quantization parameter (QP) settings shall be applied, though a one-time change of the quantization parameter from value QP to value QP+1 may be applied in order to meet the defined target bit rates. The quantization parameter settings applied for the anchors will be reported.

The second anchor uses the Joint Exploration Test Model (JEM) software package. The Joint Video Exploration Team (JVET) maintains the JEM software package in order to study coding tools in a coordinated test model [6]. JEM anchor bitstreams will be generated using this software package. It is planned that the JEM 7.0 software package will be used to generate the anchors, though a later version may be used if available.

The purpose of the anchors is to facilitate testing in accordance with BT.500,[2] providing useful reference points demonstrating the behavior of well-understood configurations of current technology, obeying the same constraints as imposed on the proposals. The anchors will be among the encodings used in the testing process; however, the purpose of the test is to compare the quality of video for proposals to each other rather than to the anchors.

Note: It is anticipated that the configuration files "encoder_randomaccess_main10.cfg" and "encoder_lowdelay_main10.cfg" provided in the HM 16.16 software package will be used to generate the HM anchor bitstreams corresponding to Constraint Set 1 and Constraint Set 2, respectively. Furthermore, it is anticipated that the configuration files "encoder_randomaccess_jvet10.cfg" and "encoder_lowdelay_jvet10.cfg" provided in the JEM7.0 software package will be used to generate the JEM anchor bitstreams corresponding to Constraint Set 1 and Constraint Set 2, respectively. As described in the footnote below, the configuration files will be made available at the same location as the anchor bit-stream for the Final Call for Proposals.

1.3 High Dynamic Range

1.3.1 Video Test Sequence Formats and Frame Rates

All test material is progressively scanned and uses 4:2:0 color sampling with 10 bits per sample.

One class of video sequences is defined:

Class HDR-HD: 1920 × 1080p 50 fps: "Market3", "Hurdles", "Starting"
1920 × 1080p 25 fps: "ShowGirl2"
1920 × 1080p 24 fps: "Cosmos1"

1.3.2 Coding Conditions of Submissions

The coding conditions and submissions shall apply to the HDR test category with the following exception:

- The quantization settings do not need to be kept static. Instead, a static temporal quantizer structure shall be used. The quantizer may then

---

[2]https://www.itu.int/rec/r-rec-bt500

Table 1.5 Combinations of classes and constraint sets for the High Dynamic Range category

	Class HDR-HD
Constraint set 1	X
Constraint set 2	–

Table 1.6 Target bit rate points not to be exceeded for the High Dynamic Range category[3]

Sequences	Target Bit Rates [kbit/s]			
	Rate 1	Rate 2	Rate 3	Rate 4
Market3	600	900	1500	2400
ShowGirl2	350	500	750	1200
Hurdles	450	800	1500	2000
Starting	450	600	850	1400
Cosmos1	500	900	1500	3000

Note: It is anticipated that the bit rates for Rate 3 of ShowGirl2 and Starting may be reduced in the Final Call for Proposals. The bit rates for other sequences may be modified as well.

be adjusted within a frame as a function of the local, average luma value and/or the local, average chroma value. A one-time change of the temporal quantizer structure is also allowed to meet the target bit rate. If either a local adjustment or a one-time change is used, a description of the adjustment scheme shall be provided in the descriptive document submission.

Submissions shall be made for the test cases (combinations of classes and constraint sets) as listed in Table 1.5.

– Submissions for Classes HDR-HD will be evaluated by means of a formal subjective assessment and BD PSNR and rate criteria.
– Submissions for Class HDR-HD will further be evaluated by the following metrics: weighted PSNR values (at least average of frame wPSNR for each video sequence and encoding point, separate for luma and chroma components), deltaE100 and PSNR-L100, as well as the Bjøntegaard Delta-Rate and Delta-PSNR for each metric.

Submissions to this Call shall, for each of the test cases defined above, submit results for the target bit rate points as listed in Table 1.6. The submitted results shall not exceed the target bit rate points.

1.3.3 Anchors

The anchor conditions and descriptions shall apply to the HDR test category with the following changes:

---

[3] 1 kbit/s means 10^3 bits per second, and 1 Mbit/s means 10^6 bits per second.

1. The generation of the anchor does not use a static quantization parameter (QP) setting. Instead, the configuration allows the QP value to vary spatially, where the variation is an explicit function of the average, local luma value. A one-time change of the quantization parameter from value QP to value QP+1 may also be applied in order to meet the defined target bit rates. The quantization parameter settings applied for the anchors will be reported.

1.4 360° Video

1.4.1 Video Test Sequence Formats and Frame Rates

All test material is progressively scanned and uses 4:2:0 color sampling with 8 or 10 bits per sample.

The classes of video sequences are:

Class 360: 8192×4096 30 fps 10 bits: "ChairliftRide"
8192×4096 30 fps 8 bits: "KiteFlite", "Harbor", "Trolley"
6144×3072 60 fps 8 bits: "Balboa"
6144×3072 30 fps 8 bits: "Landing"

Note: It is planned that the number of sequences in Class 360 will be reduced to five in the Final Call for Proposals.

1.4.2 Coding Conditions of Submissions

The coding conditions and submissions shall apply to the 360° video test category, with the following exceptions:

Submissions shall be made for the test cases (combinations of classes and constraint sets) as listed in Table 1.7.

- Submissions for Class 360 will be evaluated by means of a formal subjective assessment and BD PSNR and rate criteria.
- Submissions for Class 360 will further be evaluated using the following objective metrics: E2E WS-PSNR, E2E S-PSNR-NN, cross-format CPP-PSNR, cross-format S-PSNR-NN, codec WS-PSNR and codec S-PSNR-NN.

Table 1.7 Combinations of classes and constraint sets for the 360° video category

	Class 360
Constraint set 1	X
Constraint set 2	–

1.5 Test Sequences

Various test sequences are listed in Tables 1.8 through 1.13.

Table 1.8 SDR test sequences

Sequence Name	Resolution	Frame Count	Frame Rate	Chroma Format	Bit Depth
FoodMarket4	3840 × 2160	720	60	4:2:0	10
CatRobot1	3840 × 2160	600	60	4:2:0	10
DaylightRoad2	3840 × 2160	600	60	4:2:0	10
ParkRunning3	3840 × 2160	500	50	4:2:0	10
CampfireParty2	3840 × 2160	300	30	4:2:0	10
MountainBay2	3840 × 2160	300	30	4:2:0	10
BQTerrace	1920 × 1080	600	60	4:2:0	8
RitualDance	1920 × 1080	600	60	4:2:0	10
MarketPlace	1920 × 1080	600	60	4:2:0	10
BasketballDrive	1920 × 1080	500	50	4:2:0	8
Cactus	1920 × 1080	500	50	4:2:0	8

Table 1.9 SDR test sequence md5sums

Sequence Name	MD5Sum	Source (or Owner)
FoodMarket4	a378b34190f54f688d048a9a8b46a8ac	Netflix
CatRobot1	03a89792693fd9ecfd72ef2590025e97	B<>COM
DaylightRoad2	bf1d22643afb41b1521749d2749fb5f0	Huawei
ParkRunning3	e7a1d1ebff269767ec4bffd2998d1154	Huawei
CampfireParty2	63d3d9f9e4e8b5c344e89840e84e6428	SJTU
MountainBay2	f27b6b70244fb083baac546958fcf696	DJI
BQTerrace	efde9ce4197dd0b3e777ad32b24959cc	NTT DOCOMO Inc.
RitualDance	a3cb399a7b92eb9c5ee0db340abc43e4	Netflix
MarketPlace	dc668e7f28541e4370bdbdd078e61bba	Netflix
BasketballDrive	d38951ad478b34cf988d55f9f1bf60ee	NTT DOCOMO Inc.
Cactus	3fddb71486f209f1eb8020a0880ddf82	EBU/RAI

Note: Copyright information is included in the sequence ZIP container.

Table 1.10 HDR test sequences

Sequence Name	Resolution	Frame Count	Frame Rate	Chroma Format	Bit Depth
Market3	1920 × 1080	400	50	4:2:0	10
ShowGirl2	1920 × 1080	339	25	4:2:0	10
Hurdles	1920 × 1080	500	50	4:2:0	10
Starting	1920 × 1080	500	50	4:2:0	10
Cosmos1	1920 × 856	240	24	4:2:0	10

Note: The capture frame rate of the HDR3 (Hurdles) and HDR4 (Starting) sequences was 100 fps. However, these sequences are treated as 50 fps sequences for the evaluation processes defined in this document.

1.6 Distribution Formats for Test Sequences and Decoded Results

Table 1.11 HDR test sequence md5sums

Sequence Name	MD5Sum	Source (or Owner)
Market3	c97abe47455fd12f6d6436cecfad7c7d	Technicolor
ShowGirl2	44f1974d68f7799c71eea29fb72b245b	Stuttgart Media University
Hurdles	bc3cba849d6f4ee74d39056600722aa5	EBU
Starting	1cbc416696cb0dfcf4da9886eeb6a4a2	EBU
Cosmos1	da4a2488c249720da0535f01c3693efa	Netflix

Note: Copyright information is included in the test sequence ZIP container.

Table 1.12 A 360° video test sequences

Sequence Name	Input Resolution	Anchor Resolution	Coded Luma Sample Count of Anchors	Frame Count	Frame Rate	Chroma Format	Bit Depth
Balboa	6144 × 3072	4096 × 2048	8388608	300	60	4:2:0	8
ChairliftRide	8192 × 4096	4096 × 2048	8388608	300	30	4:2:0	10
KiteFlite	8192 × 4096	4096 × 2048	8388608	300	30	4:2:0	8
Harbor	8192 × 4096	4096 × 2048	8388608	300	30	4:2:0	8
Trolley	8192 × 4096	4096 × 2048	8388608	300	30	4:2:0	8
Landing	6144 × 3072	4096 × 2048	8388608	300	30	4:2:0	8

Note: The sequences are omnidirectional 360° × 180° degree video and are stored in an equirectangular projection (ERP) format. The number of coded luma samples in the anchor is lower than the resolution of the input sequence.

Table 1.13 360° video test sequence md5sums

Sequence	MD5Sum	Source (or Owner)
Balboa	1457bb109ae0d47265f5a589cb3464d7	InterDigital
ChairliftRide	9126f753bb216a73ec7573ecc4a280c3	GoPro
KiteFlite	18c0ea199b143a2952cf5433e8199248	InterDigital
Harbor	aa827fdd01a58d26904d1dbdbd91a105	InterDigital
Trolley	25c1082d1e572421da2b16530718156d	InterDigital
Landing	c715e332e78ab30e477445954da66c6b	GoPro

Note: Copyright information is included in the sequence ZIP container.

1.6 Distribution Formats for Test Sequences and Decoded Results, Delivery of Bitstreams and Binary Decoders, Utilities and Cross-check Meeting

Distribution of original video material files containing test sequences is done in YUV files with extension ".yuv". Color depth is 10 bit per component for all sequences. A description of the YUV file format is available at http://www.fourcc.org/ web site, designated as "planar iyuv".

HEVC Anchor bitstreams are provided with extension ".hevc". JEM Anchor bitstreams are provided with extension ".jem". Bitstream formats of proposals can be proprietary, but must contain all information necessary to decode the sequences at a given data rate (e.g. no additional parameter files). The file size of the bitstream will be used as a proof that the bit rate limitation from Tables 1.2, 1.3, 1.6, or Table 1.8 has been observed. The file extension of a proposal bitstream shall be ".bit".

Decoded sequences shall be provided in the same ".yuv" format as originals, with the exception that the color depth shall be 10 bits per component for all sequences.

All files delivered (bitstreams, decoded sequences and binary decoders) must be accompanied by an MD5 checksum file to enable identification of corrupted files. An MD5 checksum tool that may be used for that purpose is typically available as part of UNIX/LINUX operating systems; if this tool is used, it should be run with option "−b" (binary). For the Windows operating systems, a compatible tool can be obtained from http://www.pc-tools.net/win32/md5sums/. This tool should be run with additional option "−u" to generate the same output as under UNIX.

For the 360° video category, a binary projection format convertor may be provided, to be applied to the decoded YUV files. The output of the binary decoder, or the projection format convertor, if provided, must decode and/or convert each bitstream to a 4:2:0 YUV file of the same resolution as the input sequence.

The hard disk drive should be shipped (for handling in customs) with a declaration "used hard disk drive for scientific purposes, to be returned to owner" and low value specification (e.g. 20 €). The use of a hard disk drive with substantially larger size than needed is discouraged. The hard disk drive should be a 3 1/2-inch SATA solid state drive without any additional enclosure (no case, no power supply, no USB interface etc.), NTFS file format shall be used.

Before the evaluation meeting, a one-day cross-check meeting will be held. Proponents shall bring another hard disk drive, which can be connected via USB 3.0 to a Windows PC, containing original and decoded sequences in YUV format, bitstreams, binary decoder executables and all related checksum files. An adequate computer system shall also be brought to this meeting. Proponents shall specify the computing platform (hardware, OS version) on which the binary can be run. Should such a computing platform not be readily available, the proponent shall provide a computer adequate for decoder verification at this meeting.

1.7 Testing Environment and Methodology

The test procedure foreseen for the formal subjective evaluation will consider two main requirements:

- To be as much as possible reliable and effective in ranking the proposals in terms of subjective quality (and therefore adhering the existing recommendations);
- To take into account the evolution of technology and laboratory set-up oriented to the adoption of Flat Panel Display (FPD) and video server as video recording and playing equipment.

Therefore, two of the test methods described in Recommendation ITU-R BT.500 are planned to be used, applying some modification to them, relating to the kind of display, the video recording and playing equipment.

1.7.1 Selection of the Test Method

The anticipated test methods are:

1. Double Stimulus Impairment Scale (DSIS)
2. Double Stimulus Continuous Quality Scale (DSCQS)

1.7.2 DSIS

This test method is commonly adopted when the material to be evaluated shows a range of visual quality that well distributes across all quality scales.

This method will be used under the schema of evaluation of the quality (and not of the impairment); for this reason a quality rating scale made of 11 levels will be adopted, ranging from "0" (lowest quality) to "10" (highest quality). The test will be held in three different laboratories located in countries speaking different languages: This implies that it is better not to use categorical adjectives (e.g. excellent good fair etc.) to avoid any bias due to a possible different interpretation by naive subjects speaking different languages.

All the video material used for these tests will consist of video clips of 10 seconds duration.

The structure of the Basic Test Cell (BTC) of DSIS method is made by two consecutive presentations of the video clip under test; at first the original version of the video clip is displayed, immediately afterwards the coded version of the video clip is presented; then a message displays for 5 seconds asking the viewers to vote (see Figure 1.1).

14 Beyond High Efficiency Video Coding (HEVC)

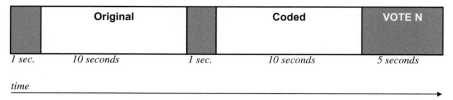

Figure 1.1 DSIS BTC.

1.7.3 DSCQS

This test method is typically selected when the range of visual quality presented to the viewing subjects is in the upper part of the quality scale.

In this method, the original and the coded samples of a video clip are presented two times.

This allows the viewing subjects to evaluate the sequences they are seeing more accurately.

An important aspect of DSCQS is that the viewing subject is asked to evaluate the two original and the coded clips separately. Furthermore, the viewer does not know which of the two clips is the original or the coded one.

The position of the original and the coded one is randomly changed for each BTC.

The structure of the Basic Test Cell of the DSCQS method therefore contains two consecutive pairs of presentations.

The first presentation is announced by a mid-grey screen with the letter "A" in the middle displaying for one second; the second presentation is announced by another mid grey screen with the letter "B"; these are repeated during the second pair of presentations changing A and B into A* and B*; the message "VOTE" displays on the screen for 5 seconds after the second pair of presentations is done (see Figure 1.2).

Figure 1.2 DSCQS BTC.

1.7.4 How to Express the Visual Quality Opinion with DSIS

The viewers will be asked to express their vote by placing a mark on a scoring sheet.

The scoring sheet for a DSIS test is made of a section for each BTC; each section is made of a column of 11 vertically arranged boxes, associated to a number from 0 to 10 (see Figure 1.3).

The viewers have to put a check mark on one of the 11 boxes; checking the box "10" the subject will express an opinion of "best" quality, while checking the box "0" the subject will express an opinion of the "worst'" quality.

The vote has to be written when the message "Vote N" appears on the screen. The number "N" is a numerical progressive indication on the screen aiming to help the viewing subjects to use the appropriate column of the scoring sheet.

1.7.5 How to Express the Visual Quality Opinion with DSCQS

The viewers will be asked to express their vote putting two marks on a scoring sheet.

The scoring sheet for a DSCQS test is made of a section for each BTC; each section is made of two continuous columns with 100 horizontal marks, (see Figure 1.4).

The viewers have to put a check mark for each of the two vertical lines; checking the upper side of the bar the subject will express an opinion of "best"

Figure 1.3 Example of DSIS test method scoring sheet.

16 *Beyond High Efficiency Video Coding (HEVC)*

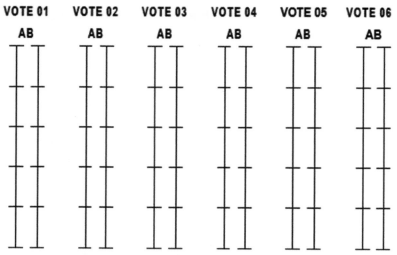

Figure 1.4   Example of DSCQS test method scoring sheet.

quality, while checking the lower side of the bar the subject will express an opinion of the "worst" quality.

The vote has to be written when the message "Vote N" appears on the screen. The number "N" is a numerical progressive indication on the screen aiming to help the viewing subjects to use the appropriate column of the scoring sheet.

1.7.6 Training and Stabilization Phase

The outcome of a test is highly dependent on a proper training of the test subjects.

For this purpose, each subject has to be trained by means of a short practice (training) session.

The video material used for the training session must be different from those of the test, but the impairments introduced by the coding have to be as much as possible similar to those in the test.

The stabilization phase uses the test material of a test session; three BTCs, containing one sample of best quality, one of worst quality and one of medium quality, are duplicated at the beginning of the test session. By this way, the test subjects have an immediate impression of the quality range they are expected to evaluate during that session.

1.7 Testing Environment and Methodology 17

The scores of the stabilization phase are discarded. Consistency of the behavior of the subjects will be checked inserting in the session a BTC in which original is compared to original.

1.7.7 The Laboratory Setup

The laboratory for a subjective assessment is planned to be set up following Recommendation ITU-R BT.500 [5], except for the selection of the display and the video play server.

High quality displays with controllable color rendition will be used.

When a video clip is shown at a resolution lower than the native resolution of the display itself, the video has to be presented in the center of the display; the active part of the display (i.e. that is actually showing the video signal) must have a dimension equal in rows and columns to the raster of the video; the remaining part of the screen has to be set to a mid-grey level (128 in 0–255 range). This constraint guarantees that no interpolation or distortion artefacts of the video images will be introduced.

The video play server, or the PC used to play video has to be able to support the display of both and UHD video formats, at 24, 30, 50 and 60 frames per second, without any limitation, or without introducing any additional temporal or visual artefacts.

1.7.8 Viewing Distance, Seats and Monitor Size

The viewing distance varies according to the physical dimensions of the active part of the video; this will lead to varying viewing distance from 2H to 4H, where H is equal to the height of the active part of the screen.

The number of subjects seating in front of the monitor is a function of the monitor size and type; for example, a monitor equal or superior to 30″ is expected to allow the seating of two or three subjects at the same time; monitors with 24″ would allow two subjects; monitors of 21″ would allow just one subject. Monitors with diagonal lower than 21″ should not be used. In any case monitors must support a wide screen aspect ratio without any picture adaptation (e.g. letter box etc.); i.e. they must give native support for the 16:9 aspect ratio.

1.7.9 Viewing Environment

The test laboratory has to be carefully protected from any external visual or audio pollution. Internal general light has to be low (just enough to allow the viewing subjects to fill out the scoring sheets) and a uniform light has to be placed behind the monitor; this light must have an intensity as specified in Recommendation ITU-R BT.500. No light source has to be directed to the screen or create reflections; ceiling, floor and walls of the laboratory have to be made of non-reflecting material (e.g. carpet or velvet) and should have a color tuned as close as possible to D65.

References

[1] 360 degree virtual reality video coding, https://github.com/Samsung/360tools.
[2] http://mpeg.chiariglione.org/standards/exploration/future-video-coding
[3] Gisle Bjontegaard, "Calculation of Average PSNR Differences between RD curves", ITU-T SG16/Q6 VCEG 13th meeting, Austin, Texas, USA, April 2001, Doc. VCEG-M33 (available at http://wftp3.itu.int/av-arch/video-site/0104_Aus/).
[4] Gisle Bjontegaard, "Improvements of the BD-PSNR model", ITU-T SG16/Q6 VCEG 35th meeting, Berlin, Germany, 16–18 July, 2008, Doc. VCEG-AI11 (available at http://wftp3.itu.int/av-arch/video-site/0807_Ber/).
[5] International Telecommunication Union – Radio Communication Sector, *Methodology for the subjective assessment of the quality of television pictures*, Recommendation ITU-R BT.500-13 (available at https://www.itu.int/rec/R-REC-BT.500).
[6] "Algorithm Description of Joint Exploration Test Model 7 (JEM7)", Joint Video Exploration Team (JVET) of ITU-T VCEG (Q6/16) and ISO/IEC MPEG (JTC 1/SC 29/WG 11), 7th Meeting, Torino, Italy, July 2017, Doc. JVET-G1001 (available at http://phenix.it-sudparis.eu/jvet/).

2

HEVC Encoder

2.1 Joint Collaborative Team on Video Coding (JCT-VC)

The Joint Collaborative Team on Video Coding is a group of video-coding experts from the ITU-T Study Group 16 (VCEG) and ISO/IEC JTC 1/SC 29/WG 11 (MPEG) created to develop a new-generation video-coding standard that will further reduce by 50% the data rate required for high-quality video coding, as compared to the earlier state-of-the-art advanced video coding (AVC) standard (ITU-T Rec. H.264 | ISO/IEC 14496-10). This new coding standardization initiative is being referred to as HEVC. In ISO/IEC, it is called MPEG-H Part2. VCEG is the acronym for the video coding experts group and MPEG for the moving picture experts group. Tan et al., [SE4] have presented the subjective and objective results of a verification test in which HEVC is compared with H.264/AVC. The conclusions in [SE4] have been repeated here: "These tests show that the bit rate savings of 59% on average can be achieved by HEVC for the same subjective quality compared with H.264/AVC. The PSNR-based BD-rate average over the same sequences was calculated to be 44%. This confirms that the subjective quality improvements of HEVC are typically greater than the objective quality improvements measured by the method that was primarily used during the standardization process of HEVC. It can therefore be concluded that the HEVC standard is able to deliver the same subjective quality as AVC, while on average (and in the vast majority of typical sequences) requiring only half or even less than half of the bit rate used by AVC". Various profiles/extensions of HEVC have been finalized starting in 2013. The joint video experts team (JVET) has called for proposals that can provide significant improvements in compression efficiency over HEVC with the new video coding standard to be finalized before the end of 2020. This is called versatile video coding (VVC). Refer the section on JVET provided in the end.

ITU-T Rec. H.264 | ISO/IEC 14496-10, commonly referred to as H.264/MPEG-4-AVC, H.264/AVC, or MPEG-4 Part 10 AVC has been

20 HEVC Encoder

Figure 2.1 Evolution of video coding standards (Alliance for Open Media).

developed as a joint activity within the joint video team (JVT). The evolution of the various video-coding standards is shown in Figure 2.1.

Footnote: H.265 and recent developments in video coding standards (Seminar presented by Dr. Madhukar Budagavi on 21 Nov. 2014 in the Dept. of Electrical Engineering, Univ. of Texas at Arlington, Arlington, Texas, USA).

Abstract: Video traffic is dominating both wireless and wireline networks. Globally, IP video is expected to be 83% of all IP traffic in 2019, up from 66% in 2013. On wireless networks, video is 70% of the global mobile data traffic in 2013 (Cisco VNI forecast). Movie studios, broadcasters, streaming video providers, TV, and consumer electronics device manufacturers are working towards providing immersive "real life" "being there" video experience to consumers using features such as increased resolution (Ultra HD 4K / 8K), higher frame rate, higher dynamic range (HDR), wider colour gamut (WCG), and 360 degrees video. These new features along with the explosive growth in video traffic are driving the need for an increased compression. This talk will cover the basics of video compression and then give an overview of the recently standardized HEVC video coding standard that provides around 50% higher compression than the current state-of-the-art H.264/AVC video coding standard. It will also highlight the recent developments in the video coding standards body related to HEVC extensions, HDR/WCG, and discussions on post-HEVC next-generation video coding.

It is expected that IP video traffic will be 82% of all consumer internet traffic by 2021 (Cisco visual networking index: Forecast and methodology, 2016–2021)

Seminar by Dr. Madukar Budagavi, Samsung Research America, 15 Sept. 2017.

EE Dept., Univ. of Texas at Arlington, Arlington, TX 76019, USA

Title: Video compression standards and recent developments

Abstract: Video traffic is dominating both the wireless and wireline networks. Movie studios, broadcasters, streaming video providers, TV and consumer electronics device manufacturers are working towards providing immersive "real life" "being there" video experience to consumers by using features such as increased resolution (Ultra HD 4K / 8K), higher frame rate, higher dynamic range (HDR), wider colour gamut (WCG), and 360 degrees video. These new features along with the explosive growth in video traffic are driving the need for increased compression. This talk will cover basics of video compression and then give an overview of the HEVC video coding standard that provides around 50% higher compression than its predecessor H.264/AVC video coding standard. The talk will also highlight recent developments in the video coding standards body related to HEVC extensions, HDR/WCG, and the ITU-T/ISO/IEC Joint Video Exploration Team (JVET) which is investigating compression techniques beyond HEVC.

Cisco VNI Global IP Traffic Forecast 2014–2019 predicts the following:

(a) Globally, IP video traffic will be 82% of all IP traffic (both business and consumer) by 2021;
(b) by 2019, more than 30% of connected flat-panel TV sets will be 4K; and
(c) traffic from wireless and mobile devices exceeded traffic from wired devices in 2016.

HEVC is designed to address the increased video traffic and increased resolution such as the 4K and 8K videos.

Latest forecast by Cisco: Video will be 78% of all mobile traffic by 2021.

Figure 2.1 Video coding standardization (Courtesy: Dr. Nam Ling, Sanfilippo family chair professor, Dept. of Computer Engineering, Santa Clara University, Santa Clara, CA, USA) [E23], AOM: Alliance for open Media.

22 HEVC Encoder

The JCT-VC is co-chaired by Jens-Rainer Ohm and Gary Sullivan, whose contact information is provided below,

ITU-T Contact for JCT-VC	Meetings
Mr Gary SULLIVAN Rapporteur, Visual coding Question 6, ITU-T Study Group 16 Tel: +1 425 703 5308 Fax: +1 425 936 7329 E-mail: garysull@microsoft.com Mr. Thomas WIEGAND Associate Rapporteur, Visual coding Question 6, ITU-T Study Group 16 Tel: +49 30 31002 617 Fax: +49 30 392 7200 E-mail: thomas.wiegand@microsoft.com	Future meetings Geneva, Switzerland, October 2013 (tentative) Vienna, Austria, 27 July – 2 August 2013 (tentative) Incheon, Korea, 20–26 April 2013 (tentative) Geneva, Switzerland, 14–23 January 2013 (tentative)
ISO/IEC contacts for JCT-VC	
Mr Jens- Rainer OHM Rapporteur, Visual coding Question 6, ITU-T Study Group 16 Tel: +49 241 80 27671 E-mail: ohm@ient.rwth-aachen.de	Mr Gary SULLIVAN Rapporteur, Visual coding Question 6, ITU-T Study Group 16 Tel: +1 425 703 5308 Fax: +1 425 936 7329 E-mail: garysull@microsoft.com

Additional information can be obtained from http://www.itu.int/en/ITU-T/studygroups/com16/video/Pages/jctvc.aspx

JCT-VC has issued a joint call for proposals in 2010 [E7]:
- 27 complete proposals submitted (some multi-organizational).
- Each proposal was a major package: several encoded videos, extensive documentation, extensive performance metric submissions, sometimes software, etc.

Table 2.1 Test Classes and Bit Rates (constraints) used in the CfP [E7]

Class	Bit Rate 1	Bit Rate 2	Bit Rate 3	Bit Rate 4	Bit Rate 5
A: 2560 × 1600p30	2.5 Mbit/s	3.5 Mbit/s	5 Mbit/s	8 Mbit/s	14 Mbit/s
B1: 1080p24	1 Mbit/s	1.6 Mbit/s	2.5 Mbit/s	4 Mbit/s	6 Mbit/s
B2: 1080P.50–60	2 Mbit/s	3 Mbit/s	4.5 Mbit/s	7 Mbit/s	10 Mbit/s
C: WVGAp30–60	384 kbit/s	512 kbit/s	768 kbit/s	1.2 Mbit/s	2 Mbit/s
D: WQVGAp30–60	256 kbit/s	384 kbit/s	512 kbit/s	850 kbit/s	1.5 Mbit/s
E: 720p60	256 kbit/s	384 kbit/s	512 kbit/s	850 kbit/s	1.5 Mbit/s

- Extensive subjective testing (three test labs, four 200 video clips evaluated, 850 human subjects, 300, 000 scores).
- Quality of proposal video was compared to AVC (ITU-T Rec. H.264| ISO/IEC 14496-10) anchor encodings.
- Test report issued JCTVC-A204/N11775.
- In a number of cases, comparable quality at half the bit rate of AVC (H.264).
- Source video sequences grouped into five classes of video resolution from quarter WVGA (416 × 240) to size 2560 × 1600 cropped from 4k × 2k ultra HD (UHD) in YC_bC_r 4:2:0 format progressively scanned with 8bpp.
- Testing for both "random access" (1 sec) and "low delay" (no picture reordering) conditions.

Figures 2.2 and 2.3 show results averaged over all of the test sequences; in which the first graph (Figure 2.2) shows the average results for the random access constraint conditions, and the second graph (Figure 2.3) shows the average results for the low delay constraint conditions.

The results were based on an 11-grade scale, where zero represents the worst and 10 represents the best individual quality measurements. Along with each mean opinion score (MOS) data point in the figures, a 95% confidence interval (CI) is shown.

A more detailed analysis performed after the tests shows that the best-performing proposals in a significant number of cases showed similar quality as the AVC anchors (H.264/AVC) at roughly half the anchor bit rate [E25, E61, E99]. More recently, Tan et al., [SE4] present the subjective and objective results of a verification test in which the performance of HEVC mail profile is compared with H.264/AVC high profile using video test sequences at various bit rates and resolutions. They conclude that the HEVC

24 HEVC Encoder

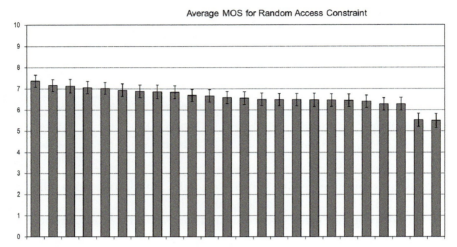

Figure 2.2 Overall average MOS results over all Classes for Random Access coding conditions [E7].

Figure 2.3 Overall average MOS results over all Classes for Low Delay coding conditions [E7].

is able to deliver the same subjective quality as H.264/AVC while on average requiring only half or even less than half the bit rate used by the latter. [SE4] describes in detail, the methodology, test procedures, test sequences, comparison metrics, and other factors in arriving at the conclusions.

2.1 Joint Collaborative Team on Video Coding (JCT-VC)

The technical assessment of the proposed technology was performed at the first JCT-VC meeting held in Dresden, Germany, 15–23 April, 2010. It revealed that all the proposed algorithms were based on the traditional hybrid coding approach, combining motion-compensated prediction between video frames with intra-picture prediction, closed loop operation with in-loop filtering, 2D transform of the spatial residual signals, and advanced adaptive entropy coding.

As an initial step towards moving forward into a collaborative work, an initial Test Model under Consideration (TMuC) document was produced, combining the identified key elements from a group of seven well-performing proposals. This first TMuC became the basis of a first software implementation, which after its development has begun to enable more rigorous assessment of the coding tools that it contains as well as additional tools to be investigated within a process of "Tool Experiments (TE)" as planned at the first JCT-VC meeting.

P.S.: Detailed subjective evaluations in mobile environments (smart phone/iPad, Tablet), however, have shown that the user (observer) experience is not significantly different when comparing H.264/AVC and HEVC compressed video sequences at low bit rates (200 and 400 kbps) and small screen sizes [E124, E210]. Advantages of HEVC over H.264/AVC appear to increase dramatically at higher bit rates and high resolutions such as HDTV and UHDTV.

One of the most beneficial elements for higher compression performance in high-resolution video comes due to the introduction of larger block structures with flexible mechanisms of sub-partitioning. For this purpose, the TMuC defines *coding units* (CUs) that define a sub-partitioning of a picture into rectangular regions of equal or (typically) variable size. The coding unit replaces the macroblock structure (H.264) and contains one or several *prediction unit(s)* (PUs) and *transform units* (TUs). The basic partition geometry of all these elements is encoded by a scheme similar to the quad-tree segmentation structure. At the level of PU, either intra-picture or inter-picture prediction is selected.

The paper "Block partitioning structure in the HEVC standard", by I.-K. Kim et al., [E93] explains the technical details of the block partitioning structure and presents the results of an analysis of the coding efficiency and complexity.

- Intra-picture prediction is performed from samples of already decoded adjacent PUs, where the different modes are DC (flat average), horizontal, vertical, or one of up to 28 angular directions (number depending

on block size), plane (amplitude surface) prediction, and bilinear prediction. The signalling of the mode is derived from the modes of adjacent PUs.
- Inter-picture prediction is performed from region(s) of the already decoded pictures stored in the reference picture. This allows selection among multiple reference pictures, as well as bi-prediction (including weighted averaging) from two reference pictures or two positions in the same reference picture. In terms of the usage of the motion vector (quarter pixel precision), merging of adjacent PUs is possible, and non-rectangular sub-partitions are also possible in this context. For efficient encoding, skip and direct modes similar to the ones of H.264/AVC (Chapter 4) are defined, and the derivation of motion vectors from those of adjacent PUs is made by various means such as median computation or a new scheme referred to as *motion vector competition*.

At the TU level (which typically would not be larger than the PU), an integer spatial transform similar in concept to the DCT is used, with a selectable block size ranging from 4×4 to 64×64. For the directional intra modes, which usually exhibit directional structures in the prediction residual, special *mode-dependent directional transforms* (MDDT) [E8, E51, E17, E301], [DCT19]] are employed for block sizes 4×4 and 8×8. Additionally, a *rotational transform* (See P.2.13) can be used for the cases of block sizes larger than 8×8. Scaling, quantization, and scanning of transform coefficient values are performed in a similar manner as in AVC.

At the CU level, it is possible to switch on an *adaptive loop filter* (ALF) that is applied in the prediction loop prior to copying the frame into the reference picture buffer. This is an FIR filter designed with the goal to minimize the distortion relative to the original picture (e.g., with a least-squares or Wiener filter optimization). Filter coefficients are encoded at the slice level. In addition, a deblocking filter (similar to the deblocking filter design in H.264/AVC) [E73] is operated within the prediction loop. The display output of the decoder is written to the decoded picture buffer after applying these two filters. Please note that the ALF has been dropped in the HEVC standard [E235, E61, E99]. In the updated version, in-loop filtering consists of deblocking [E73, E209, E283, E357, E378, E425, E428] and sample adaptive offset (SAO) filters (Figure 2.4). See [E87, E111, E373, E433] about SAO in the HEVC standard. Recently, the [E456] convolution neural networks for in-loop filtering [E436] after DB and SAO filters in the HEVC encoder (Figure 2.5) have been proposed. The authors demonstrate that the proposed

2.1 Joint Collaborative Team on Video Coding (JCT-VC) 27

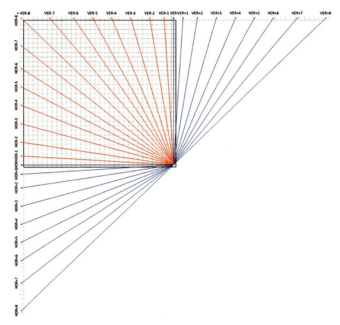

Figure 2.4 Available prediction directions in the unified intra prediction in HM 1.0.

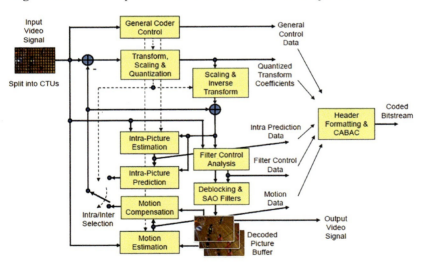

Figure 2.5 HEVC encoder block diagram [E59]. In the Joint Exploration Model (JEM) – bilateral filter before DF and adaptive loop filter after SAO are adopted.

method is able to not only raise the PSNR of reconstructed frame but also prominently reduce the bit-rate compared with HEVC reference software. Several papers related to in-loop filtering have been cited at the end.

The TMuC defines two context-adaptive entropy coding schemes, one for operation in a lower-complexity mode and one for higher-complexity mode.

A software implementation of the TMuC has been developed. On this basis, the JCT-VC is performing a detailed investigation about the performance of the coding tools contained in the TMuC package, as well as other tools that have been proposed in addition to those. Based on the results of such *Tool Experiments* (TE), the group will define a more well-validated design referred to as a Test Model (TM) as the next significant step in HEVC standardization. Specific experiments have been planned relating to a tool-by-tool evaluation of the elements of the current TMuC, as well as the evaluation of other tools that could give additional benefit in terms of compression capability or complexity reduction in areas such as intra-frame and inter-frame prediction, transforms, entropy coding and motion vector coding. Various *ad hoc groups* (AHGs) have been set up to perform additional studies on issues such as complexity analysis, as listed below:

Ad Hoc Coordination Groups Formed
- JCT-VC project management
- Test model under consideration editing
- Software development and TMuC software technical evaluation
- Intra-prediction
- Alternative transforms
- MV precision
- In-loop filtering
- Large block structures
- Parallel entropy coding

In summary, the addition of number of tools within the motion-compensated (MC) transform prediction hybrid coding framework (adaptive intra/inter-frame coding, adaptive directional intra prediction, multiple block size motion estimation, SAO filter [E87, E111, E373], in-loop deblocking filter [E73, E209, E283, E302, E357, E378, E398, E425, E428, E436, E443, JPEG14] entropy coding (CABAC, see [E67, E68], RE390), multiple frame MC-weighted prediction, integer transforms from 4×4 to 32×32 [E8, E74], Hadamard transform coding of the dc coefficients in intra-frame coding introduced in H.264/AVC and various other tools as enumerated below have shown further gains in the coding efficiency, reduced bit rates, higher PSNR, etc. compared to H.264/AVC. Thus, the HEVC holds promise and potential in a number of diverse applications/fields and is expected to eventually overtake H.264/AVC.

a) Directional prediction modes (up to 34) for different PUs in intra prediction (Figure 2.4) [E49, E104]
b) Mode-dependent directional transform (MDDT)$^+$ besides the traditional horizontal/vertical scans for intra frame coding [E51, E17, E301], [DCT15, DCT19, DCT32]
c) Rotational transforms for block sizes larger than (8 × 8) (see P.2.13)
d) Large size transforms up to 32 × 32 [E8, E74]
e) In-loop deblocking filter [E73, E209, E283, E289, E302, E357, E378, E398, E425, E428, E436, E443, JPEG14] (See also J. Erfurt et al., "Multiple feature-based classifications of adaptive lop filter", PCS2018, San Francisco, CA, June 2018.) and SAO filter [E87, E111, E373, E398, E433, E444]. See also Q. Han and W.-K. Cham, "High performance loop filter for HEVC", IEEE ICIP 2015
f) Large size blocks for ME/MC
g) Non-rectangular motion partition [E61]

Note b) and c) have been dropped as they contribute very little to the coding efficiency at the cost of a substantial increase in complexity. Other directional transforms that were proposed, such as directional DCT [E112] and mode-dependent DCT/DST [E113], were also dropped due to the same reason. Recently, steerable DCT (SDCT) and its integer approximation INTSDCT that can be steered in any direction have been proposed [E392]. The authors state that "SDTC allows to rotate in a flexible way pairs of basis vectors and enables precise matching of directionality in each image block, achieving improved coding efficiency". They have implemented SDCT in image coding and compared with DCT and DDCT. They also suggest a possible implementation of SDCT and INTSDCT in a video compression standard such as HEVC. See the projects P.2.262 through P.2.267. Both in-loop deblocking and SAO filters reduce the blocking and ringing artefacts. However, in HD and UHD videos, false contour is a major artefact due to unbalanced quantization in smooth areas. Huang et al., [E393] have identified the cause of false contours and have developed an effective method for detection and removal of false contours in both compressed images and videos while preserving the true edges and textures.

It is interesting to note that some of these tools such as the rotational transforms are reconsidered in NGVC (beyond HEVC) [BH2]. Other tools considered in NGVC (proposed earlier in HEVC) are as follows:

30   HEVC Encoder

- Large blocks (both CU and TU sizes).
- Fine granularity intra-prediction angles.
- Bi-directional optical flow [9].
- Secondary transform, both implicit and explicit [10, 11].
- Multi-parameter Intra prediction [12].
- Multi-hypothesis CABAC probability estimation [13].

References and figures cited here are from [BH2].
P.S: The introduction to JCT-VC is based on the paper titled by "Recent developments in standardization of high efficiency video coding (HEVC)" [E7] by G.J. Sullivan and J.-R. Ohm published in applications of digital image processing XXXIII, proc. of SPIE vol. 7798, pp. 7798V–1 through 7798V–7, 2010.

For the recent developments in HEVC, the reader is referred to an excellent review paper:
G.J. Sullivan et al., "Overview of high efficiency video coding (HEVC) standard", IEEE Trans. CSVT, vol. 22, pp. 1649–1668, Dec. 2012 [E61]. Also keynote speeches on HEVC [E23, E25, E99]. Also tutorials on HEVC (see the section on tutorials). Also see HEVC text specification draft 8 [E60]. An updated paper on HEVC is G.J. Sullivan et al., "Standardized extensions of high efficiency video coding (HEVC)", IEEE Journal of selected topics in signal processing, vol. 7, pp. 1001–1016, Dec. 2013 [E160]. Another valuable resource is the book: V. Sze, M. Budagavi and G.J. Sullivan (editors) "High efficiency video coding (HEVC): algorithms and architectures", Springer, 2014 [E202]. Another valuable book is
M. Wien, "High Efficiency Video Coding: Coding Tools and Specification", Springer, 2015. (See section on books on HEVC),[+] For intra mode, an alternative transform derived from DST is applied to 4 × 4 luma blocks only. For all other cases integer DCT is applied.

2.2 Analysis of Coding Tools in HEVC Test Model, HM 1.0 – Intra Prediction

In HM 1.0, unified intra-prediction provides up to 34 directional prediction modes for different PUs. With the PU sizes of 4 × 4, 8 × 8, 16 × 16, 32 × 32, and 64 × 64, there are 17, 34, 34, 34, and five prediction modes available, respectively. The prediction directions in the unified intra prediction have the angles of $+/- [0, 2, 5, 9, 13, 17, 21, 26, 32]/32$. The angle is given by displacement of the bottom row of the PU and the reference row above the PU in case of vertical prediction, or displacement of the rightmost column of the PU and the reference column left from the PU in case of horizontal prediction. Figure 2.4 shows an example of prediction directions for 32x32 block size. Instead of different accuracies for different sizes, the reconstruction of the pixel uses the linear interpolation of the reference top or left samples at 1/32th pixel accuracy for all block sizes.

More details on unified intra prediction in HM 1.0 are available in http://www.h265.net/2010/12/analysis-of-coding-tools-in-hevc-test-model-hm-intra-prediction.html

The working draft (WD) of the HEVC has gone through several updates/revisions, and the final draft international standard (FDIS) was released in January 2013. This refers to the Main, Main 10, and Main Intra profiles. In August 2013, five additional profiles Main 12, Main 4:2:2 12, Main 4:4:4 10, and Main 4:4:4 12 were released [E160]. Other range extensions include increased emphasis on high quality coding, lossless coding, and screen content coding (SCC). See the references section on screen content coding. Scalable video coding (spatial, temporal, quality, hybrid, and colour gamut scalabilities) and multiview video coding were finalized in July 2014 and standardized in October 2014. See the section on special issues on HEVC.

The IEEE Journal on Emerging and Selected Topics in Circuits and Systems (JETCAS) published the special Issue on Screen Content Video Coding and Applications in December 2016. See the overview paper on SCC [OP11 and SCC48]. This special issue opened up a number of research areas in SCC. (See P.2.85 and P.2.85a).

Scalability extensions [E160, E325, E326, and E410] and 3D video extensions that enable stereoscopic and multiview representations and consider newer 3D capabilities such as the depth maps and view-synthesis techniques have been finalized in 2015. For work on 3D video topics for multiple standards including 3D video extensions, the JCT-VC formed a team known as JCT on 3D Video (JCT-3V) in July 2012 [E160]. The overall objective is to reduce the bit rate and increase the PSNR significantly compared to H.264/AVC with a reasonable increase in encoder/decoder complexity. These three overview papers [E325, E326, and E160] give a detailed development and description of these extensions. See also references listed under overview papers.

2.3 HEVC Encoder

IEEE Trans. CSVT vol. 22, Dec. 2012 is a special issue on emerging research and standards in next generation video coding [E45]. This special issue provides the latest developments in HEVC related technologies, implementations and systems with focus on further research. As the HEVC development is ongoing, this chapter concludes with a number of projects related to HEVC with appropriate references and the info on the KTA and HEVC software [E97]. Hopefully these projects can provide additional insights to the tools and techniques proposed and provide a forum for their modifications leading to further improvements in the HEVC encoder/decoder. Figure 2.5 describes the HEVC encoder block diagram [E61]. This does not show the coded bit stream representing the various modes (intra/inter, CU/PU/TU sizes, intra angular prediction directions/modes, MV prediction, scaling, and quantization of transform coefficients and other modes are shown in the decoder block diagram – see Figure 2.6).

For directional intra modes, an alternative transform related to DST is applied to 4 × 4 luma prediction residuals. For all other cases, integer DCT

Figure 2.6 HEVC decoder block diagram[1].

---

[1] The decoder block diagram is adopted from C. Fogg, "Suggested figures for the HEVC specification", ITU-T / ISO-IEC Document: JCTVC-J0292r1, July 2012.

2.3 HEVC Encoder

is applied (Intra/Inter chroma and Inter luma). This change was adopted in Stockholm in July 2012.

In entropy coding, only the context adaptive binary arithmetic coding (CABAC) [E67, E68, E390] is adopted unlike two (CAVLC and CABAC) in H.264/AVC. Details on context modelling, adaptive coefficient scanning, and coefficient coding are provided in [E58]. Chapter 2 – Entropy coding in HEVC – authored by Sze and Marpe in [E202] describes the functionality and design methodology behind CABAC entropy coding in HEVC. In the conclusions section of this Chapter Sze and Marpe state "The final design of CABAC in HEVC shows that by accounting for implementation cost and coding efficiency when designing entropy coding algorithms results in a design that can maximize processing speed and minimize area cost, while delivering high coding efficiency in the latest video coding standard". The references cited at the end of this chapter are extensive and valuable.

Mode-dependent directional transform [E8, E17, E51, E301] is not adopted. Only INTDCT (separable 2-D) for all cases other than 4×4 intra luma is used.

While the HEVC follows the traditional (also proven) block-based motion-compensated prediction followed by transform, quantization, and variable length coding (also adaptive intra mode), it differs significantly by adopting a flexible quad tree coding block partitioning structure. This recursive tree-coding structure is augmented by large block size transforms (up to 32×32), advanced motion prediction, sample adaptive offset (SAO) [E87, E111, 373, E398] besides the deblocking filter [E73, E209, E283, E302, E357, E378, E398]. The large multiple sizes recursive structure is categorized into a coding unit (CU), prediction unit (PU), and a transform unit (Figure 2.7a) [E44]. For details about transform coefficient coding in HEVC, see [E74].

In HEVC, the coding tree unit (CTU) is the basic unit of the quadtree coding method, and prediction and transform are performed at each coding unit (CU), that is, a leaf node of the tree. The sizes of CTUs can be 16×16, 32×32, or 64×64 pixels, and CU sizes can be 8×8, 16×16, 32×32, or 64×64 pixels. Figure 2.7b shows CU formations in a 64×64 CTU and each CU can be partitioned into prediction units (PUs), as shown in Figure 2.7c. There are a number of modes, such as skip/merge $2N \times 2N$, inter $2N \times 2N$, inter $N \times N$, inter $N \times 2N$, inter $2N \times N$, inter $2N \times nD$, inter $2N \times nU$, inter $nL \times 2N$, inter $nR \times 2N$, intra $2N \times 2N$, and intra $N \times N$. For each CU, the mode having the minimum rate–distortion (RD) cost is called the best mode, which is used as a competitor to decide

34　HEVC Encoder

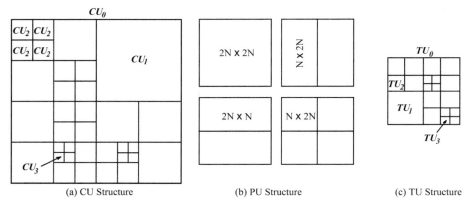

(a) CU Structure　　(b) PU Structure　　(c) TU Structure

Figure 2.7a Recursive block structure for HEVC, where k indicates the depth for CU_k and TU_k [E44].

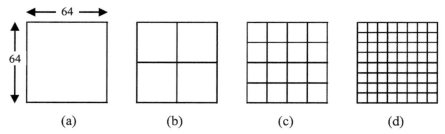

Figure 2.7b Quadtree partitioning of a 64 × 64 CTU. (a) 64 × 64 CU at depth 0. (b) Four CUs of 32 × 32 at depth 1. (c) 16 CUs of 16 × 16 at depth 2. (d) 64 CUs of 8 × 8 at depth 3.

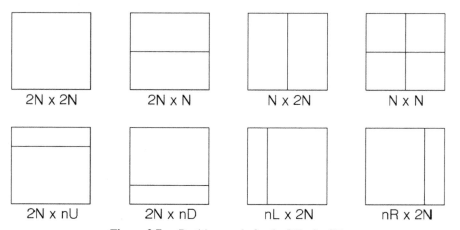

Figure 2.7c Partition mode for the PU of a CU.

the quadtree partition structure of the CTU. The decision process of the best mode includes prediction, transform, quantization, and entropy coding, which usually requires a high computational cost.

Complexity of the HEVC encoder is further increased by introducing an intra-adaptive angular direction prediction, mode-dependent context sample smoothing, adaptive motion parameter prediction, and in-loop filtering (deblocking filter and sample adaptive offset-SAO). Details on these two in-loop filters and their effectiveness in improving the subjective quality are described in Chapter 7 In-loop filters in HEVC authored by Norkin et al., see [E202]. These and other tools contribute to 50% improved coding efficiency over H.264 at the cost of substantial increase in encoder complexity. The decoder complexity, however, is similar to that of H.264/AVC [E23, E61, E105]. Several techniques for reducing the intra-prediction encoder complexity (see [E42] and various references cited at the end) are suggested. See also [E108]. Zhang and Ma [E42, E147] are also exploring the reduction of inter-prediction complexity. They suggest that these two areas (intra- and inter-prediction modes) can be combined to reduce the overall HEVC encoder complexity (see projects P.2.16 through P.2.19). This is a fertile ground for research. A summary of the tools included in main and high efficiency 10 [HE10] is shown in Table 2.2. Details of these tools are described in Test Model encoder description [E59]. See also the review papers [E61, E99, E107]. The paper [E61] states "To assist industry community in learning how to use the standard, the standardization effort not only includes the development of a text specification document (HM8) but also reference software source code (both encoder/decoder)" [E97]. This software can be used as a research tool and as the basis of products. This paper also states "A standard test data suite is also being developed for testing conformance to the standard".

2.3.1 Intra Prediction

Figure 2.8 shows the 33 intra prediction angle directions [E59, E61, E125] corresponding to the VER and HOR described in Figure 2.4. Figure 2.9 shows the 33 intra prediction mode directions. The mapping between the intra prediction mode directions and angles is shown in Table 2.3 [E125]. See also [E102]. These intra prediction modes contribute significantly to the improved performance of HEVC. Statistical analysis of the usage of the directional prediction modes for all intra case has shown that besides planar (mode 0) and dc (mode 1), horizontal (mode 10), and vertical (mode 26) are at the top

Table 2.2 Structure of tools in HM9 configuration

Main	High Efficiency 10 (HE10)
High-level Structure:	
High-level support for frame rate temporal nesting and random access	
Clean random access (CRA) support	
Rectangular tile-structured scanning	
Wavefront-structured processing dependencies for parallelism	
Slices with spatial granularity equal to coding tree unit	
Slices with independent and dependent slice segments	
Coding units, Prediction units, and Transform units:	
Coding unit quadtree structure square coding unit block sizes	
$2N \times 2N$, for $N = 4, 8, 16, 32$ (i.e. up to 64×64 luma samples in size)	
Prediction units (for coding unit size	
$2N \times 2N$: for Inter, $2N \times 2N$, $2N \times N$, $N \times 2N$, and, for $N > 4$, also $2N \times (N/2 + 3N/2)$ & $(N/2 + 3N/2) \times 2N$; for Intra, only $2N \times 2N$ and, for $N = 4$, also $N \times N$)	
Transform unit tree structure within coding unit (maximum of 3 levels)	
Transform block size of 4×4 to 32×32 samples (always square)	
Spatial Signal Transformation and PCM Representation:	
DCT-like integer block transform;	
for Intra also a DST-based integer block transform (only for Luma 4×4)	
Transforms can cross prediction unit boundaries for Inter; not for Intra	
Skipping transform is allowed for 4×4 transform unit	
PCM coding with worst-case bit usage limit	
Intra-picture Prediction:	
Angular intra prediction (35 modes including DC and Planar)	
Planar intra prediction	
Inter-picture Prediction:	
Luma motion compensation interpolation: 1/4 sample precision, 8×8 separable with 6 bit tap values for 1/2 precision, 7×7 separable with 6 bit tap values for 1/4 precision	
Chroma motion compensation interpolation: 1/8 sample precision, 4×4 separable with 6 bit tap values	
Advanced motion vector prediction with motion vector "competition" and "merging"	
Entropy Coding:	
Context adaptive binary arithmetic entropy coding (CABAC)	
Rate-distortion optimized quantization (RDOQ)	
Picture Storage and Output Precision:	
8 bit-per-sample storage and output	10 bit-per-sample storage and output
In-Loop Filtering:	
Deblocking filter	
Sample-adaptive offset filter (SAO)	

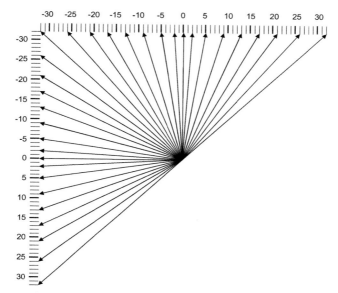

Figure 2.8 Intra prediction angle definition [E59, E125].

of this ranking [E104]. The authors in [E104] by developing a new angular table have demonstrated improved coding gains for video sequences with large amounts of various textures. Each intra-coded PU shall have an intra prediction mode for luma and another for chroma components. All TUs within a PU shall use the same associated mode for each component. The encoder then selects the best luma intra prediction mode from the 35 directions (33 plus planar and DC). Due to the increased number of directions (compared to those in H.264/AVC) HEVC considers three most probable modes (mpm) compared to one mpm in H.264/AVC. For chroma of intra PU, the encoder then selects the best chroma prediction mode from 5 modes including planar, DC, horizontal, vertical and direct copy of the intra prediction mode for luma. Details about mapping between intra prediction direction and mode # for chroma are given in [E59, E125].

A detailed description of the HEVC encoder related to slices/tiles, coding units, prediction units, transform units, inter prediction modes, special coding modes, MV estimation/ prediction, interpolation filters for fractional pixel MV resolution, weighted prediction, transform sizes (4×4, 8×8, 16×16 and 32×32) [E74], scanning the transform coefficients – Figure 2.10 – (see [E74] for details), scaling/quantization, loop filtering (deblocking filter [E73, E209, E283, E357, E378, E398] and SAO [E87, E111, E373, E398]), entropy

38 HEVC Encoder

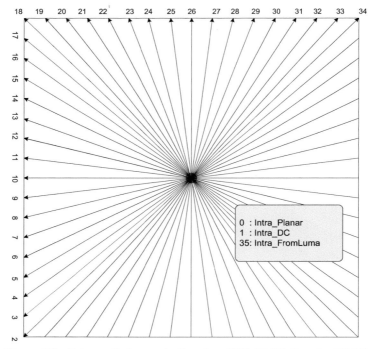

Figure 2.9 Intra prediction mode directions [E59, E125]. See also [E104].

coding (CABAC [E67, E68, E390]) and myriad other functionalities are provided in [E61, E99, E107]. An excellent resource is V. Sze, M. Budagavi and G.J. Sullivan (Editors), "High efficiency video coding: Algorithms and architectures", Springer 2014. [E202]. Another excellent resource is the book M. Wien, "High efficiency video coding: Coding tools and specification", Springer 2015. In [E202], various aspects of HEVC are dealt within different chapters contributed by various authors who have been involved in all phases of the HEVC development as an ITU-T and ISO/IEC standard. Popular zigzag scan is not adopted.

P.S. Hsu and Shen [E357] have developed the VLSI architecture and hardware implementation of a highly efficient deblocking filter for HEVC that can achieve 60 fps video (4K × 2K) under an operating frequency of 100 MHz. They also list several references related to deblocking filter (H.264/AVC and H.265) and SAO filter (H.265).

Both visual and textual descriptions of these directional modes, besides the filtering process of the reference samples as predictors, post-processing

2.3 HEVC Encoder

Table 2.3 Mapping between intra prediction mode direction (shown in Figure 2.9) and intra prediction angles (shown in Figure 2.8) [E125]

predModeIntra	1	2	3	4	5	6	7	8	9	10	11	12	13	14	15	16	17
intraPredAngle	–	32	26	21	17	13	9	5	2	0	–2	–5	–9	–13	–17	–21	–26
predModeIntra	18	19	20	21	22	23	24	25	26	27	28	29	30	31	32	33	34
intraPredAngle	–32	–26	–21	–17	–13	–9	–5	–2	0	2	5	9	13	17	21	26	32

of predicted samples and myriad other details are described in Chapter 4 Lainema and Han, "Intra-picture prediction in HEVC", [E202].

2.3.2 Transform Coefficient Scanning

The three transform coefficient scanning methods, diagonal, horizontal, and vertical, adopted in HEVC for a 8×8 transform block (TB) are shown in Figure 2.10 [E74]. The scan in a 4×4 transform block is diagonal. Horizontal

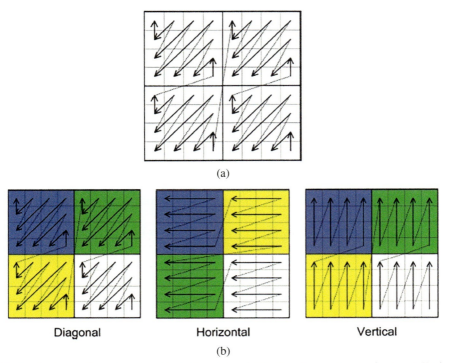

Figure 2.10 (a) Diagonal scan pattern in an 8×8 TB. The diagonal scan of a 4×4 TB is used within each 4×4 sub-block of larger blocks. (b) Coefficient groups for 8×8 TB [E74].

40 HEVC Encoder

and vertical scans may also be applied in the intra case for 4 × 4 and 8 × 8 transform blocks.

Chapter 6, Budagavi, Fuldseth, and Bjontegaard, "HEVC transform and quantization", in [E202] describe the 4 × 4 to 32 × 32 integer 2-D DCTs including the embedding process (small-sized transforms are embedded in large-sized transforms) default quantization matrices for transform block sizes of 4 × 4 and 8 × 8, and other details are addressed. The embedding feature allows for different transform sizes to be implemented using the same architecture thereby facilitating hardware sharing. An extensive list of references related to integer DCT architectures is provided in [E381]. An extensive list of references related in integer DCT architectures is provided in [E381].

An alternate 4 × 4 integer transform derived from DST for intra 4 × 4 luma blocks is also listed.

2.3.3 Luma and Chroma Fractional Pixel Interpolation

Integer ($A_{i,j}$) and fractional pixel positions (lower case letters) for luma interpolation are shown in Figure 2.11 [E61, E109]. See [E69] for generalized interpolation.

Unlike a two-stage interpolation process adopted in H.264, HEVC uses separable 8-tap filter for $1/2$ pixels and 7-tap filter for $1/4$ pixels (Table 2.4) [E61, E71, E109]. Similarly, 4-tap filter coefficients for chroma fractional (1/8 accuracy) pixel interpolation are listed in Table 2.5. Lv et al., [E109] have conducted a detailed study of performance comparison of fractional-pel interpolation filters in HEVC and H.264/AVC and conclude that the filters in HEVC increase the BD rates [E81, E82, E198] by more than 10% compared to those in H.264/AVC (Chapter 4) at the cost of increased implementation complexity.

M. Wien, "High Efficiency Video Coding: Coding Tools and Specification", Springer, 2014. See Section 7.3.3

Determination of the interpolation filter (IF) coefficients (pages 194–196) for the theory behind how these IF coefficients are derived using DCT.

Table 2.4 Filter coefficients for luma fractional sample interpolation [E61]

Index	−3	−2	−1	0	1	2	3	4
hfilter[i]	−1	4	−11	40	40	−11	4	1
qfilter[i]	−1	4	−10	58	17	−5	1	

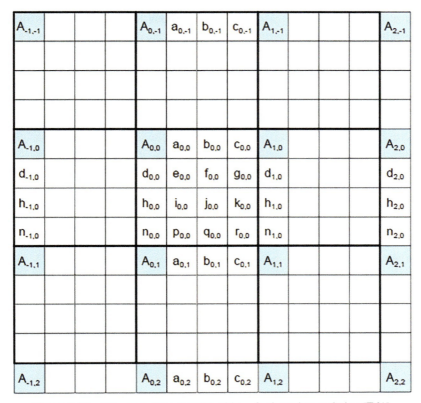

Figure 2.11 Integer and fractional positions for luma interpolation [E61].

2.3.4 Comparison of Coding Tools of HM1 and HEVC Draft 9

Coding tools of HEVC test model version 1 (HM1) and draft 9 [E59] are summarized in Table 2.6 [E59]. The overview paper on HEVC by Sullivan et al., [E61] is an excellent resource which clarifies not only all the inherent functionalities but also addresses the history and standardization process

Table 2.5 Filter coefficients for chroma fractional sample interpolation [E61]

Index	−1	0	1	2
filter1[i]	−2	58	10	−2
filter2[i]	−4	54	16	−2
filter3[i]	−6	46	28	−4
filter4[i]	−4	36	36	−4

leading to this most efficient standard. In the long run, the HEVC (including the additions/extensions/profiles) has the potential/prospects/promise to overtake all the previous standards including H.264/AVC (Chapter 4).

2.4 Extensions to HEVC

As with H.264/AVC, additions/extensions to HEVC include 4:2:2 and 4:4:4 formats, higher bit depths (10 and 12), scalable video coding (SVC), (References are listed in Section 2.9 Summary) 3D/stereo/multiview coding and screen content coding. See [E23, E25, E34, E39, E260, E326, SCC48]. Several proposals related to SVC have been presented at the Shanghai meeting of HEVC (ITU-T/ISO, IEC) held in October 2012. The reader is referred to the poster session [E27] MA.P2 "High efficiency video coding", IEEE

Table 2.6 Summary of coding tools of high efficiency configuration in HM1 and HEVC [E61]

Functionality	HM 1 High Efficiency	HEVC (draft 9)
CTU structure	Tree structure from 8×8 to 64×64	Tree structure from 8×8 to 64×64
PU structure	Square and symmetric	Square, symmetric and asymmetric (only square for intra)
TU structure	Tree structure of square TUs	Tree structure of square TUs
Core transform	Integer transforms from 4 to 32 points (full factorization)	Integer transforms from 4 to 32 points (partially factorable)
Alternative transform	n/a	Integer DST type for 4×4
Intra prediction	33 angular modes with DC mode	33 angular modes with planar and DC modes
Luma interpolation	12-tap separable	8-tap/7-tap separable
Chroma interpolation	Bilinear	4-tap separable
MV prediction	AMVP	AMVP
MC region merging	Spatial CU merging	PU merging
Entropy coding	CABAC	CABAC
Deblocking filter	Nonparallel	Parallel
Sample adaptive offset	n/a	Enabled
Adaptive loop filter	Multiple shapes	n/a
Dedicated tools for parallel processing	Slices	Slices, tiles, wavefronts, and dependent slice segments

ICIP 2012, Orlando FL, Sept.,-Oct. 2012 and the special issue on emerging research and standards in next generation video coding" IEEE Trans. CSVT, vol. 22, Dec. 2012 [E45]. Several papers from this special issue are cited as references here [E61–E63, E67, E73, E74, E76–E79, E83–E85, E87–E93, E120]. See also [E321, E325, E326].

These papers cover not only the various aspects of HEVC, but also ways to reduce the implementation complexity and the reasons behind the various tools and techniques adopted in this standard. The justification for the initial consideration of some tools such as adaptive loop filter, MDDT, ROT, etc. and the subsequent elimination of these tools in the standard is also provided in some of these papers.

Some of these tools are reconsidered in the NGVC, beyond HEVC. Also, the IEEE Journal of selected topics of signal processing has published a special issue on video coding: HEVC and beyond, vol. 7, December 2013 [E80].

These additions/extensions have been standardized in 2015.

2.5 Profiles and Levels

Three profiles (Main, Main10, and Main Still Picture – intra frame only) are listed in Annex A in the final draft international standard (FDIS) (January 2013) [E125]. The ITU-T study group 16 has agreed to this first stage approval formally known as Recommendation H.265 or ISO/IEC 23008-2. Main profile is limited to YC_bC_r 4:2:0 format, 8-bit depth, progressive scanning (non-interlaced) with spatial resolutions ranging from QCIF (176 × 144) to 7640 × 4320 (called 8K × 4K). Figure 2.12 [E23] lists the spatial resolutions ranging from SD (NTSC) to super Hi-Vision/ultra HD video. In the main profile, 13 levels are included in the first version (Table 2.7) [E61, E99].

2.6 Performance and Computational Complexity of HEVC Encoders

Correa et al., [E84] have carried out a thorough and detailed investigation of the performance of coding efficiency versus computational complexity of HEVC encoders. This investigation focuses on identifying the tools that most affect these two vital parameters (efficiency and complexity). An invaluable outcome is the trade-off between the efficiency and complexity useful for implementing complexity-constrained encoders.

44 HEVC Encoder

Figure 2.12 Future visual applications and demands. (NHK and Mitsubishi have developed a SHV-HEVC hardware encoder).

Additionally, the development of low-complexity encoders that achieve coding efficiency comparable to high-complexity configurations is a valuable resource for industry. This combination of low complexity and high efficiency can be achieved by enabling Hadamard ME, asymmetric motion partitions,

Table 2.7 Level limits for the main profile in HEVC [E61]

Level	Max Luma Picture Size (samples)	Max Luma Sample Rate (samples/sec)	Main Tier Max Bit Rate (1000 bits/s)	High Tier Max Bit Rate (1000 bits/s)	Min Comp. Ratio
1	36,864	552,960	128	–	2
2	122,880	3,686,400	1,500	–	2
2.1	245,760	7,372,800	3,000	–	2
3	552,960	16,588,800	6,000	–	2
3.1	983,040	33,177,600	10,000	–	2
4	2,228,224	66,846,720	12,000	30,000	4
4.1	2,228,224	133,693,440	20,000	50,000	4
5	8,912,896	267,386,880	25,000	100,000	6
5.1	8,912,896	534,773,760	40,000	160,000	8
5.2	8,912,896	1,069,547,520	60,000	240,000	8
6	33,423,360	1,069,547,520	60,000	240,000	8
6.1	33,423,360	2,005,401,600	120,000	480,000	8
6.2	33,423,360	4,010,803,200	240,000	800,000	6

and the loop filters instead of computationally demanding tools such as ME. Their analysis included the three coding tools (non-square transform, adaptive loop filter, and LM chroma), which have been subsequently dropped from the draft standard [E60].

2.7 System Layer Integration of HEVC

Schierl et al., describe the system layer integration of HEVC [E83]. The integration of HEVC into end-to-end multimedia systems, formats, and protocols (RTP, MPEG-2 TS, ISO File Format, and dynamic adaptive streaming over HTTP – DASH) is discussed. Error resiliency tools in HEVC are also addressed. Many error resilience tools of H.264/AVC such as FMO, ASO, redundant slices, data partitioning, and SP/SI pictures (Chapter 4) have been removed due to their usage. They suggest that the use of error concealment in HEVC should be carefully considered in implementation and is a topic for further research. They conclude with a description of video transport and delivery such as broadcast, IPTV, internet streaming, video conversation and storage as provided by the different system layers. See reference 24 listed in this paper i.e., T. Schierl, K. Gruenberg and S. Narasimhan, "Working draft 1.0 for transport of HEVC over MPEG-2 Systems", ISO/IEC SC29/WG11, MPEG99/N12466, February 2012. See other references related to various transport protocols.

2.8 HEVC Lossless Coding and Improvements [E86]

Zhou et al., [E85] have implemented lossless coding mode of HEVC main profile (bypassing transform, quantization and in-loop filters – Figure 2.13 [E85]) and have shown significant improvements (bit rate reduction) over current lossless techniques such as JPEG2000, JPEG-LS, 7-Zip and WinRAR. (See Appendix F) They improve the coding efficiency further by introducing the sample based intra angular prediction (SAP).

Cai et al., [E114] also conducted the lossless coding of HEVC intra mode, H.264 High 4:4:4 Profile intra mode, MJPEG 2000 and JPEG-LS using a set of recommended video sequences during the development of HEVC standard. Their conclusion is that the performance of HEVC has matched that of H.264/AVC and is comparable to JPEG-LS and MJPEG 2000. Similar tests on lossy intra coding show that HEVC high10, H.264/AVC HP 4:4:4 and HEVC MP have similar performance. However, MJPEG 2000 outperforms the former three in low bit scenario, while this advantage is gradually

46   HEVC Encoder

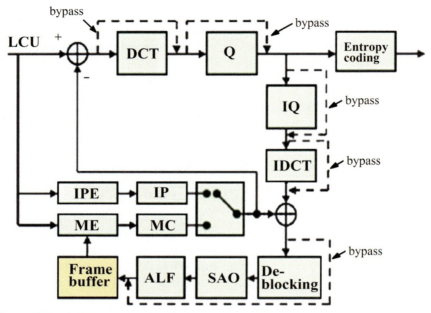

Figure 2.13 HEVC encoder block diagram with lossless coding mode [E85] (See Figure 2.5). See also [E194, E196].

compensated and finally surpassed by the former three as the bit rate increases. Several other interesting papers on the performance comparison of these and other standards are listed in [E114]. Wige et al., [E147] have implemented HEVC lossless coding using pixel-based averaging predictor. Heindel, Wige, and Kaup [E194] have developed lossless compression of enhancement layer (SELC) with base layer as lossy. Wang et al., [E196] have implemented lossy to lossless image compression based on reversible integer DCT.

Horowitz et al., [E66] have conducted the informal subjective quality (double blind fashion) comparison of HEVC MP (reference software HM 7.1) and H.264/AVC HP reference encoder (JM 18.3) for low delay applications. Table 2.8 describes the comparison results for test sequences encoded using the HM and JM encoders. They conclude that HEVC generally produced better subjective quality compared with H.264/AVC at roughly half the bit rate with viewers favouring HEVC in 73.6% trials.

To reinforce these results, ×264 the production quality H.264/AVC encoder is compared with eBrisk Video (production quality HEVC implementation). See Table 2.9 for the comparison results. The viewers favoured

Table 2.8 Subjective viewing comparison results for sequences encoded using the HM and JM encoders [E66]

	Subjective Viewing for eBrisk vs. ×264								
Sequence	eBrisk Bit Rate (kbps)	eBrisk QP	×264 Bit Rate (kbps)	×264 QP	eBrisk: ×264 Bit Rate	Votes Favoring eBrisk	Votes Favoring ×264	% Favoring HEVC	Total Number of Votes
KristenAndSara	332	36	657	33	51%	12	13	48%	25
Vidyo1	363	36	773	32	47%	10	15	40%	25
Old Town Cross	904	35	1716	34	53%	22	3	88%	25
Kimono1	1384	35	2670	32	52%	17	8	68%	25
toys_and_calendar	729	36	1553	33	47%	17	8	68%	25
				Average	50%	78	47	62.4%	125

Table 2.9 Subjective viewing comparison results for sequences encoded using the eBrisk and ×264 encoders [E66]

	Subjective Viewing for HM vs. JM								
Sequence	HM Bit Rate (kbps)	HM QP	JM Bit Rate (kbps)	JM QP	HM: JM Bit Rate	Votes Favoring HM	Votes Favoring JM	% Favoring HEVC	Total Number of Votes
KristenAndSara	149	38	302	37	49%	10	15	40%	25
Vidyo1	190	37	367	36	52%	14	11	56%	25
OldTownCross	408	37	879	37	46%	22	3	88%	25
Kimono1	682	36	1404	35	49%	21	4	84%	25
toys_and_calendar	347	37	734	38	47%	25	0	100%	25
				Average	49%	92	33	73.6%	125

eBrisk-encoded video at roughly half the bit rate compared to ×264 in 62.4% trials. These tests confirm that HEVC yields similar subjective quality at half the bit rate of H.264/AVC. Chapter 9 – Compression performance analysis in HEVC – by Tabatabai et al., in [E202] describes in detail the subjective and objective quality comparison of HEVC with H.264/AVC. Tan et al., [4] have presented the subjective and objective comparison results of HEVC with H.264/AVC and conclude that the key objective of the HEVC has been achieved i.e., substantial improvement in compression efficiency compared to H.264/AVC.

2.9 Summary

In the family of video coding standards (Figure 2.1), HEVC has the promise and potential to replace/supplement all the existing standards (MPEG and

H.26x series including H.264/AVC). While the complexity of the HEVC encoder is several times that of the H.264/AVC, the decoder complexity [E163] is within the range of the latter. Researchers are exploring about reducing the HEVC encoder/decoder complexity [E63, E79, E84, E88, E89, E106, E108, E166–E17, E187, E195R, E199, E204, E205, E208, E221, E230, E232, E240, E242, E247, E248, E252, E257, E258, E259, E269, E270, E274, E276, E278, E290, E308, E312, E332, E348, E368, E369, E374, E383, E388, E390, E414, E423, E424, E440, E454, E458, E460, E463, E465, E468]. Bin and Cheung [E247] have developed a fast decision algorithm for the HEVC Intra coder and report encoding time is reduced by 52% on average compared with HM 10.0 reference software with negligible losses. Similarly, Kim and Park [E248] developed a fast CU partitioning algorithm resulting in reducing the computational complexity of the HEVC encoder by 53.6% on the average with negligible BD bit rate loss. Chi et al., [E246] have developed an SIMD optimization for the entire HEVC decoder resulting in up to 5x speed up. Min, Xu and Cheung [E382] have developed a fully pipelined architecture for intra prediction of HEVC that achieves a throughput of 4 pixels in one clock cycle and is capable to support 3840 × 2160 video at 30 fps. Correa et al., [E383] have developed (part of the abstract is repeated here).

"A set of Pareto-efficient encoding configurations, identified through rate-distortion-complexity analysis. This method combines a medium-granularity encoding time control with a fine-granularity encoding time control to accurately limit the HEVC encoding time below a predefined target for each GOP. The encoding time can be kept below a desired target for a wide range of encoding time reductions, e.g., up to 90% in comparison with the original encoder. The results also show that compression efficiency loss (Bjøntegaard delta-rate) varies from negligible (0.16%) to moderate (9.83%) in the extreme case of 90% computational complexity reduction"

Index Terms—Encoding time limitation, High Efficiency

In particular, refer to [E242], wherein significant encoder complexity reduction is achieved with a negligible increase in BD-rate by combining early termination processes in CTU, PU and RQT structures. See also P.2.139 related to this. Chapter 4 – Intra-picture prediction in HEVC – and Chapter 5 – inter-picture prediction in HEVC – and the references at the end of these chapters [E202] provide insights related to encoder complexity reduction. Another approach in reducing the encoder complexity is based on zero block detection [E166, E171, E186]. Kim et al., [E168] have shown

2.9 Summary

that motion estimation (ME) occupies 77–81% of HEVC encoder implementation. Hence, the focus has been in reducing the ME complexity [E137, E152, E168, E170, E179, E205, E221, E239, E275, E351, E359, E370, E371, E412, E430, E434, E449, E450, E470]. See also various references related to fast/simplified ME algorithms listed in [E470]. Fan et al., [E412] have developed a hardware solution that supports 4k × 2k@ 30 fps videos with a working frequency of no more than 500 MHz and compared with several other integer ME-based HEVC encoders. PPT slides related to fast ME algorithm and architecture design are presented in [Tut10]. Machine learning techniques [E259, E263, E317, E359, E440] have been applied effectively resulting in significant reduction of HEVC encoder complexity. Recently, Liu et al., [E440] have developed an adaptive CU size decision algorithm for HEVC intra prediction based on the complexity comparison using machine learning and show that their proposed algorithm can achieve around 60% encoding time reduction for various video test sequences with a negligible BDBR increase. They also summarize earlier papers related to complexity reduction in intra prediction of HEVC encoder. Several researchers have implemented performance comparison of HEVC with other standards such as H.264/AVC (Chapter 4 in [Book4]), MPEG-4 Part 2 visual, H.262/MPEG-2 Video [E183], H.263, and VP9 (Chapter 6 in [Book4]) and also with the image coding standards such as JPEG2000, JPEG-LS, and JPEG-XR [E22, E47, E62, E63, E65, E66, E114, E177, E191, E245]. Similar comparisons with MPEG-4 Part 10 (AVC) and VP9 are reported in [E191]. See also papers listed under VP8, VP9, and VP10. Most significant of all these is the detailed performance comparison (both objective and subjective) described in [E62]. Special attention may be drawn to references 39, 40, and 42-47 cited in [E62]. Chapter 9 – Comparison performance analysis in HEVC – in [E202] provides both objective/subjective comparison between HM 12.1 (HEVC) and JM 18.5 (AVC – H.264). See Table 9.4 in [E202] for description of the test conditions.

However, the payoff is several tests have shown that HEVC provides improved compression efficiency up to 50% bit rate reduction [E61, E62, E99] for the same subjective video quality compared to H.264/AVC (Chapter 4 in [Book4]). See Tables 2.8 and 2.9. Besides addressing all current applications, HEVC is designed and developed to focus on two key issues: increased video resolution – up to 8k × 4k – and an increased use of parallel processing architecture. A detailed description of the parallelism in HEVC is outlined in Chapter 3 – H. Schwarz, T. Schierl, and D. Marpe, "Block structures and parallelism features in HEVC ", in [E202]. This chapter also demonstrates that more than half of the average bit –rate savings compared to H.264 are due to increased flexibility of block partitioning for prediction and

50   HEVC Encoder

transform coding. Brief description of the HEVC is provided. However, for details and implementation, the reader is referred to the JCT-VC documents [E55, E188], overview papers (See section on overview papers).

[E321] is an overview paper describing the range extensions (Rext) of the HEVC. Beyond the Main, Main 10, and Main Intra profiles finalized in version1, Rext (version 2) focuses on monochrome, 4:2:4 and 4:4:4 sampling formats and higher bit depths. The new coding tools integrated into version 2 are: Cross-component prediction (CCP) [E355], Adaptive chroma, Version 1 tools are: Filtering for smoothing of samples in intra-picture prediction, transform skip and transform quantizer bypass modes, truncated Rice binarization, internal accuracy and k-th order Exp-Golomb (EGk) binarization, and CABAC bypass decoding. These tools and modifications are meant to address the application scenarios such as: content production in a digital video broadcasting delivery chain, storage, and transmission of video captured by a professional camera, compression of high dynamic range content, improved lossless compression, and coding of screen content. This overview paper describes in detail not only the justification for introducing these additional tools, modifications, and profiles but also the improved performance. Using a spectrum of standard video test sequences, the authors have compared the performance of HEVC version II (HM 16.2 software) with H.264/AVC (JM 18.6) using BD bit rate as a metric. This comparison extends to lossless coding for different temporal structures called as All-Intra (AI), Random Access (RA) and Low Delay (LD). In the overview paper on SCC [E322], implementation complexity as a performance measure is conspicuous by its absence. An extensive list of references related to SCC are cited at the end. See also the overview paper on SCC [SCC48].

Keynote speeches [E23, E99], tutorials (See section on tutorials), panel discussions [E98, E366, E380], poster sessions [E27, E139], workshop [E257], special issues (See special issues on HEVC) test models (TM/HM) [E54], web/ftp sites [E48, E53–E55], open source software (See section on open source software), test sequences, file formats [E327], and anchor bit streams [E49]. Also researchers are exploring transcoding between HEVC and other standards such as MPEG-2 (See section on transcoders). Further extensions to HEVC [E158] are scalable video coding (SHVC) [E76, E119, E126–E129, E141, E158, E159, E208, E213, E214, E234, E238, E264, E284, E325] (note that E325 is an overview paper on SHVC), 3D video/multiview video coding [E34, E39, E201, E227, E271, E326] (note that E326 is an overview paper on the 3D video/multiview coding) and range extensions [E155, E321] which include screen content coding. See the references on

2.9 Summary

screen content coding besides the overview paper [E322] bit depths larger than 10 bits and colour sampling of 4:2:2 and 4:4:4 formats.

In an overview paper, Boyce et al., [E325] describe the scalable extensions of the HEVC (SHVC), which include temporal, spatial, quality (SNR), bit depth, and colour gamut functionalities and as well their combination (hybrid). They compare the performance of SHVC with the scalable extensions of H.264/AVC called SVC and conclude using various test sequences (Table 5 in [E325]), that SHVC is 50% to 63% more efficient than SVC depending on the scalability type and the use scenario. In view of the superior performance, various standardization bodies such as 3GPP and ATSC for broadcasting, and video streaming and IMTC for videoconferencing are evaluating SHVC as potential codec choice. Investigation of bit rate allocation in SHVC is addressed in [E350].

[E322, SCC48] are overview papers on the HEVC SCC extension finalized in 2016. They describe several new modules/tools added to the original HEVC framework and also the new coding tools introduced during the development of HEVC – SCC. [E322] concludes with the performance analysis of the new coding tools relative to HEVC – Rext and H.264/AVC.

Screen Content Coding (SCC) [E322, SCC48] in general refers to computer-generated objects and screenshots from computer applications (both images and videos) and may require lossless coding. HEVC SCC extension reference software is listed in [E256]. Some aspects of SCC are addressed in [E352, E353]. They also provide a fertile ground for R & D. Screen Content Coding has been finalized in 2016. [SCC15] is the special issue on screen content video coding and applications published in Dec. 2016 and reports briefly on ongoing standardization activities. This special issue covers on all aspects of SCC besides opening the avenue for a number of projects/problems. The overview paper on screen content video coding (SCVC) by Peng et al., [OP11] addresses the recent advances in SCC with an emphasis on two state-of-the-art standards: HEVC/H.265 Screen Content Coding Extensions (HEVC-SCC) by ISO/IEC MPEG and Display Stream Compression (DSC) by VESA. They also explore future research opportunities and standards developments in order to give a comprehensive overview of SCVC. Iguchi et al., [E164] have already developed a hardware encoder for super hi-vision (SHV) i.e., ultra HDTV at 7680 × 4320 pixel resolution. Shishikui [E320] has described the development of SHV broadcasting in Japan and states that 8K SHV broadcasting to begin in 2016 during the Rio de Janeiro Summer Olympics and for full pledged broadcasting to begin in 2020 with the Tokyo Summer Olympics. He has also presented a keynote speech

"8K super hi-vision: science, engineering, deployment and expectations", in IEEE PCS 2015 in Cairns, Australia.

Tsai et al., [E222] have designed a 1062 Mpixels/s 8192 × 4320 HEVC encoder chip. See also [E165] that describes real-time hardware implementation of an HEVC encoder for 1080p HD video. NHK is planning SHV experimental broadcasting in 2016. See also [E226]. A 249-Mpixel/s HEVC video decoder chip for 4k Ultra-HD applications has already been developed [E182]. Bross et al., [E184] have shown that real-time software decoding of 4K (3840 × 2160) video with HEVC is feasible on current desktop CPUs using four CPU cores. They also state that encoding 4K video in real time on the other hand is a challenge. AceThought, among many other companies, provides high-quality and efficient software implementation of HEVC Main Profile, Main 10 Profile, and Main Still Picture Profile decoder algorithms on various handheld and desktop platforms [E189]. Cho et al., [E219] have developed CTU level parallel processing for single-core and picture partition-level multi-core parallel processing for 4K / 8K UHD high-performance HEVC hardware decoder implementation. They concluded that their multi-core decoder design is suitable for 4K / 8K UHD real time decoding of HEVC bit streams when implemented on SoC. Kim et al., [E220] have extended this to scalable HEVC hardware decoder and the decoder design is estimated to be able to decode 4K@60 fps HEVC video. Decoder hardware architecture and encoder hardware architecture for HEVC are addressed in Chapters 10 and 11 respectively in the book Sze, Budagavi and Sullivan (editors) "High efficiency video coding (HEVC): algorithms and architectures", Springer, 2014 [E202]. For 4K video, software decoder has been developed [E329]. A hardware architecture of a CABAC decoder for HEVC that can lead to real time decoding of video bit streams at levels 6.2 (8K UHD at 10 fps) and 5.1 (4K UHD at 60 fps) has been developed by Chen and Sze [E249]. See also [E390]. Zhou et al., also have developed similar VLSI architecture for H.265/HEVC CABAC encoder [E250]. Wang et al., [E359] have presented an improved and highly efficient VLSI implementation of HEVC MC that supports 7680 × 4320@60fps bidirectional prediction. The authors [E359] claim that the proposed MC design not only achieves 8x throughput enhancement but also improves hardware efficiency by at least 2.01 times in comparison with prior arts. Zooming and displaying region of interest (ROI) from a high resolution video, i.e., ROI-based streaming in HEVC has also been developed by Umezaki and Goto [E206]. The authors suggest future work to reduce the decoding cost. A novel wavefront-based high parallel solution for HEVC Encoding integrating data level, optimal Single Instruction – Multiple – Data

2.9 Summary

(SIMD) algorithms are designed by Chen et al., [E272], in which the overall WHP solution can bring up to 57.65x, 65.55x and 88.17x speedup for HEVC encoding of WVGA, 720p and 1080p standard test sequences while maintaining the same coding performance as with Wavefront Parallel Processing (WPP). Abeydeera et al., [E338] have developed a 4K real-time decoder on FPGA. Chiang et al., [E345] have developed a real-time HEVC decoder capable of decoding 4096 × 2160@30 fps at 270 MHz. Open source code for HEVC encoder is developed by Kvazaar [OSS22]. Transport of HEVC video over MPEG-2 systems is described in [E197]. Encode/decode and multiplex/demultiplex HEVC video/AAC audio while maintaining lip synch has been implemented in software [E203]. This, however, is limited to HEVC main profile intra coding only. Extension to inter prediction and also to other profiles are valid research areas. Kim, Ho, and Kim [E254] have developed a HEVC-complaint perceptual video coding scheme based on the JND models that have significant bit-rate reductions with a negligible subjective quality loss at a slight increase in the encoder complexity. Several companies and research institutes have developed various products related to HEVC (see the booths and exhibits at NAB 2016, held in Las Vegas 16–21 April 2016 www.nabshow.com). One highlight is the single chip #8K #video decoder developed by Socionext US. Another activity is the realization of low-cost embedded HEVC encoder on TI Keysone multicore DSP [E448]. Several companies have developed or developing hardware encoders/decoders (chips or SOC) that can process up to 4K and 8K video [TUT10]. Details on software codecs are also provided. This tutorial concludes with the future challenges for next-generation video encoders that relate to extensions such as HSVC, SCC, Multiview, 3D, etc. A new standard called MPEG-4 internet video coding (IVC) has been approved as ISO/IEC 14496–33 in June 2015 [E399]. This standard is highly beneficial for video services in the internet domain. The HEVC standard and all its extensions are expected to go through the various revisions and refinements meeting the growth in consumer electronics, internet, VLSI, security, defence, medical, broadcasting, and related areas over the next several years. The next major standard is projected to be around 2020 designed to meet 5G specification. [BH1] and [BH2] address video coding techniques beyond HEVC. References [1] through [8] in [BH2] describe the algorithms and techniques proposed for NGVC beyond HEVC. Sze and Budagavi and also Grois et al., (see tutorials) have suggested directions for post HEVC activities. While the main emphasis in developing HEVC has been on efficient video compression and coding, it is interesting to note that the features in its compressed domain have been adopted in video saliency detection [E400].

The projects listed below range from graduate 3 credit hours to research at the M.S. and doctoral levels. These projects are designed to provide additional insights into the tools and techniques and provide a forum for their modifications leading to further improvements in the HEVC encoder/decoder). Besides the references related to HEVC, references are listed under various categories such as, books, tutorials, transcoders, online courses (OLC), overview papers, open source software, video test sequences, SSIM, screen content coding, file format, JVT reflector, MSU codec comparison, general, beyond HEVC etc. for easy access. References and projects related to JPEG, JPEG-LS, JPEG-2000, JPEG-XR, JPEG XS, JPEG-XT, JPEG-pleno, PNG, WebP, WebM, DIRAC and DAALA are added at the end. Keynote address by T. Ebrahimi "JPEG PLENO: Towards a new standard for plenoptic image compression", IEEE DCC, Snow Bird, Utah, 2016 along with the abstract is included. This developing new standard while maintaining backward compatibility with legacy JPEG opens up thought provoking and challenging research (JP-P1, through JP-P5). Panel presentations by Sullivan and others [E366] "Video coding: Recent developments for HEVC and future trends' during IEEE DCC 2016 opens up research areas for next generation video coding (NGVC). Slides related to high dynamic range, wide colour gamut video, 360 video, and royalty-free video standards are of special interest providing fertile ground for research. Access to HEVC SCC extensions [OP11], new profiles, and further compression exploration beyond HEVC are also provided. Several tests [E372] have shown superior performance of HEVC intra compared to JPEG, JPEG LS, JPEG 2000, JPEG XT, JPEG XR, and VP9 intra in terms of PSNR and SSIM. Related problems are described in P.2.237 and P.2.238. However, implementation complexity of HEVC intra is higher compared to all the others. A brief introduction to AV1 codec developed by Alliance for Open Media (AOM) is included. This is followed by brief description of real-Media HD (RMHD) codec developed by Real Networks. Special sessions on recent developments in video coding, genomic compression and quality metrics/perceptual compression held in IEEE DCC, April 2017 are of interests to researchers, academia and industry [E386]. Projects related to demosaicking of mosaicked images are listed in P.2.254 through P.2.257 (See [E384] and [E387]). In IEEE DCC 2017 [E386] special sessions on "Video coding", "Genome Compression", and "Quality Metrics and Perceptual Compression" can lead to number of R&D projects. Of particular interest is perceptual compression which is based on psycho visual perception (vision science). Another interesting topic is the keynote address "Video Quality Metrics" by Scott Daly, Senior Member Technical Staff, Dolby Laboratories. Besides compression and coding, some features of HEVC in compressed domain are used in developing video saliency detection [E400]. The authors in [E400] claim that their method is superior to other state-of-the-art saliency detection techniques. Recently, Kim et al., [E355] have demonstrated significant coding gains for both natural and screen content video in HEVC based on cross component prediction (CCP). IEEE ICCE, Las Vegas, Jan. 2017 has also many papers related to HEVC. The domain of the video coding standards, in general, is limited to encoder/decoder. Both preprocessing (RGB to colour space conversion, quantization and chroma subsampling) and post processing (corresponding inverse processes) are trends for future video codecs. For example, HDR and WCG include both pre and post processing. This implies end-to-end pipeline video capture and display in RGB format. The special session "Toward a next-generation international video coding standard" in IEEE DCC, 27–30 March 2018 [E426] – Dr. G.S. Sullivan and Dr. Y. Ye as the session coordinators is of special interest related to research directions in NGVC. Another special session in this conference [E426] is "The emerging AV1 video codec' – AV1 by alliance for open media (AOM) with J. Han and I. Sodagar as the session coordinators. AV1 codec has been finalized and is released as a royalty free open source software by AOM. See the section AV1 codec (alliance for open media) towards the end–http://aomedia.org. Joint Exploration Model (JEM) being developed by Joint Video Experts Team (JVET) is another major development in the video coding arena. [E406] See the section on Joint Video Experts Team towards the end. Performance comparison of AV1, JEM, VP9 and HEVC coders is explored in [E445]. Some of these developments are categorized as next generation video coding (NGVC), beyond HEVC, future video coding (FVC), versatile video coding (VVC) etc. The invited talk by Ohm and Sullivan on versatile video coding – towards the next generation of video compression (IEEE PCS, San Francisco, CA, June 2018) presents an optimistic/realistic gain in compression efficiency (50% bit rate reduction with the same subjective quality compared to HEVC) with the standard projected for 2020. This would also address 360 degree video, 8K video, HDR video etc. [OP16]. Some more recent topics such as MPEG-DASH and HDR image/video coding are also briefly described.

In summary several improved/advanced video codecs are developed/finalized by various organizations/alliances/groups with no end in sight.

IEEE DCC 2019 (26–29 March) is setting up two special sessions:

Advances in Video Coding and its Applications"
Yuriy Reznik, *Brightcove, Inc.*
Gary Sullivan, *Microsoft Research*
Jiangtao Wen, *Tsinghua University*
Yan Ye, *Interdigital, Inc.*
Session Coordinators

"Plenoptic Image Compression"
Thomas Richter, *Fraunhofer IIS*
Peter Schelkens, *Vrije Universiteit Brussel*
Session Coordinators

T. Sugimoto, Y. Nakayama and T. Komori, "22.2 channel audio encoding/decoding hardware system based on MPEG-4 AAC", IEEE Trans. Broadcasting, vol. 63, pp. 426–432, June 2017. MPEG-4 AAC uses MDCT for psycho analysis in the audio coder. 8K Super hi-vision (UHDTV) 7680 × 4320 broadcasting (program production equipment, coding system, multiplexer, transmitter/receiver) Launched in public in 2018, widespread use by 2020 in time for summer Olympics in Tokyo, Japan. NHK developed a world-first 22.2 channel audio encoding/decoding hardware system capable of real time encoding/decoding (22.2 channel audio codec). 8K (video resolution 7680 × 4320) video at 59.94 Hz satellite test broadcasting has been used in Japan since August 2016 [E462]. This complies with the Japanese 8K broadcasting standard ARIB STD-B32 – See P.2.385. This standard is based on HEVC/H.265 Main 20 profile. Fujitsu has already developed a commercial HEVC/H.265 Main 4:2:2 10 Profile real-time video codec. See reference 3 in [E462]. Zhou et al., in Waseda University, Japan have developed a 4 GPixels/s 8/10b H. 265/HEVC video decoder chip for 8K ultra HD applications (IEEE ISSCC, pp. 266–268, San Francisco, CA, 31 Jan. – 4 Feb. 2016). Researchers are looking at higher frame rates i.e., 119.88 (120 Hz).

INFO ON AOM DEMO.

AOMedia Presents and Demos AV1 at the NAB Streaming Smmit, NYC, Oct. 17–18, 2018

Join AOMedia members Bitmovin, Google and Mozilla at the AV1 Happy Hour on Wednesday, Oct. 17, 2018, 5:00-6:00 p.m. at the CMIIP Theater

New York, USA - Oct. 16, 2018 – The Alliance for Open Media (AOMedia) today announced that members will be leading presentations and demos focusing on the AOMedia Video Codec (AV1) from Oct. 17-18, 2018 during the NAB Show's Streaming Summit in New York at the Jacob K. Javits Center. Steaming Media attendees are also invited to the AV1 Happy Hour on Wednesday, Oct. 17, 2018, 5:00–6:00 p.m. in the CMIIP Theater, sponsored by AOMedia, Bitmovin, Mozilla and Google.

Co-produced by streaming industry expert Dan Rayburn, the Streaming Summit at NAB will feature presentations, round-table discussions and fireside chats that will address trends and technologies that are having a direct impact on the continued development, proliferation and profitability of online video.

Track C: AV1, A Win for the Open Web: Demos by Google and Mozilla

When: Wednesday, Oct. 17, 2018, 3:15 p.m.–4:00 p.m.

Where: Location: 3D10

What: Presenters from AOMedia members Google and Mozilla will share a look at AV1, the royalty-free video codec using Chrome and Firefox. AV1 offers a generational improvement in coding efficiency over any web codec and comes backed by nearly 40 companies, including some of the largest in the internet, tech, and video industries. This talk will cover why Google and Mozilla invest in web codecs, the benefits of AV1, where the codec stands going into 2019, and demos of AV1 already available in browsers today as well as on different platforms and for different use cases.

- PRESENTERS:
    - Jai Krishnan, Product Manager, Google
    - Nathan Egge, Video Codec Engineer, Mozilla

AV1 Happy Hour - Sponsored by AOMedia, Bitmovin, Mozilla, and Google

When: Wednesday, Oct. 17, 2018, 5:00–6:00 p.m.

Where: Location CMIIP Theater

What: Attendees are invited to the AV1 Happy Hour. AOMedia members Bitmovin, Mozilla, and Google will deliver a quick general interest update on AV1 and refreshments will be served.

Track B: AV1: How the Codec of the Future is Being Used Today

When: Thursday, Oct. 18, 2018, 10:15 a.m.–11:15 a.m.

Where: Location: 3D09

What: The next-generation video codec AV1 was standardized in the Alliance for Open Media earlier this year, and it is coming to the web in a hurry. Content and encoding providers such as YouTube, Ateme, Harmonic, Netflix, and Bitmovin presented AV1 demos this fall, and both Chrome and Firefox have decode support in browser releases. Join panelists from Google, Mozilla, and Bitmovin to talk about why their companies have invested in AOM, the early gains shown by AV1, and what to expect in 2019 and beyond.

- MODERATOR:
    - John Luther, SVP, Technology, JW Player
- PANELISTS:
    - Nathan Egge, Video Codec Engineer, Mozilla
    - Steve Geiger, Director, Solutions, Bitmovin
    - Jai Krishnan, Product Manager, Google
    - Mike Coward, Director of Engineering, Facebook

Track C: AV1 Benchmarks, 6 Months Post-Launch

Launched in late March 2018 by the Alliance for Open Media (AOMedia), AV1 marks a significant shift in the balance of power in online entertainment. While HEVC and VP9 offer the best compatibility across devices and platforms, the newest standard promises a much higher compression efficiency and thereby also a better user experience. How does this new codec compare to the incumbents in compression and speed? This presentation will showcase the latest performance benchmarks and key encoding techniques that make AV1 an interesting choice for the future of your video delivery.

- PRESENTER
    - Steve Geiger, Director of Solutions, Bitmovin

To set up a press or analyst briefing with an AOMedia spokesperson, please email jhennion@virtualinc.com.

About the Alliance for Open Media

Launched in 2015, the Alliance for Open Media (AOMedia) was formed to define and develop media technologies to address marketplace demand for an open standard for video compression and delivery over the web. Founding Members include Amazon, Apple, Arm, Cisco, Facebook, Google, IBM, Intel, Microsoft, Mozilla, Netflix, and NVIDIA. AOMedia is a Joint Development Foundation Projects, LLC series. The Joint Development Foundation is an independent nonprofit organization that provides the corporate and legal infrastructure to enable groups to establish and operate standards and source code development collaborations.

Alliance for Open Media, 401 Edgewater Place, Suite 600, Wakefield, MA 01880

2.10 Projects

There are some projects that refer to the EE Dept., University of Texas at Arlington (UTA), Arlington, Texas, 76019, USA. For details please go to www.uta.edu/faculty/krrao/dip. Click on courses and then click on Graduate courses followed by EE5359 Multimedia Processing. Scroll down to theses and also to projects.

P.2.1 Deng et al., [E17] have added further extensions to H.264/AVC FRExt such as larger MV search range, larger macroblock, skipped block sizes and 1-D DDCT. They compared its performance with motion JPEG 2000 using high resolution (HR) (4096 × 2160) video sequences and showed significant improvement of the former in terms of PSNR at various bit rates. Implement the extended H.264/AVC and Motion JPEG 2000 and confirm that the former has a superior performance using HR test sequences. C. Deng et al., "Performance analysis, parameter selection and extension to H.264/AVC FRExt for high resolution video coding", J. VCIR, vol. 22, pp. 687–760, Feb. 2011.

P.2.2 Karczewicz et al., [E10] have proposed a hybrid video codec superior to H.264/AVC (Chapter 4) codec by adding additional features such as extended block sizes (up to 64 × 64), mode dependent directional transforms (MDDT) for intra coding, luma and chroma high precision filtering, adaptive coefficient scanning, extended block size partition, adaptive loop filtering, large size integer transform, etc. By using several test sequences at different spatial resolutions, they have shown that the new codec out performs the traditional H. 264/AVC codec in terms of both subjective quality and objective metrics. Also, this requires only moderate increase in complexity of both the encoder and decoder. Implement this new codec and obtain results similar to those described in this paper, consider SSIM [SSIM1] also as another metric in all the simulations. Use the latest JM software for H.264/AVC. M. Karczewicz et al., "A hybrid video coder based on extended macroblock sizes, improved interpolation, and flexible motion representation", IEEE Trans. CSVT, vol. 20, pp. 1698–1708, Dec. 2010.

P.2.3 Ma and Segall [E20] have developed a low-resolution (LR) decoder for HEVC. The objective here is to provide a low power decoder within a high resolution bit stream for handheld and mobile devices. This is facilitated by adopting hybrid frame buffer compression, LR intra prediction, cascaded motion compensation and in loop deblocking [E73], within the HEVC framework. Implement this low power HEVC decoder. Also port these tools in the HEVC reference model (HM9.0) [E56] and evaluate the performance. Z. Ma and A. Segall, "Low resolution decoding for high efficiency video coding", IASTED SIP 2011, pp. Dallas, TX, Dec. 2011.

P.2.4 Joshi, Reznik and Karczewicz [E8] have developed scaled integer transforms which are numerically stable, recursive in structure and are orthogonal. They have also embedded these transforms in H.265/JMKTA framework. Specifically develop the 16-point scaled transforms and implement in H.265 using JMKTA software. Develop 32 and 64 point scaled transforms. R. Joshi, Y.A. Reznik and M. Karczewicz, "Efficient large size transforms for high-performance video coding", Applications of Digital Image Processing XXXIII, Proc. of SPIE, vol. 7798, 77980W-1 through 77980W-7, 2010.

P.2.5 Please access S. Subbarayappa's thesis (2012) from EE 5359, MPL web site, "Implementation and Analysis of Directional Discrete Cosine Transform in Baseline Profile in H.264". Obtain the basis images for all the directional modes related to (4 × 4) and (8 × 8) DDCT[2]. Modes 4, 6, 7, and 8 can be obtained from modes 3 and 5 as shown in Figures 13–16 (project). See also [E112]. Use this approach for obtaining the basis images.

P.2.6 Please access the web site http://www.h265.net/ and go to analysis of coding tools in HEVC test model (HM 1.0) – intra prediction. It describes that up to 34 directional prediction modes for different PUs can be used in intra prediction of H.265. Implement these modes in HM 1.0 and evaluate the H.265 performance using TMuC HEVC software [E97]. (HM: HEVC test model, TMuC – Test Model under Consideration).

P.2.7 Using TMuC HEVC software [E95], implement HM1.0 considering various test sequences at different bit rates. Compare the performance of HEVC (h265.net) with H.264/AVC (use JM software) using SSIM [SSIM1], bit rates, PSNR, BD measures [E81, E82, E96, E198] and computational time as the metrics. Use WD 8.0 [E60].

P.2.8 In the document JCTVC-G399 r2, Li has compared the compression performance of HEVC WD4 with H.264/AVC high profile. Implement this comparison using HEVC WD8 and the latest JM software for H.264/AVC based on several test sequences at different bit rates. As before, SSIM [SSIM1], PSNR, bit rates, BD measures [E81, E82, E96, E198] and implementation complexity are the metrics. JCT-VC, 7th meeting, Geneva, CH, 21–30, Nov. 2011. (comparison of compression performance of HEVC working draft 4 with H.264/AVC High profile)

P.2.9 Please access J.S. Park and T. Ogunfunmi, "A new approach for image quality assessment", ICIEA 2012, Singapore, 18–20 July 2012. They have developed a subjective measure (similar to SSIM) for evaluating video quality based on (8 × 8) 2D-DCT. They suggest that it is much simpler to implement compared to SSIM [SSIM1] while in performance it is close to SSIM. Evaluate this based on

---

[2]Please access: http://www.h265.net/2009/9/mode-dependent-directional-transform-mddt-in-jmkta.html

various artefacts. Also consider (4 × 4) and (16 × 16) 2D-DCTs besides the (8 × 8). Can this concept be extended to integer DCTs? Can DCT be replaced by DST (discrete sine transform)?

P.2.10 Please access J. Dong and K.N. Ngan, "Adaptive pre-interpolation filter for high efficiency video coding", J. VCIR, vol. 22, pp. 697–703, Nov. 2011. Dong and Ngan [E16] have designed an adaptive pre-interpolation filter (APIF) followed by the normative interpolation filter [E69]. They have integrated the APIF into VCEG's reference software KTA 2.6 and have compared with the non-separable adaptive interpolation filter (AIF) and adaptive loop filter (ALF). Using various HD sequences, they have shown that APIF outperforms either AIF or ALF and is comparable to AIF+ALF and at much less complexity. Implement the APIF and confirm their conclusions.

P.2.11 Please access W. Ding et al., "Fast mode dependent directional transform via butterfly-style transform and integer lifting steps", J. VCIR, vol. 22, pp. 721–726, Nov. 2011 [E17]. They have developed a new design for fast MDDT through integer lifting steps. This scheme can significantly reduce the MDDCT complexity with negligible loss in coding performance. Develop the fast MDDT with integer lifting steps for (4 × 4) and (8 × 8) and compare its performance (see Figures 6–10) with the DCT and BSTM (butterfly style transform matrices) using video test sequences.

P.2.12 Please access B. Li, G.J. Sullivan and J. Xu, "Compression performance of high efficiency video coding (HEVC) working draft 4", IEEE ISCAS, pp. 886–889, Seoul, Korea, May 2012 [E22]. They have compared the performance of HEVC (WD4) with H.264/AVC (JM 18.0) using various test sequences. They have shown that WD4 provides a bit rate savings (for equal PSNR) of about 39% for random access applications, 44% for low-delay use and 25% for all intra use. Verify these tests.

P.2.13 Please access the paper E. Alshina, A. Alshin and F.C. Fernandez, "Rotational transform for image and video compression", IEEE ICIP, pp. 3689–3692, 2011. See also [BH2].

Alshina, Alshin, and Fernandez have applied ROT 4 to 4 × 4 blocks and ROT8 to upper left sub matrix in all other cases (see Figures 2 and 3 in the paper) and have shown a BD-rate gain of 2.5% on average for [E81, E82, E96, E198] all test sequences (see Table 4 in the paper). Implement this technique using the test sequences and confirm the results (ROT: rotational transform).

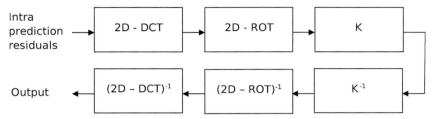

Figure 2.14 Block diagram for DCT/ROT applied to intra prediction residuals only.

P.2.14 Please access the document JCTVC-C108, Oct. 2010 submitted by Saxena and Fernandez, (Title: Jointly optimal prediction and adaptive primary transform). They have compared TMuC 0.7 between the proposed adaptive DCT/DST as the primary transform and the DCT in intra prediction for 16×16, 32×32 and 64×64 block sizes for two cases i.e., secondary transform (ROT) is off or on. Implement this scheme and verify the results shown in Tables 2 and 3 of this document. Use TMuC 0.7.

P.2.15 In Stockholm, Sweden JCT-VC meeting, adaptive DCT/DST has been dropped. Directional DCT [E112] (to the residuals of adaptive intra directional prediction) and also the rotational secondary transform (See P.2.13) are not considered. Only a transform derived from DST for 4×4 size luma intra prediction residuals and integer DCT for all other cases (both intra and inter) have been adopted. The DDCT and ROT (rotational transform) contribute very little to image quality but at the cost of significant increase in implementation complexity.

See the paper by A. Saxena and F.C. Fernandez, "On secondary transforms for prediction residuals", IEEE ICIP 2012, Orlando, FL, 2012 [E26]. They have implemented the HEVC using mode dependent DCT/DST to (4×4) sizes for both intra and inter prediction residuals. For all other cases, (i.e., both intra and inter block sizes other than 4×4), they have applied a secondary transform to the top left (low frequency) coefficients after the primary 2D-DCT. This has resulted in BD rate gains (see Tables 1–3) [E81, E82, E96, E198, E431] for various test sequences compared to the case where no secondary transform is implemented. Implement this scheme and show results similar to Tables 1–3.

P.2.16 Please access H. Zhang and Z. Ma, "Fast intra prediction for high efficiency video coding", Pacific Rim Conf.

on Multimedia, PCM 2012, Singapore, Dec. 2012 [E44], (http://cement.ntu.edu.sg/pcm2012/index.html)

Zhang and Ma [E44] (see also [E149] have proposed a novel intra prediction approach at the PU level and achieved a significant reduction in HEVC encoding time at the cost of negligible increase in bit rate and negligible loss in PSNR. Please implement this. They suggest that their source code is an open source and can be used for research purposes only. (http://vision.poly.edu/~zma03/opensrc/sourceHM6.zip)

P.2.17 Please see P.2.16. The authors also suggest that similar approaches by other researchers (see Section 2 of this paper) can be combined with their work to further decrease the encoding time. See also [E43] and the references at the end of this paper. Explore this.

P.2.18 Please see P.2.17. The authors Zhang and Ma [E44, E149] also plan to explore the possibility of reducing the complexity of inter prediction modes. Investigate this.

P.2.19 Please see P.2.16 through P.2.18. Combine both the complexity reduction techniques (intra/inter prediction modes) that can lead to practical HEVC encoders and evaluate the extent of complexity reduction in HEVC encoder with negligible loss in its compression performance.

Note that P.2.17 through P.2.19 are research-oriented projects leading to M.S. theses and Ph.D. dissertations.

P.2.20 Please access M. Zhang, C. Zhao and J. Xu, "An adaptive fast intra mode decision in HEVC", IEEE ICIP 2012, Orlando, FL, Sept.-Oct. 2012 [E43]. By utilizing the block's texture characteristics from rough mode decision and by further simplification of residual quad tree splitting process, their proposed method saves average encoding times 15% and 20% in the all intra high efficiency and all intra low complexity test conditions respectively with a marginal BD-rate increase [E81, E82, E96, E198]. Confirm these test results by implementing their approach.

P.2.21 See the paper by J. Nightingale, Q. Wang and C. Grecos, "HEVStream; A framework for streaming and evaluation of high efficiency video coding (HEVC) content in loss-prone networks', IEEE Trans. Consumer Electronics, vol. 59, pp. 404–412, May 2012 [E57]. They have designed and implemented a comprehensive streaming and evaluation framework for HEVC-encoded video streams and tested its performance under a varied range of network conditions. Using some of the recommended test conditions (See Table III) the effects of applying bandwidth, packet loss, and path latency constraints on the quality (PSNR) of received video streams are reported. Implement and verify these tests. Besides PSNR, use SSIM [SSIM1] and BD rates [E81, E82, E 96, E198] as benchmarks for comparison purposes.

P.2.22 See P.2.21. In terms of future work, the authors propose to focus on the development of suitable packet/NAL unit prioritization schemes for use in selective dropping schemes for HEVC. Explore this as further research followed by conclusions.

P.2.23 See the paper D. Marpe et al., "Improved video compression technology and the emerging high efficiency video coding standard", IEEE International Conf. on Consumer Electronics, pp. 52–56, Berlin, Germany, Sept. 2011 [E58]. The authors on behalf of Fraunhofer HHI have proposed a newly developed video coding scheme leading to about 30% bit rate savings compared to H.264/AVC HP (high profile) at the cost of significant increase in computational complexity. Several new features that contribute to the bit rate reduction have been explored. Implement this proposal and verify the bandwidth reduction. Explore the various techniques that were successfully used in reducing the complexity of H.264/AVC encoders. Hopefully these and other approaches can result in similar complexity reduction of HEVC encoders.

P.2.24 See the paper, M. Budagavi and V. Sze, "Unified forward + inverse transform architecture for HEVC', IEEE ICIP 2012, Orlando, FL, Sept.-Oct., 2012 [E35]. They take advantage of several symmetry properties of the HEVC core transform and show that the unified implementation (embedding multiple block size transforms, symmetry between forward and inverse transforms etc.) results in 43–45% less area than separate forward and inverse core transform implementations. They show the unified forward + inverse 4-point and 8-point transform architectures in Figures 2 and 3 respectively. Develop similar architectures for the unified forward + inverse 16-point and 32-point transforms. Note that this requires developing equations for the 16 and 32 point transforms similar to those described in equations 10–17 of this paper.

P.2.25 See P.2.24. The authors claim that the hardware sharing between the forward and inverse transforms has enabled an area reduction of over 40%. Verify this.

P.2.26 In the transcoding arena, several researchers have developed, designed, tested, and evaluated transcoders among H.264/AVC, AVS China, DIRAC, MPEG-2 and VC-1. Develop a transcoding system between H.264/AVC and HEVC (main profile) [E181]. Use HM9. [See E93].

P.2.27 Repeat P.2.26 for transcoding between MPEG-2 and HEVC (main profile), See [E98]

P.2.28 Repeat P.2.26 for transcoding between DIRAC and HEVC (main profile).

P.2.29 Repeat P.2.26 for transcoding between VC-1 and HEVC (main profile).

P.2.30 Repeat P.2.26 for transcoding between AVS China and HEVC (main profile).

P.2.31 As with H.264/AVC (Chapter 4), HEVC covers only video coding. To be practical and useful for the consumer, audio needs to be integrated with an HEVC-encoded video. Encode HEVC video along with audio coder such as AAC, HEAAC etc. following the multiplexing the coded bit streams at the transmitter. Demultiplexing the two bit streams, followed by decoding the audio and video while maintaining the lip sync is the role of the receiver. Implement these schemes for various video spatial resolutions and multiple channel audio. This comprises several research areas at M.S. and doctoral levels. Such integrated schemes have been implemented for H.264/AVC, DIRAC and AVS China video with audio coders.

P.2.32 Similar to H.264/AVC for high video quality required within the broadcast studios (not for transmission/distribution), HEVC intra frame coding only can be explored. Compare this (HEVC intra frame coding only) with H.264/AVC intra frame coding only and JPEG 2000 at various bit rates using different test sequences. Use MSE/PSNR/SSIM/BD rates [E81, E82, E96, E198] and implementation complexity as comparison metrics.

P.2.33 In [E62], Ohm et al., compared the coding efficiency of HEVC at different bit rates using various test sequences with the earlier standards such as H.262/MPEG-2 video, H.263, MPEG-4 Visual (part 2) and H.264/AVC using PSNR and subjective quality as the metrics. They also indicate that the software and test sequences for reproducing the selected results can be accessed from ftp://ftp.hhi.de/ieee-tcsvt/2012/ Repeat these tests and validate their results. Note that the DSIS used for measuring the subjective quality requires enormous test

2.10 Projects   67

facilities and subjects (novices and experts) and may be beyond the availability of many research labs.

P.2.34 Repeat P.2.33 using SSIM [SSIM1] and BD-rates [E81, E82, E96, E198, E431] as the performance metric and evaluate how these results compare with those based on PSNR.

P.2.35 Horowitz et al., [E66] compared the subjective quality (subjective viewing experiments carried out in double blind fashion) of HEVC (HM7.1) – main profile/low delay configuration – and H.264/AVC high profile (JM18.3) for low delay applications as in P.2.31 using various test sequences at different bit rates. To compliment these results, production quality H.264/AVC encoder known as x264 is compared with a production quality HEVC implementation from eBrisk Video (VideoLAN x264 software library, http://www.videolan.org/developers/x264.html version core 122 r2184, March 2012). They conclude that HEVC generally produced better subjective quality compared with H.264/AVC for low delay applications at approximately 50% average bit rate of the latter. Note that the x264 configuration setting details are available from the authors on request. Several papers related to subjective quality/tests are cited in [E46]. Repeat these tests using PSNR, BD rate [E81, E82, E96, E198] and SSIM [SSIM1] as the performance metrics and evaluate how these metrics can be related to the subjective quality.

P.2.36 Bossen et al., [E63] present a detailed and comprehensive coverage of HEVC complexity (both encoders and decoders) and compare with H.264/AVC high profile. They conclude for similar visual quality HEVC encoder is several times more complex than that of H.264/AVC. The payoff is HEVC accomplishes the same visual quality as that of H.264/AVC at half the bit rate required for H.264/AVC. The HEVC decoder complexity, on the other hand, is similar to that of H.264/AVC. They claim that hand held/mobile devices, lap tops, desk tops, tablets, etc. can decode and display the encoded video bit stream. Thus, real-time HEVC decoders are practical and feasible. Their optimized software decoder (no claims are made as to its optimality) does not rely on multiple threads and without any parallelization using ARM and X64 computer. Implement this software for several test sequences at different bit rates and explore additional avenues for further optimization. See also [E192].

P.2.37 One of the three profiles in HEVC listed in FDIS (Jan. 2013) is intra frame (image) coding only. Implement this coding mode in HEVC

68   HEVC Encoder

P.2.38 and compare with other image coding standards such as JPEG, JPEG2000, JPEG-LS, JPEG-XR and JPEG using MSE/PSNR, SSIM [SSIM1] and BD-rate [E81, E82, E96, E198, E431] as the metrics. As before, perform this comparison using various test sequences at different spatial resolutions and bit rates. See P.2.47.

P.2.38 Besides multiview/3D video [E34, E39], scalable video coding (temporal, spatial SNR-quality and hybrid) is one of the extensions/additions to HEVC [E325]. Scalable video coding (SVC) at present is limited to two layers (base layer and enhancement layer). SVC is one of the extensions in H.264/AVC and a special issue on this has been published [E69]. Software for SVC is available online http://ip.hhi.de/omagecom_GI/savce/downloads/SVC-Reference-software.htm [E70]. Design, develop and implement these three different scalabilities in HEVC.

P.2.39 Sze and Budagavi [E67] have proposed several techniques in implementing CABAC (major challenge in HEVC) resulting in higher throughput, higher processing speed and reduced hardware cost without affecting the high coding efficiency. Review these techniques in detail and confirm these gains.

P.2.40 In [E73] details of the deblocking filter in HEVC are explained clearly. They show that this filter has lower computational complexity and better parallelization on multi cores besides significant reduction in visual artefacts compared to the deblocking filter in H.264/AVC. They validate these conclusions by using test sequences based on three configurations; 1) All-intra, 2) Random access and 3) Low delay. Go through this paper and related references cited at the end and confirm these results by running the simulations.

In [E209], deblocking filter is implemented in Verilog HDL.

P.2.41 Lakshman, Schwartz, and Wiegand [E71] have developed a generalized interpolation framework using maximal-order interpolation with minimal support (MOMS) for estimating fractional pels in motion compensated prediction. Their technique shows improved performance compared to 6-tap and 12-tap filters [E109], especially for sequences with fine spatial details. This however may increase the complexity and latency. Develop parallel processing techniques to reduce the latency.

P.2.42 See P.2.41. Source code, complexity analysis and test results can be downloaded from

H. Lakshman et al., "CE3: Luma interpolation using MOMS", JCT-VC D056, Jan. 2011". http://phenix.int-evry.fr/jct/doc_end_user/documents/4_Deagu/wg11/CTVC-D056-v2.zip. See [E71]. Carry out this complexity analysis in detail.

P.2.43 Correa et al., [E84] have investigated the coding efficiency and computational complexity of HEVC encoders. Implement this analysis by considering the set of 16 different encoding configurations.

P.2.44 See P.2.43 Show that the low complexity encoding configurations achieve coding efficiency comparable to that of high complexity encoders as described in draft 8 [E60].

P.2.45 See P.2.43 Efficiency and complexity analysis explored by Correa et al., [E84] included the tools (non-square transform, adaptive loop filter, and LM luma) which have been subsequently removed in the HEVC draft standard [E60]. Carry out this analysis by dropping these three tools.

P.2.46 Schierl et al., in their paper "System layer integration of HEVC" [E83] suggest that the use of error concealment in HEVC should be carefully considered in implementation and is a topic for further research. Go through this paper in detail and explore various error resilience tools in HEVC. Please note that many error resilience tools of H.264/AVC such as FMO, ASO, redundant slices, data partitioning, and SP/SI pictures have been removed due to their very rare deployment in real-world applications.

P.2.47 Implement the lossless coding of HEVC main profile (Figure 2.13) proposed by Zhou et al., [E85] and validate their results. Also compare with current lossless coding methods such as JPEG-2000 etc. based on several test sequences at different resolutions and bit rates. Comparison metrics are PSNR/MSE, SSIM, BD-rates [E81, E82, E96, E198, E431], etc. Consider the implementation complexity also in the comparison.

Cai et al., [E112] have also compared the performance of HEVC, H.264/AVC, JPEG2000 and JPEG-LS for both lossy and lossless modes. For lossy mode, their comparison is based on $PSNR_{avg} = (6 \times PSNR_y + PSNR_u + PSNR_v)/8$ only. This is for 4:2:0 format and is by default. Extend this comparison based on SSIM, BD-rate [81, E82, E96, E198, E431] and implementation complexity. Include also JPEG-XR which is based on HD-Photo of Microsoft in this comparison. They have provided an extensive list of references related to performance comparison of intra coding of several standards. See also P.2.37. See also [E145].

P.2.48 See [E95]. An efficient transcoder for H.264/AVC to HEVC by using a modified MV reuse has been developed. This also includes complexity scalability trading off RD performance for complexity reduction. Implement this. Access references [4–7] related to transcoding overview papers cited at the end of [E95]. See also [E98], [E146], [E148] and [E333].

P.2.49 See P.2.48. The authors in [E95] suggest that more of the H.264/AVC information can be reused in the transcoder to further reduce the transcoder complexity as future work. Explore this in detail and see how the transcoder complexity can be further reduced. The developed techniques must be justified based on the comparison metrics (See P.2.47).

P.2.50 See P.2.48 and P.2.49. Several other transcoders can be developed i.e.,

A) Transcoder between MPEG-2 and HEVC (there are still many decoders based on MPEG-2). Please access [E98] T. Shanableh, E. Peixoto and E. Izquierdo, "MPEG-2 to HEVC video transcoding with content-based modeling", IEEE Trans. CSVT, vol. 23, pp. 1191–1196, July 2013. The authors have developed an efficient transcoder based on content-based machine learning. In the conclusions section, they have proposed future work. Explore this. In the abstract they state "Since this is the first work to report on MPEG-2 to HEVC video transcoding, the reported results can be used as a benchmark for future transcoding research". This is a challenging research in the transcoding arena.

B) Transcoder between AVS China and HEVC.

C) Transcoder between VC-1 and HEVC.

D) Transcoder between VP9 and HEVC.

Implement these transcoders. Note that these research projects are at the M.S. theses levels.

You can access the theses related to transcoders that have been implemented as M.S. theses from the web site http://www.uta.edu/faculty/krrao/dip, click on courses and then click on EE5359. Or access directly http://www.uta.edu/faculty/krrao/dip/Courses/EE5359/index.html

2.10 Projects   71

P.2.51 Please access [E72]. This paper describes low complexity-high performance video coding proposed to HEVC standardization effort during its early stages of development. Parts of this proposal have been adopted into TMuC. This proposal is called Tandberg, Ericsson and Nokia test model (TENTM). Implement this proposal and validate the results. TENTM proposal can be accessed from reference 5 cited at the end of this paper.

P.2.52 Reference 3 (also web site) cited in [E72] refers to video coding technology proposal by Samsung and BBC [Online]. Implement this proposal.

P.2.53 Reference 4 (also web site) cited in [E72] refers to video coding technology proposal by Fraunhoff HHI [Online]. Implement this proposal.

P.2.54 Please access [E106] M.S. Thesis by S. Gangavathi entitled, "Complexity reduction of H.264 using parallel programming" M.S. Thesis, EE Dept., University of Texas at Arlington, Arlington, Texas, Dec. 2012. http://www.uta.edu/faculty/krrao/dip click on courses and then click on EE5359 Scroll down and go to Thesis/Project Title and click on S. Gangavathi.

By using CUDA he has reduced the H.264 encoder complexity by 50% in the baseline profile. Extend this to Main and High profiles of H.264.

P.2.55 See P.2.54. Extend Gangavathi's approach to HEVC using several test sequences coded at different bit rates. Show the performance results in terms of encoder complexity reduction and evaluate this approach based on SSIM [SSIM1], BD-PSNR, BD- bit rates [E81, E82, E96, E198] and PSNR as the metrics.

UTA/EE5359 course web site: http://www-ee.uta.edu/Dip/Courses/EE5359/index.html

P.2.56 Zhang, Li and Li [E108] have developed a gradient-based fast decision algorithm for intra prediction in HEVC. This includes both prediction unit (PU) size and angular prediction modes. They claim a 56.7% savings of the encoding time in intra HE setting and up to 70.86% in intra low complexity setting compared to the HM software [E97]. Implement this and validate their results.

P.2.57 Please see P.2.56 In the section Conclusion the authors suggest future work on how to obtain the precise coding unit partition for the complex texture picture combined with RDO technique used in HEVC. Explore this.

P.2.58 Wang et al., [E110] present a study of multiple sign bit hiding scheme adopted in HEVC. This technique addresses the joint design of quantization transform coefficient coding using the data hiding approach. They also show that this method consistently improves the rate-distortion performance for all standard test images resulting in overall coding gain in HEVC. In terms of future work, they suggest that additional gains can be expected by applying the data hiding technique to other syntax elements. Explore this.

P.2.59 Please see P.2.58. The authors comment that the general problem of joint quantization and entropy coding design remains open. Explore this.

P.2.60 Lv et al., [E116] have developed a fast and efficient method for accelerating the quarter-pel interpolation for ME/MC using SIMD instructions on ARM processor. They claim that this is five times faster than that based on the HEVC reference software HM 5.2. See Section V acceleration results for details. Using NEON technology verify their results.

P.2.61 Shi, Sun and Wu [E76] have developed an efficient spatially scalable video coding (SSVC) for HEVC. Using two layer inter prediction schemes. Using some test sequences they demonstrate the superiority of their technique compared with other SSVC schemes. Implement this and validate their results. In the conclusion section, they suggest future work to further improve the performance of their scheme. Explore this in detail. Review the papers related to SVC listed at the end in references. See H. Schwarz, D. Marpe, and T. Wiegand, "Overview of the Scalable Video Coding Extension of the H.264/AVC Standard", *IEEE Trans. on Circuits and Systems for Video Technology,* vol. 17, pp. 1103–1120, Sept. 2007. This is a special issue on SVC. There are many other papers on SVC.

P.2.62 Zhou et al., [E85, E123] have implemented HEVC lossless coding for main profile by simply bypassing transform, quantization and in-loop filters and compared with other lossless coding methods such as JPEG-2000, ZIP, 7-Zip, WinRAR etc. Implement this and also compare with JPEG, JPEG-XR, PNG, etc. Consider implementation complexity as another metric.

P.2.63 References [E126–E129, E158–E160, E325] among others address scalable video coding extensions to HEVC. Review these and implement spatial/quality (SNR)/temporal scalabilities. See also P.2.61.

2.10 Projects   73

P.2.64 Please access [E66]. In this paper Horowitz et al., demonstrate that HEVC yields similar subjective quality at half the bit rate of H.264/AVC using both HM 7.1 and JM 18.3 softwares. Similar conclusions are also made using eBrisk and x264 softwares. Using the latest HM software, conduct similar tests on the video test sequences and confirm these results. Consider implementation complexity as another comparison metric.

P.2.65 Kim et al., [E134] developed a fast intra-mode (Figures 2.8 and 2.9) decision based on the difference between the minimum and second minimum SATD-based (sum of absolute transform differences) RD cost estimation and a fast CU-size (Figure 7) decision based on RD cost of the best intra mode. Other details are described in this paper. Based on the simulations conducted on class A and class B test sequences (Table 2.1), they claim that their proposed method achieves an average time reduction (ATR) of 49.04% in luma intra prediction and an ATR of 32.74% in total encoding time compared to the HM 2.1 encoder. Implement their method and confirm these results. Use the latest HM software [E56]. Extend the luma intra prediction to chroma components also.

P.2.66 Flynn, Martyin-Cocher, and He [E135] proposed to JCT-VC best effort decoding a 10 bit video bit stream using an 8-bit decoder. Simulations using several test sequences based on two techniques a) 8-bit decoding by adjusting inverse transform scaling and b) hybrid 8-bit–10-bit decoding using rounding in picture construction process were carried out. HM-10 low-B main-10 sequences (F. Bossen, "Common HM test conditions and software reference configurations", JCT-VC-L1100, JCT-VC, Jan. 2013, E178) were decoded and the PSNR measured against the original input sequences. PSNR losses averaged 6 dB and 2.5 dB respectively for the two techniques compared against the PSNR of the normal (10 bit) decoder. Implement these techniques and confirm their results. Explore other techniques that can reduce the PSNR loss.

P.2.67 Pan et al., [E140] have developed two early terminations for TZSearch algorithm in HEVC motion estimation and have shown that these early terminations can achieve almost 39% encoding saving time with negligible loss in RD performance using several test sequences. Several references related to early terminations are listed at the end of this paper. Review these references and simulate the techniques proposed by Pan et al., and validate their conclusions. Extend these simulations using HDTV and ultra HDTV test sequences.

P.2.68 The joint call for proposals for scalable video coding extension of HEVC was issued in July 2012 and the standardization started in Oct. 2012 (See E76 and E126 through E129, E158–E160, E325). Some details on scalable video codec design based on multi-loop and single-loop architectures are provided in [E141]. The authors in [E141] have developed a multi-loop scalable video coder for HEVC that provides a good complexity and coding efficiency trade off. Review this paper and simulate the multi-loop scalable video codec. Can any further design improvements be done on their codec? Please access the web sites below.

JSVM-joint scalable video model reference software for scalable video coding. (online) http://ip.hhi.de/imagecom_GI/savce/downloads/SVC-Reference-software.htm

JSVM9 Joint scalable video model 9 http://ip.hhi.de/imagecom_GI/savce/dowloads

See H. Schwarz, D. Marpe, and T. Wiegand, "Overview of the Scalable Video Coding Extension of the H.264/AVC Standard", *IEEE Trans. on Circuits and Systems for Video Technology,* vol. 17, pp. 1103–1120, Sept. 2007. This is a special issue on SVC. There are many other papers on SVC.

P.2.69 Tohidpour, Pourazad, and Nasiopoulos [E142] proposed an early-termination interlayer motion prediction mode search in HEVC/SVC (quality/fidelity scalability) and demonstrate complexity reduction up to 85.77% with at most 3.51% bit rate increase (almost same PSNR). Simulate this approach. Can this be combined with the technique developed in [E141] – See P.2.68.

P.2.70 In [E143], Tan, Yeo, and Li proposed a new lossless coding scheme for HEVC and also suggest how it can be incorporated into the HEVC coding framework. In [E83], Zhou et al., have implemented a HEVC lossless coding scheme. Compare these two approaches and evaluate their performances in terms of complexity, SSIM, BD-rates [E81, E82, E96, E198, E431] and PSNR for different video sequences at various bit rates. See also P.2.62.

P.2.71 See the paper Y. Pan, D. Zhou, and S. Goto, "An FPGA-based 4K UHDTV H.264/AVC decoder", IEEE ICME, San Jose, CA, July

2.10 Projects 75

2013. Can a similar HEVC decoder be developed? (Sony already has a 4K UHDTV receiver on the market).

P.2.72 In [E13], Asai et al., proposed a new video coding scheme optimized for high resolution video sources. This scheme follows the traditional block based MC+DCT hybrid coding approach similar to H.264/AVC, HEVC and other ITU-T/VCEG and ISO/IEC MPEG standards. By introducing various technical optimizations in each functional block, they achieve roughly 26% bit-rate savings on average compared to H.264/AVC high profile at the same PSNR. They also suggest improved measures for complexity comparison. Go through this paper in detail and implement the codec. Consider BD-PSNR and BD-Rate [E81, E82, E96, E198, E431] and SSIM [SSIM1] as the comparison metrics. For evaluating the implementation complexity, consider both encoders and decoders. Can this video coding scheme be further improved? Explore all options.

P.2.73 See P.2.72 By bypassing some functional blocks such as transform/quantization/in loop deblocking filter, Zhou et al., [E85, E123] have implemented a HEVC lossless coding scheme (Figure 2.13). Can a similar lossless coding scheme be implemented in the codec proposed by Asai et al., [E13]. If so, compare its performance with other lossless coding schemes. See P.2.62.

P.2.74 In [E145], the authors propose HEVC intra coding acceleration based on tree level mode correlation. They claim a reduction by up to 37.85% in encoding processing time compared with HM4.0 intra prediction algorithm with negligible BD-PSNR loss [E81, E82, E96, E198, E431]. Implement their algorithm and confirm the results using various test sequences. Use the latest HM instead of HM 4.0.

P.2.75 See P.2.48 In [E146], Peixoto et al., have developed a H.264/AVC to HEVC video transcoder which uses machine learning techniques to map H.264/AVC macroblocks into HEVC coding units. Implement this transcoder. In the conclusions, the authors suggest ways and methods by which the transcoder performance can be improved (future work). Explore these options. See [E263], where different methods such as mode mapping, machine learning, complexity scalable and background modelling are discussed and applied to H.264/AVC to HEVC transcoder. (Also see [E267])

P.2.76 Lainema et al., [E78] give a detailed description of intra coding of the HEVC standard developed by the JCT-VC. Based on different test sequences, they demonstrate significant improvements in both subjective and objective metrics over H.264/AVC and also carry out a complexity analysis of the decoder. In the conclusions, they state "Potential future work in the area includes e.g., extending and tuning the tools for multiview/scalable coding, higher dynamic range operation and 4:4:4 sampling formats". Investigate this future work thoroughly.

P.2.77 Please access the paper [E175], Y. Tew and K.S. Wong, "An overview of information hiding in H.264/AVC compressed video", IEEE Trans. CSVT, vol. 24, pp. 305–319, Feb. 2014. This is an excellent review paper on information hiding especially in H.264/AVC compressed domain. Consider implementing information hiding in HEVC (H.265) compressed domain.

Abstract is reproduced below:

Abstract—Information hiding refers to the process of inserting information into a host to serve specific purpose(s). In this article, information hiding methods in the H.264/AVC compressed video domain are surveyed. First, the general framework of information hiding is conceptualized by relating state of an entity to a meaning (i.e., sequences of bits). This concept is illustrated by using various data representation schemes such as bit plane replacement, spread spectrum, histogram manipulation, divisibility, mapping rules and matrix encoding. Venues at which information hiding takes place are then identified, including prediction process, transformation, quantization and entropy coding. Related information hiding methods at each venue are briefly reviewed, along with the presentation of the targeted applications, appropriate diagrams and references. A timeline diagram is constructed to chronologically summarize the invention of information hiding methods in the compressed still image and video domains since year 1992. Comparison among the considered information hiding methods is also conducted in terms of venue, payload, bitstream size overhead, video quality, computational complexity, and video criteria. Further perspectives and recommendations are presented to provide a better understanding on the current trend of information hiding and to identify new opportunities for information hiding in compressed video.

Implement similar approaches in HEVC – various profiles. The authors suggest this, among others, as future work in VII Recommendation and further research direction and VIII Conclusion sections. These two sections can lead to several projects/theses.

P.2.78 Please access X.-F. Wang and D.-B. Zhao, "Performance comparison of AVS and H.264/AVC video coding standards", J. Comput. Sci. & Tech., vol. 21, pp. 310–314, May 2006. Implement similar performance comparison between AVS China and HEVC (various profiles).

P.2.79 See [E152] about combining template matching prediction and block motion compensation. Go through several papers on template matching listed in the references at the end of this paper and investigate/evaluate thoroughly effects of template matching in video coding.

P.2.80 See P.2.79 in [E152], the authors have developed the inter frame prediction technique combining template matching prediction and block motion compensation for high efficiency video coding. This technique results in 1.7–2.0% BD-rate [E81, E82, E96, E198, E431] reduction at a cost of 26% and 39% increase in encoding and decoding times respectively based on HM-6.0. Confirm these results using the latest HM software. In the conclusions the authors state "These open issues need further investigation". Explore these.

P.2.81 In [E120] detailed analysis of decoder side motion vector derivation (DMVD) and its inclusion in call for proposals for HEVC is thoroughly presented. See also S. Kamp, *Decoder-Side Motion Vector Derivation for Hybrid Video Coding*, (Aachen Series on Multimedia and Communications Engineering Series). Aachen, Germany: Shaker-Verlag, Dec. 2011, no. 9. The DMVD results in very moderate bit rate reduction, however, offset by increase in decoder side computational resources. Investigate this thoroughly and confirm the conclusions in [E120].

P.2.82 Implement, evaluate and compare Daala codec with HEVC. Daala is the collaboration between Mozilla foundation and Xiph.org foundation. It is an open source codec. Access details on Daala codec from Google. "The goal of the DAALA project is to provide a free to implement, use and distribute digital media format and reference implementation with technical performance superior to H.265".

P.2.83 Please access the thesis 'Complexity Reduction for HEVC intraframe Luma mode decision using image statistics and neutral networks', by D.P. Kumar from UTA–MPL web site EE5359. Extend this approach to HEVC interframe coding using neural networks. Kumar has kindly agreed to help any way he can on this. This extension is actually a M.S. thesis.

P.2.84 Please see P.2.83. Combine both interframe and intraframe HEVC coding to achieve complexity reduction using neural networks.

P.2.85 HEVC range extensions include screen content (text, graphics, icons, logos, lines etc.) coding [E322]. Combination of natural video and screen content has gained importance in view of applications such as wireless displays, automotive infotainment, remote desktop, distance education, cloud computing, video walls in control rooms etc. These papers specifically develop techniques that address screen content coding within the HEVC framework. Review and implement these techniques and also explore future work. IEEE Journal on Emerging and Selected Topics in Circuits and Systems (JETCAS) has called for papers for the special Issue on Screen Content Video Coding and Applications. Many topics related to screen content coding (SCC) and display stream compression (see P.2.85a) are suggested. These can lead to several projects and theses. This special issue was published in Dec. 2016 [SCC15].

P.2.85a The Video Electronics Standards Association also recently completed a Display Stream Compression (DSC) standard for next-generation mobile or TV/Computer display interfaces. The development of these standards introduced many new ideas, which are expected to inspire more future innovations and benefit the varied usage of screen content coding. Review this standard in detail and implement DSC for next-generation mobile or TV/Computer display interfaces.

P.2.86 Access the M.S Thesis (2011) by P. Ramolia, "Low complexity AVS-M using machine learning algorithm C4.5", from UTA web site. Explore and implement similar complexity reduction in HEVC – all profiles – using machine learning.

P.2.87 The papers cited in [E161-E163] deal with DSP based implementation of HEVC9.0 decoder. They state that with an optimization process HEVC decoders for SD resolution can be achieved with a single DSP. Implement this.

P.2.88 See P.2.87. The authors state that for HD formats multi-DSP technology is needed. Explore this.

P.2.89 In [E168], the authors state that the motion estimation (ME) for ultra-high definition (UHD) HEVC requires 77–81% of computation time in the encoder. They developed a fast and low complexity ME algorithm and architecture which uses less than 1% of the number of operations compared to full search algorithm at negligible loss

2.10 Projects   79

in BD bit rate [E81, E82, E96, E198, E431]. They claim that the proposed ME algorithm has the smallest hardware size and the lowest power consumption among existing ME engines for HEVC. Review this paper along with the references (listed at the end) related to ME algorithms. Implement their algorithm and verify the results and conclusions based on various UHD test sequences. Consider also BD-PSNR as another performance metric. Review the paper, M. Jakubowski and G. Pastuzak, "Block based motion estimation algorithms – a survey", Opto-Electronics review, vol. 21, no. 1, pp. 86–102, 2013.

P.2.90 See P.2.89 Implement the fast and low complexity ME algorithm and architecture in HEVC encoder using super hi-vision test sequences. See [E164].

P.2.91 In Section 2.9 Summary some references related to reducing the implementation complexity of HEVC encoder (includes both intra and inter predictions) are cited. Review these and other related papers and develop a technique that can significantly reduce HEVC encoder complexity with negligible loss in BD-PSNR and BD bit rates [E81, E82, E96, E198, E431]. Demonstrate this based on various test sequences at different bit rates and resolutions.

P.2.92 Fast early zero block detection is developed for HEVC resulting in substantial savings in encoder implementation. Go through [E166, E171] and related papers and implement this technique in HEVC encoder for all profiles and levels. Compare the complexity reduction with HM software using BD bit rates, BDPSNR, [E81, E82, E96, E198, E431]. SSIM [SSIM1] and computational time as performance metrics.

P.2.93 Nguyen et al., [E173] present a detailed overview of transform coding techniques in HEVC focusing on residual quadtree structure and entropy coding structure. Using the standard test sequences and test conditions they show the resulting improved coding efficiency. Implement their techniques and confirm the results (Tables I – VI). Use the latest HM. If possible try also 4k × 2k test sequences. This project is highly complex and can be subdivided into various sections.

P.2.94 See P.2.85. Naccari et al., [E157] have specifically developed residual DPCM (RDPCM) for lossless screen content coding in HEVC and demonstrated that their algorithm achieves up to 8% average bit rate reduction with negligible increase in decoder complexity using the screen content coding set. Confirm their findings by implementing the RDPCM.

P.2.95 See P.2.94 The authors state that future work involves extension of the inter RDPCM tool for the lossy coding at very high bit rates leading to visually lossless coding. Explore this in detail.

P.2.96 Lv et al., [E109] carried out a detailed analysis of fractional pel interpolation filters both in H.264/AVC and in HEVC. They also conducted complexity analysis of these filters. By comparing the contribution of these filters to the compression performance they conclude that the interpolation filters contribute 10% performance gain to HEVC [E45] compared to H.264/AVC at the cost of increased complexity. They also demonstrate that the frequency responses of quarter pel filters in HEVC are superior to those in H.264/AVC. Their conclusions are based on reference software HM 5.2 using low resolution (416 × 240) test sequences. Carry out this analysis using HM 15.0 software and high resolution test sequences.

P.2.97 See P.2.96. Go through the analysis of how DCTIF (discrete cosine transform interpolation filters) have been developed [E109]. By plotting the frequency responses of the filters in H.264/AVC and in HEVC [E45], they show that in the passband the filters in HEVC are much flatter and have much smaller ripples than those in H.264/AVC (See Figure 5 in [E107]). Confirm these conclusions.

P.2.98 Na and Kim [E176] developed a detailed analysis of no-reference PSNR estimation affected by deblocking filtering in H.264/AVC bit streams. Go through this paper in detail and develop similar analysis for HEVC. Note that in HEVC there are two in-loop filters, deblocking, and SAO filters.

P.2.99 Grois et al., [E 177] conducted a detailed performance comparison of H.265/MPEG-HEVC, VP9 and H.264/MPEG-AVC encoders. Using various test sequences and common HM test conditions [E176]. They conclude that H.265/MPEG-HEVC provides average bit-rate savings of 43.3% and 39.35% relative to VP9 and H.264/MPEG-AVC encoders respectively. However H.265/MPEG-HEVC reference software requires 7.35 times that of VP9 encoding time. On the other hand VP9 requires 132.6 times that of x264 encoding time. These results are based on equal $PSNR_{YUV}$. Implement these performance comparison tests using the HM15.0 software. See the references cited at the end of [E177]. Please see P.2.33, P.2.36 and P.2.64.

2.10 Projects

P.2.100 See P.2.99. Review the Tables IV and VI in [E 177]. Develop bit-rate savings between VP9 and x264 (Table IV) for all the test sequences. Develop encoding run times between H.265/MPEG-HEVC and x264 (Table VI) for all the test sequences. Both these parameters need to be based on equal $PSNR_{YUV}$.

P.2.101 Abdelazim, Masri, and Noaman [E179] have developed ME optimization tools for HEVC. With these tools they are able to reduce the encoder complexity up to 18% compared to the HM software at the same PSNR and bit rate using various test sequences. Implement their scheme and confirm the results.

P.2.102 Ohm et al., [E62] have compared the HEVC coding efficiency with other video coding standards such as H.264/MPEG-4 AVC, H.264/MPEG-4 visual (part 2), H.262/MPEG-2 part 2 video [E181] and H.263. This comparison is based on several standard test sequences at various bit rates. The comparison metrics include PSNR and mean opinion score (MOS) [E180] based on double stimulus impairment scale (DSIS) both for interactive and entertainment applications. The comparison, however, does not include implementation complexity. Implement this metric for all the video coding standards and draw the conclusions.

P.2.103 See P.2.102 Extend the coding efficiency comparison between HEVC and DIRAC. Software and data for reproducing selected results can be found at ftp://ftp.hhi.de/ieee-tcsvt/2012

P.2.104 See P.2.102 Extend the coding efficiency comparison between HEVC and VC-1.

P.2.105 See P.2.102 Extend the coding efficiency comparison between HEVC and AVS China.

P.2.106 See P.2.102 Extend the coding efficiency comparison between HEVC and VP9.

P.2.107 See [E188]. Rerabek and Ebrahimi have compared the compression efficiency between HEVC and VP9 based on image quality. Implement this comparison based on BD-PSNR, BD-bit rate [E81, E82, E96, E198, E431], implementation complexity, etc., using the UHDTV sequences.

P.2.108 See P.2.107. The authors [E188], in the conclusion, suggest extending this comparison to Internet streaming scenarios. Implement this.

P.2.109 Grois et al., [E191] have made a comparative assessment of HEVC/H.264/VP9 encoders for low delay applications using

1280 × 720 60 fps test sequences. Implement this. Extend this comparison to UHDTV sequences. Consider implementation complexity as another metric.

P.2.110 Heindel, Wige, and Kaup [E193] have proposed lossy to lossless scalable video coding. This is based on lossy base layer (BL) HEVC encoder and lossless enhancement layer (EL) encoder. They claim that this approach has advantages over the scalable extension of the HEVC. Implement this technique and confirm their results. They have suggested future work. Explore this.

P.2.111 See P.2.110. Heindel, Wige, and Kaup [E193] have investigated different prediction schemes for the EL beyond [E192] and compared the performance with single layer JPEG-LS (lossless compression) (See references 8 and 9 in [E193]). Implement this.

P.2.112 Bross et al., [E192] have demonstrated that real time HEVC decoding of 4K (3840 × 2160) video sequences is feasible in current desktop CPUs using four CPU cores. Implement this using the EBU HD-1 TEST SET (available http://tech.ebu.ch/test sequences/uhd-1).

P.2.113, Vanne, Viitanen, and Hamalainen [E195] have optimized inter prediction mode decision schemes for HEVC encoder and have reduced the encoder complexity (compared to HM 8.0) by 31%–51% at a cost of 0.2%–1.3% for random access (similar ranges for low delay). They have used the test sequences (Table VI) specified in F. Bossen, "Common test conditions and software reference configurations", Document JCTVC-J1100, Incheon, S. Korea, April 2013. Their mode decision schemes are described in Section V – proposed mode decision schemes. Implement their complexity reduction techniques in the HEVC encoder.

P.2.113 See P.2.112 Extend this to ultra HDTV (4K resolution) and to super-hi vision (8K resolution) [E164] test sequences.

P.2.114 See P.2.113. In Section VII Conclusion the authors propose to combine the proposed techniques with the other existing approaches (see the references at the end) without compromising the RD performance or losing the ability for parallelization and hardware acceleration. Explore this.

P.2.115 Several lossless image compression techniques have been developed-HEVC, H.264, JPEG, JPEG-LS, JPEG 2000, 7-Zip, WinRAR, etc. [E85, E114, E147, E194, E196]. Compare these techniques in terms of implementation complexity using several test images at different resolutions.

P.2.116 Several techniques [E164, E171, E186] for detection of zero blocks in HEVC have been investigated with a view to reduce the encoder complexity. Review these in detail and also the techniques developed for H.264 – Chapter 4 – (See the references at the end of these papers) and develop an optimal approach for zero block detection in HEVC. The optimal approach needs to be justified in terms of encoder complexity and comparison metrics such as BD-PSNR, BD-bitrate [E81, E82, E96, E198, E431] and visual quality using several test sequences.

P.2.117 Chapter 6, Budagavi, Fuldseth, and Bjontegaard, "HEVC transform and quantization", in [E202] describes the 4×4, 8×8, 16×16 and 32×32 integer DCTs including the embedding process (small size transforms are embedded in large size transforms) Review the references listed at the end of this chapter. The inverse integer DCTs are transposes of the corresponding integer DCTs. Compute the orthogonal property of these integer DCTs. Hint; Matrix multiply the integer DCTS with their corresponding transposes.

See J. Sole et al., "Transform coefficient coding in HEVC", IEEE Trans. CSVT, vol. 22, pp. 1765–1777, Dec. 2012. See also M. Budagavi et al., "Core transform design in the high efficient video coding standard (HEVC)", IEEE J. of selected topics in signal processing, vol. 7, pp. 1029–1041, Dec. 2013. The philosophy in designing these integer DCTs (core transform) specially meeting some properties can be observed in this chapter. Review this and justify why alternative proposals to the core transform design were not selected by the JCT-VC.

P.2.118 A fast inter-mode decision algorithm for HEVC is proposed by Shen, Zhang, and Li [E199]. This algorithm is based on jointly using the inter-level correlation of quad-tree structure and spatiotemporal correlation. Using several test sequences, they show that the proposed algorithm can save 49% to 52%computational complexity on average with negligible loss of coding efficiency (see Tables X through XIII and Figures 5 through 8). They also show the superiority of their algorithm compared to other algorithms (see references at the end) in terms of coding time saving (Figures 9a and 10a) with negligible BDBR increase (see Figures 9b and 10b). Implement this algorithm and validate their results. Explore this algorithm for further improvements in reducing the encoder complexity.

P.2.119 Peng and Cosman [E200] have developed a weighted boundary matching error concealment (EC) technique for HEVC and have demonstrated its superiority over other EC methods (see references 4–6 cited at the end). For simulation two video sequences BQMall and Drill (832 × 480) at 50 fps are used. The results are based on QP 28 using HM 11. Consider various other test sequences including higher resolutions (HDTV and ultra HDTV), using more QPs and HM15. Show the results in terms of PSNR versus error rate (Figure 3) and the frames (original, loss pattern and error concealed (Figure 4)).

P.2.120 Warrier [E203] has implemented encode/decode, multiplex/demulti plex HEVC video/AAC audio while maintaining lip synch. She used main profile/intra prediction in HEVC. Extend this to inter prediction and also to other profiles.

P.2.121 Y. Umezaki and S. Goto [E206] have developed two techniques – partial decoding and tiled encoding – for region of interest (ROI) streaming in HEVC. They have compared these two methods using HM10.0 in terms of decoding cost, bandwidth efficiency and video quality. Regarding future work, they suggest combining tiled encoding and partial decoding to further reduce the decoding cost. Explore this future work and evaluate the reduction in the decoding cost. Use the latest HEVC reference software HM 15.0.

P.2.122 Mehta [E204] has effectively introduced parallel optimization of mode decision for intra prediction in HEVC and was able to reduce the encoding time by 35% to 40% on average with negligible loss in image quality. In future work, he has suggested that there are many other effective techniques to implement parallelism to different sections of the HM software leading to further reduction in the encoding time. Review the thesis and future work in detail and implement these techniques.

P.2.123 See [E208] and [E207] Scalable HEVC (SHVC) test model 6 is described in [E206]. Overview of the scalable extensions of the H.265/HEVC video coding standard is described in [E206]. Implement the SHVC using test sequences for scalabilities; Temporal, SNR, bit-depth, spatial, interlaced-to-progressive, colour gamut, hybrid and their combinations. Note that each scalability, by itself, is a project. Compare the enhancement layer output with direct encoder/decoder (no scalability) output based on the standard metrics. See the papers below:

H. Schwarz, D. Marpe, and T. Wiegand, "Overview of Scalable Video Coding Extension of the H.264/AVC Standard", IEEE Trans. on CSVT, vol. 17, no. 9, pp. 1051–1215, Sept. 2007.

H. Schwarz and T. Wiegand, "The Scalable Video Coding Amendment of the H.264/AVC Standard", csdn.net, world pharos, blog. [online]. Available: http://blog.csdn.net/worldpharos/article/details/3369933 (accessed on June 20th 2014).

Also access Karuna Gubbi S.S. Sastri, "Efficient intra-mode decision for spatial scalable extension of HEVC", M.S. Thesis, EE Dept., University of Texas at Arlington, Arlington, Texas, Aug. 2014. http://www.uta.edu/faculty/krrao/dip click on courses and then click on EE5359 Scroll down and go to Thesis/Project Title and click on Karuna Gubbi S.S. Sastri

P.2.124 Whereas Mehta [E204] was able to reduce the encoding time in intra frame coding, Dubhashi [E205] was able to reduce the time taken for the motion estimation process in inter frame coding in HEVC. Combine these two techniques to reduce the encoding time taken by the HEVC encoder. Consider various test sequences at different spatial and temporal resolutions.

P.2.125 Projects related to ATSC3.0 (ADVANCED TELEVISION SYSTEMS COMMITTEE) www.atsc.org

ATSC (ESTABLISHED IN 1982) is a consortium of broadcasters, vendors and trade groups. ATSC 3.0

The Advanced Television Systems Committee is an international, non-profit organization developing voluntary standards for digital television. The ATSC member organizations represent the broadcast, broadcast equipment, motion picture, consumer electronics, computer, cable, satellite, and semiconductor industries. For more information, visit www.atsc.org

The overall intent is to enable seamless transmission of HD, 4K video, 22.2 kHz audio and other data streams to fixed, mobile, and handheld devices in all types of terrains. Industry, research institutes, and academia have submitted proposals and are submitting proposals for different layers. Hence, this is a fertile ground for R & D. Access these proposals and explore possible research topics. (See M.S. Richter et al., "The ATSC digital television system", Proc. IEEE, vol. 94, pp. 37–43, Jan. 2006. Special issue on global digital television technology, Proc. IEEE, vol. 94, Jan. 2006. In future, this may be extended to 8K video.

See J.M. Boyce, "The U.S. digital television broadcasting transition", IEEE SP Magazine, vol. 29, pp. 108–112, May 2012.
ATSC APPROVES MOBILE & HANDHELD 20 July 2009
CANDIDATE STANDARD
ATSC DTV Moves into High Gear with Mobile and Handheld Specifications WASHINGTON, December 1 – The Advanced Television Systems Committee, Inc. (ATSC) has elevated its specification for Mobile Digital Television to Candidate Standard status. The new Mobile DTV Candidate Standard provides the technical capabilities necessary for broadcasters' to provide new services to mobile and handheld devices using their digital television transmissions. ATSC Mobile DTV includes a highly robust transmission system based on vestigial sideband (VSB) modulation coupled with a flexible and extensible IP based transport, efficient MPEG AVC (H.264) video and HE AAC v2 audio (ISO/IEC 14496–3) coding. (HE-AAC High efficiency advanced audio coder).

It seems ATSC is exploring HEVC for these services. This can lead to number of R&D projects.

P.2.126 In [E160], it is stated that ATSC also is conducting using reference software and experimental evaluation methodology for the 3D-TV terrestrial broadcasting. Review [E160, E211] and implement the real-time delivery of 3D-TV terrestrial broadcasting. Access also www.atsc.org.

Please do extensive literature survey on ATSC before writing the thesis proposal.
Please access documents related to ATSC 3.0:

> Report on ATSC 3.0:
> Link: http://www.atsc.org/cms/pdf/pt2/PT2-046r11-Final-Report-on-NGBT.pdf
> Presentation slides on ATSC 3.0:
> Link: https://mentor.ieee.org/802.18/dcn/12/18-12-0011-00-0000-nab-presentation-on-atsc-3.pdf
> List of ATSC standards and an overview of ATSC:
> Link: http://www.atsc.org/cms/pdf/ATSC2013_PDF.pdf
> Link to ATSC standards
> Link: http://www.atsc.org/cms/index.php/standards/standards?layout=default

P.2.128 Wang, Zhou, and Goto [E212] have proposed a motion compensation architecture that can support real-time decoding 7680x4320@30fps at 280 MHz. Develop a similar architecture for ME/MC that can support real time encoder for ultra HDTV. See also M. Tikekar et al., "Decoder architecture for HEVC", Chapter 10 in [E202].

P.2.129 VESA/DSC Please see below:

The Video Electronics Standards Association (VESA) also recently completed a Display Stream Compression (DSC) standard for next-generation mobile or TV/Computer display interfaces. The development of these standards introduced many new ideas, which are expected to inspire more future innovations and benefit the varied usage of screen content coding

In contrast with other image or video compression standards, the proposed Display Stream Compression Standard targets a relatively low compression ratio and emphasizes visually-lossless performance, high data throughput, low latency, low complexity, and includes special considerations geared for future display architectures.

VESA's DSC standard version 1.0 enables up to 66% data rate reduction, extending battery life in mobile systems and laptops, while simplifying the electrical interface requirements for future 4K and 8K displays. The standard enables a single codec for system chips that have multiple interfaces

As display resolutions continue to increase, the data rate across the video electrical interface has also increased. Higher display refresh rates and colour depths push rates up even further. For example, a 4K display at 60 frames per second with a 30 bit colour depth requires a data rate of about 17.3 gigabits per second, which is the current limit of the Display Port specification. Higher interface data rates demand more power, can increase the interface wire count, and require more shielding to prevent interference with the device's wireless services. These attributes increase system hardware complexity and weight and are undesirable for today's sleek product designs.

http://www.prweb.com/releases/real-time-video/compression-VESA-DSC/prweb11732917.htm

Implement the DSC standard using various test sequences. (www.vesa.org) For non VESA members the DSC standard is available for $350.00.

P.2.130 Yan et al., [E221] have achieved significant speed up by implementing efficient parallel framework for HEVC motion estimation on multi core processors compared with serial execution. In the conclusions they state that they like to find efficient parallel methods for other processing stages in the encoder and to find an efficient parallel framework for HEVC encoder (Figure 2.5). Explore this.

P.2.131 See [E223]. This can lead to several projects. Several papers related to frame recompression are listed in the references. Pixel truncation scheme for motion estimation [17], combining pixel truncation and compression to reconstruct lossless pixels for MC [9, 18], new frame recompression algorithm integrated with H.264/AVC video compression [7]. See also [9] and [11]. Also, the mixed lossy and lossless reference frame recompression is proposed in [E223]. These techniques can be investigated for improved compression in HEVC and as well ME/MC functions. Both [E223] and the papers listed as references need to be reviewed thoroughly.

P.2.132 See P.2.131. The frame recompression techniques proposed and applied to HEVC in [E223] (besides H.264/AVC) can also be explored in other video coding standards such as DIRAC, AVS China, VP9 and VC-1. Each of these can be a project/thesis. Investigate in detail with regards to additional gains that can be achieved by integrating frame recompression with these standards.

P.2.133 The papers presented in IEEE ICCE 2015 (Las Vegas, NV, Jan. 2015) are listed in [E224] through [E238]. In IEEE ICCE, in general, each paper (extended abstracts) is limited to 2 pages. The authors can be contacted by emails and full papers, if any, can be requested. These papers can lead to additional projects.

P.2.134 Zhao, Onoye, and Song [E240] have developed detailed algorithms for HEVC resulting in 54.0–68.4% reduction in encoding time with negligible performance degradation. They also show that fast intra mode decision algorithm can be easily implemented on a hardware platform with fewer resources consumed. Simulation results are based on class A through class E test sequences using HM 11. Implement these algorithms using HM 16 and extend to 8K $\times$ 4K sequences.

P.2.135 See P.2.134. The authors [E240] suggest some aspects of future work in conclusions. Explore these in detail with the goal of reducing the HEVC encoder complexity even further.

P.2.136 you, Chang and Chang [E239] have proposed an efficient ME design with a joint algorithm and architecture optimization for HEVC encoder In Table XV, this proposal is compared with other ME

2.10 Projects

designs and demonstrate that their design reduces the gate count and on-chip memory size significantly. Verify their ME design using HM 16.0 and extend to 8Kx4K sequences.

P.2.137 The theory behind development of DCT based fractional pel interpolation filters and their selection in HEVC is described clearly in [E109]. The performance of these interpolation filters in HEVC is compared with corresponding filters adopted in H.264/AVC. Also some of the fractional pel interpolation filters in H.264/AVC are replaced with those in HEVC to evaluate their effects. Go through this paper and the related references listed at the end and confirm the results shown in Tables 3–6 and Figure 6. Use HM 16.0. Replace the interpolation filters adopted in H.264/AVC completely by those specified in HEVC and evaluate its performance.

P.2.138 Umezaki and Goto [E241] have developed two methods for region of interest based streaming in HEVC and have evaluated in terms of decoding cost, bandwidth efficiency and video quality. They suggest possible future research directions to further reduce the decoding cost. Explore these directions further to achieve the desired results.

P.2.139 Correa et al., [E242] have achieved an average complexity reduction of 65% by combining three early termination schemes in CTU, PU and RQT structures with a negligible BD-rate increase of 1.36% in the HEVC encoder. Their proposed method uses data mining as a tool to build a set of decision trees that allow terminating the decision processes thus relieving the encoder of testing all encoding structure partitioning possibilities. These results are confirmed by simulations based on various test sequences (different resolutions) and are compared with earlier works. Implement this work and confirm their results. Extend this to 4K × 2K and 8K × 4K test sequences. Can the HEVC encoder complexity reduction be further Improved?

P.2.140 Ngyuen and Marpe [E245] have conducted a detailed performance comparison of HEVC main still picture (MSP) profile with other still image and video compression schemes (intra coding only) such as JPEG, JPEG 2000, JPEG XR, H.264/MPEG-4 AVC, VP8, VP9 and WebP. They used PSNR and BD-bit rate as the comparison metrics (see Figures 2–4 in [E245]). Extend this comparison using test sequences with spatial resolution higher than 1280x1600. Another image compression standard is JPEG-LS. Consider this also regarding comparison with HEVC MSP profile.

P.2.141 See ([E245] and P.2.140). Consider mean opinion score (MOS) as the metric for performance comparison of all these standards. Access reference 35 in [E245] regarding HEVC subjective evaluation. Note that this comparison requires extensive man power, detailed testing lab and other resources.

P.2.142 Min and Cheung [E247] have developed a fast CU decision algorithm that reduces the HEVC intra encoder complexity by nearly 52% with negligible R-D performance loss. In the conclusion section they state that for the blur sequences that the BD rate [E81, E82, E96, E198, E431] is worse than the sequences with sharp edges. Explore how this can be minimized for sequences with blur background.

P.2.143 Kim and Park [E248] have proposed a CU partitioning method that reduces the HEVC encoder complexity by 53.6% on the average compared with the HM 15.0 with negligible coding efficiency loss. They state that further work will focus on the optimal feature selection for CU partitioning. Explore and implement this selection.

P.2.144 Chi et al., [E246] developed SIMD optimization for the entire HEVC decoder resulting in 5x speed up. This has been substantiated by implementing the optimized HEVC decoder on 14 mobile and PC platforms covering most major architectures released in recent years. In the conclusions, they state "General purpose architectures have lately increased their floating point performance at a much faster rate than integer performance and in the latest architectures have even a higher floating point throughput. As the use of floating point numbers might also improve compression performance, it is an interesting and promising direction for future work". Implement SIMD acceleration for HEVC decoding using floating point arithmetic and evaluate any improvement in compression performance. This project is complex and may require a group of researchers.

P.2.145 Hautala et al., [E253] have developed low-power multicore coprocessor architecture for HEVC/H.265 in-loop filtering that can decode 1920 × 1080 video sequences (luma only) at 30 fps. Extend this architecture for in-loop filtering of chroma components also. They state that the in-loop filters (both deblocking and SAO) typically consume about 20% of the total decoding time.

P.2.146 See [E254]. Go through this paper in detail. Extend this scheme to 4K and 8K video sequences using the latest HM. Evaluate the bit-rate reductions, subjective quality and computational complexity and compare with the conclusions in [E254].

P.2.10 Projects 91

See P.2.145. The authors state that by using four instances of the proposed architecture, in-loop filtering can be extended to 4K sequences. Explore and implement this.

P.2.147 Zhang et al., [E259] have proposed a Machine Learning-Based Coding Unit depth decision that reduces the HEVC encoder complexity on average 51.45% compared with very little loss in BD-bit rate and BD-PSNR [E81, E82, E96, E198, E431]. Go through this paper in detail and extend this technique to 4K and 8K test sequences. Compile a comparative table similar to Table III in this paper.

P.2.148 Chen et al., [E257] have developed a New Block Based Coding method for HEVC Intra Coding, which results in a 2% BD-rate reduction on average compared with the HM 12.0 Main Profile Intra Coding. However, the encoding time has increased by 130%. In terms of future work, the authors plan on designing new interpolators to improve the R-D performance and also extend this technique to inter coding. Explore this in detail. Consider various avenues related to reducing the encoder complexity (130% increase in the complexity essentially nullifies any BD-rate reductions). Consider the latest HM.

P.2.149 Zou et al., [E255] have developed a Hash Based Intra String Copy method for HEVC based Screen Content Coding that achieves 12.3% and 11% BD-rate savings for 1080P RGB and YUV respectively in full frame Intra Block Copy condition. However, this reduction in BD rates comes at the cost of 50% increase in encode complexity (see Table 1). Can the encoder complexity be reduced by disabling the proposed mode for certain CU sizes? Explore this in detail.

P.2.150 Hu et al., [E258] have developed a Hardware-Oriented Rate-Distortion optimization Algorithm for HEVC Intra-Frame Encoder that achieves 72.22% time reduction of rate-distortion optimization (RDO) compared with original HEVC Test Model while the BD-rate is only 1.76%. Also see all the references listed at the end in [E258]. They suggest that the next step is to implement in Hardware. Design and develop an encoder architecture layout, to implement the same.

P.2.151 In [E261] several methods such as mode mapping, machine learning, complexity scalable, and background modelling for H.264/AVC to HEVC/H.265 transcoder are explained. Implement all these techniques for H.264/AVC to HEVC transcoder and compare their performances using standard metrics. See references at the end.

P.2.152 E. Peixoto [E263] et al., have developed a Fast H.264/AVC to HEVC Transcoding based on Machine Learning, which is 3.4 times faster, on average, than the trivial transcoder, and 1.65 times faster than a previous transcoding solution. Explore HEVC to H.264/AVC transcoder based on Machine Learning for various Profiles/Levels.

P.2.153 Please access the paper, D. Mukherjee, "An overview of new video coding tools under consideration for VP10: the successor to VP9," [9599–50], SPIE. Optics + photonics, San Diego, California, USA, 9–13, Aug. 2015. Implement VP10 encoder using new video coding tools under consideration and compare with HEVC (H.265) based on standard metrics. Note that VP10 effort has been dropped by Google as AV1 codec has been finalized by alliance for open media (AOM).

P.2.154 Please access the paper, D. Mukherjee, "An overview of new video coding tools under consideration for VP10: the successor to VP9," [9599–50], SPIE. Optics + photonics, San Diego, California, USA, 9–13, Aug. 2015. Develop the VP10 decoder block diagram, compare its implementation complexity with HEVC decoder.

Note that VP10 effort has been dropped by Google as AV1 codec has been finalized by alliance for open media (AOM).

P.2.155 Please see the Paper, R. G. Wang et al., "Overview of MPEG Internet Video Coding", [9599–53], SPIE. Optics + photonics, San Diego, California, USA, 9–13, Aug. 2015 and access all the references listed at the end. In response to the call for proposals for internet video coding by MPEG, three different codecs Web Video Coding (WVC), Video Coding for browsers (VCB) and Internet Video Coding (IVC). WVC and VCB were proposed by different group of companies and IVC was proposed by several Universities and its coding tools were developed from zero. Section 2 describes coding tools (key technologies) used in the current test model of IVC (ITM 12.0). Specific Internet applications and the codec requirements are listed. Besides an overview of IVC, this paper presents a performance comparison with the WVC and VCB codecs using different test sequences at various bit rates (see Table 9). Constraints are listed in Section 3.1 – Test cases and Constraints. Implement the IVC, WVC and VCB codecs and compare with AVC HP (H.264). Extend this comparison based on 4k × 2k video test sequences (see Table 8). Consider also implantation complexity as another comparison metric. Implementation complexity requirements are also described in this paper (Section 1). Please see [38–41] in references for detailed encoding settings.

2.10 Projects

Note that JVT – VC documents can be accessed as follows:

E1. JCT-VC DOCUMENTS can be found in JCT-VC document management system
http://phenix.int-evry.fr/jct (see [E185])

E2. All JCT-VC documents can be accessed. [online]. Available: http://phenix.intevry.fr/jct/doc_end_user/current_meeting.php?id_meeting=154&type_order=&sql_type=document_number_

P.2.156 Please see the Paper, R. G. Wang et al., "Overview of MPEG Internet Video Coding", [9599–53], SPIE. Optics + photonics, San Diego, California, USA, 9–13, Aug. 2015. This Paper says "VCB was proposed by Google and it is in fact VP8". Replace VP8 by the tools proposed in VP10 and implement the VCB in detail and evaluate thoroughly its performance with IVC, WVC and AVC HP. Please refer the paper, D. Mukherjee, "An overview of new video coding tools under consideration for VP10: the successor to VP9," [9599–50], SPIE. Optics + photonics, San Diego, California, USA, 9–13, Aug. 2015.

P.2.157 Please review [E266] in detail. This paper has proposed a Novel JND – based HEVC – Complaint Perceptual Video Coding (PVC) scheme that yielded a remarkable bitrate reduction of 49.10% maximum and 16.10% average with negligible subjective Quality loss. Develop and implement A VP10 – Compliant Perceptual Video Coding Scheme based on JND models for Variable Block – sized Transform Kernels and consider bit – rate reductions, implementation complexity, subjective quality etc., as performance metrics. Subjective Quality evaluation requires extensive test set up (Monitor, lighting, display) and a number of trained viewers to get Mean Opinion Score (MOS). (Also refer: D. Mukherjee, "An overview of new video coding tools under consideration for VP10: the successor to VP9," [9599–50], SPIE. Optics + photonics, San Diego, California, USA, 9–13, Aug. 2015 and Ref 11 at the end of [E266]).

P.2.158 Please see P.2.157. Repeat this project for AVS – China. (See References in AVS China Section).

P.2.159 Please see P.2.157. Repeat this project for DIRAC. (See References in DIRAC Section).

P.2.160 Please see P.2.157. Repeat this project for HEVC – IVC. (Also Refer: R. G. Wang et al., "Overview of MPEG Internet Video Coding",

[9599–53], SPIE. Optics + photonics, San Diego, California, USA, 9–13, Aug. 2015).

P.2.161 Review [E264]. This paper presents an overview of scalable extensions of HEVC where in the scalabilities include Spatial, SNR, bit depth, and colour gamut as well as combinations of any of these. In the conclusions, it is stated that ATSC is considering SHVC for broadcasting and video streaming. Go to WWW.atsc.org and extract related documents. Implement SHVC for different scalabilities compatible with ATSC requirements. Go to http://atsc.org/newsletter/2015, to get newsletters on different video coding standards.

P.2.162 Review [E264]. Implement HEVC (all profiles / levels) compatible with ATSC requirements. Go to www.atsc.org. Also refer to P.2.161. Go to http://atsc.org/newsletter/2015, to get newsletters on different video coding standards.

P.2.163 Chen et al., [E267] have proposed a novel framework for software based H.264/AVC to HEVC transcoding, integrated with parallel processing tools that are useful for achieving higher levels of parallelism on multicore processors and distributed systems. This proposed transcoder can achieve up to 60x speed up on a Quad core 8-thread server over decoding – re-encoding based on FFMPEG and the HM software with a BD-rate loss of 15%–20% and cam also achieve a speed of 720P at 30 Hz by implementing a group of picture level task distribution on a distributed system with nine processing units. Implement the same for higher resolution (8k × 4k) sequences.

P.2.164 Please see P.2.163. Develop a framework for software based HEVC (All Profiles/Levels) to H.264/AVC transcoding.

P.2.165 Please see P.2.163. Develop an algorithm for H.264 to HEVC transcoding that can reduce compression performance loss while improving/maintaining transcoding speed for 4k × 2k video sequences. Also extend to 8k × 4k video sequences.

P.2.166 Lee et al., [E268] have developed an early skip mode decision for HEVC encoder that reduces the encoder complexity by 30.1% for random access and around by 26.4% for low delay with no coding loss. Implement this for 4k and 8K video using the latest HM software.

P.2.167 Lim et al., [E269] have developed a fast PU skip and split termination algorithm that achieves a 44.05% time savings on average when $\alpha = 0.6$ and 53.52% time savings on average when $\alpha = 0.5$ while maintaining almost the same RD performance compared with HM 14.0. In the conclusion it is stated that the there are some complexity and coding efficiency tradeoff differences between class F and other class sequences in the simulation results. Explore this as future research.

P.2.168 Won and Jeon [E270] have proposed a complexity – efficient rate estimation for mode decision of the HEVC Encoder that shows an average saving in rate calculation time for all intra, low delay and random access conditions of 55.29%, 41.05% and 42.96% respectively. Implement this for 4k and 8k video sequences using the latest HM software.

P.2.169 Chen et al., [E272] have developed a novel wavefront-based high parallel (WHP) solution for HEVC encoding integrating data level and task level methods that bring up to 88.17 times speedup on 1080P sequences, 65.55 times speed up on 720P sequences and 57.65 times speedup on WVGA sequences compared with serial implementation. They also state that the proposed solution is also applied in several leading video companies in China, providing HEVC video service for more than 1.3 million users every day. Implement this for 4k and 8k video using the latest HM software. Develop tables similar to tables III through XI and graphs similar to Figures 14 and 15. If possible use the hardware platform described in Table IV.

P.2.170 Zhang, Li and Li [E273] have proposed an efficient fast mode decision method for Inter Prediction in HEVC which can save about 77% encoding time with only about 4.1% bit rate increase compared with HM16.4 anchor and 48% encoding time with only about 2.9% bit rate increase compared with fast mode decision method adopted in HM16.4. Flow chart of the proposed algorithm is shown in Figure 5. Previous Fast Mode Decision Algorithms designed for HEVC Inter prediction are also described. Review this paper. Implement the same and confirm the results. If possible use hardware platform described in Table IV.

P.2.171 Please see P.2.170. Implement the code based on the given flow chart using C++ compiler and use "clock" function to measure the run time.

P.2.172 Please see P.2.170. Implement the same project for 4k and 8k video sequences and tabulate the results.

P.2.173 Hu and Yang [E274] have developed a Fast Mode Selection for HEVC intra frame coding with entropy coding refinement. Using various test sequences, they have achieved 50% reduction in encoding time compared with HM 15.0 with negligible BD-Rate Change. Review this paper thoroughly and confirm their results. Test sequences for the simulation are class A through class F (see Table IV). Extend this simulation using 4k & 8k sequences and cite the results and conclusions.

P.2.174 Jou, Chang, and Chang [E275] have developed an efficient ME design with a joint algorithm and architecture optimization that reduces the integer Motion Estimation and fractional Motion Estimation complexity significantly. Architectural design also supports real – time encoding of 4k × 2k video at 60 fps at 270 MHz. Review this paper in detail and simulate the algorithm / architecture Motion Estimation design. Please confirm the results shown in the last column of Table XV. In the conclusion it says further optimization can be achieved by tweaking the proposed algorithms and corresponding architecture. Explore this in detail.

P.2.175 Go through the overview paper on emerging HEVC SCC extension described in [E322] and the related references listed at the end. Using the test sequences as described in tables III and IV, verify the results shown in table V using JM 18.6 (H.264/AVC), HM 16.4 (HEVC) and SCM 3.0 / SCM 4.0 (HEVC – SCC).

P.2.176 Repeat P.175 using Table VI from the overview paper on emerging HEVC SCC extension described in [E322].

P.2.177 Repeat P.175 using Table VII from the overview paper on emerging HEVC SCC extension described in [E322].

P.2.178 Repeat P.175 using Table VIII from the overview paper on emerging HEVC SCC extension described in [E322].

P.2.179 Repeat P.175 using Table XI from the overview paper on emerging HEVC SCC extension described in [E322].

P.2.180 Repeat P.175 using Table XII from the overview paper on emerging HEVC SCC extension described in [E322].

P.2.181 Repeat P.175 using Table XV from the overview paper on emerging HEVC SCC extension described in [E322].

P.2.182 Repeat P. 175 using Table XVI from the overview paper on emerging HEVC SCC extension described in [E322].

Note that the projects P.2.175 through P.2.182 require thorough understanding of the H.264/AVC, HEVC version 1 and

2.10 Projects

HEVC– SCC besides familiarity of the corresponding software JM 18.6, HM 16.4 and SCM 3.0/SCM 4.0 (respectively). Note that implementation complexity is not considered in any of these simulations. The additional tools/modules added to the HEVC version for SCC as described in [E322] probably result in increased complexity. This is a relevant comparison metric. Develop Tables showing the comparison of this complexity.

P.2.183 Using the standard test sequences, confirm the results shown in Table II [E321] for All Intra (AI), RA (Random Access) and LD (Low Delay) configurations.

P.2.184 Repeat P.2.183 for lossless performance. See Table III [E321].

P.2.185 Repeat P.2.184. Extend the performance comparison of version II with H.264/AVC (see Tables II and III) [E321] based on implementation complexity as a metric.

P.2.186 Using the test sequences in Table 5 [E325], confirm the results shown in Table 6 [E325] in terms of coding performance of SHVC and HEVC simulcast.

P.2.187 See P.2.186. Confirm the results shown in Table 7 [E325] in terms of coding performance of SHVC EL and HEVC single – layer coding equivalent to EL.

P.2.188 See P.2.186. Confirm the results shown in Table 8 [E325] in terms of coding performance of SHVC and HEVC single – layer coding equivalent to EL.

P.2.189 See P.2.186. Confirm the results shown in Table 9 [E325] in terms of coding performance of SHVC and SVC.

P.2.190 In [E325] implementation, complexity has not been included as a performance metric. Evaluate this for all cases listed in Tables 6 through 8 [E325].

P.2.191 [E348] M.S. Thesis "REDUCING ENCODER COMPLEXITY OF INTRA-MODE DECISION USING CU EARLY TERMINATION ALGORITHM' by Nishit Shah. By developing CU early termination and TU mode decision algorithm, he has reduced the computational complexity for HEVC intra prediction mode with negligible loss in PSNR and slight increase in bit rate". Conclusions and future work are reproduced here:

Conclusions

In this thesis, a CU splitting algorithm and the TU mode decision algorithm are proposed to reduce the computational complexity of the HEVC encoder, which includes two strategies, i.e. CU splitting algorithm and the TU mode decision. The results of comparative experiments demonstrate that the proposed algorithm can effectively reduce the computational complexity (encoding time) by 12–24% on average as compared to the HM 16.0 encoder [38], while only incurring a slight drop in the PSNR and a negligible increase in the bitrate and encoding bitstream size for different values of the quantization parameter based on the various standard test sequences [29]. The results of simulation also demonstrate negligible decrease in BD-PSNR [30], i.e., 0.29 dB to 0.51 dB as compared to the original HM16.0 software and negligible increase in the BD-bitrate [31].

Future Work

There are many other ways to explore in the CU splitting algorithm and the TU mode decision in the intra prediction area. Many of these methods can be combined with this method, or if needed, one method may be replaced by a new method and encoding time gains can be explored.

Similar algorithms can be developed for fast inter-prediction in which the RD cost of the different modes in inter-prediction are explored, and depending upon the adaptive threshold, mode decision can be terminated resulting in less encoding time and reduced complexity combining with the above proposed algorithm.

Another fact of encoding is CU size decisions which are the leaf nodes of the encoding process in the quad tree. Bayesian decision rule can be applied to calculate the CU size, and then this information can be combined with the proposed method to achieve further encoding time gains.

Complexity reduction can also be achieved through hardware implementation of a specific algorithm which requires much computation. The FPGA implementation can be useful to evaluate the performance of the system on hardware in terms of power consumption and encoding time.

Explore this future work.

P.2.192 Hingole [E349] has implemented the HEVC bit stream to H.264/MPEG4 AVC bit stream transcoder (see figure below)

The transcoder can possibly be significantly reduced in complexity by adaptively reusing the adaptive intra directional predictions (note HEVC has 35 intra directional predictions versus 9 for H.264/MPEG4 AVC) and MV reuse from HEVC in H.264 (again HEVC multiple block sizes, both symmetric and asymmetric, for ME and up to 1/8 fractional MV resolution versus simpler block sizes in H.264). Explore this in detail and see how the various modes in HEVC can be tailored to those in H.264/AVC such that the overall transcoder complexity can be significantly reduced. This requires thorough understanding of both HEVC and H.264 codecs besides review of the various transcoding techniques already developed.

P.2.193 Dayananda [E350] in her M.S. thesis entitled *"Investigation of Scalable HEVC & its bitrate allocation for UHD deployment in the context of HTTP streaming"*, has suggested

Further, work on exploring optimal bitrate algorithms for allocation of bits into layer of SHVC based on Game theory and other approaches can be done, considering various scalability options (such as spatial, quality and combined scalabilities). Explore this.

P.2.194 See P.2.192 Additional experiments to study the effect of scalability overhead for its modelling can be done using several test sequences. Implement this.

P.2.195 See P.2.192 Also, evaluation of SHVC for its computational complexity can be done and parallel processing techniques for encoding base and enhancement layers in SHVC can be explored. Investigate this. Note that this is a major task.

P.2.196 See [E351]. M.S. thesis by V. Vijayaraghavan, "Reducing the encoding time of motion estimation in HEVC using parallel programming". Conclusions and future work are reproduced here;

Conclusions

Through thorough analysis with the most powerful tool, Intel® vTune™ amplifier, hotspots were identified in the HM16.7 encoder. These hotspots are the most time consuming functions/loops in the encoder. The functions are optimized using optimal C++ coding techniques and the loops that do not pose dependencies are parallelized using the OpenMP directives available by default in Windows Visual Studio.

Not every loop is parallelizable. Thorough efforts are needed to understand the functionality of the loop to identify dependencies and the capability of the loop to be made parallel. Overall observation is that the HM code is already vectorized in many regions, and hence, parallel programming on top of vectorization may lead to degradation in performance in many cases. Thus, the results of this thesis can be summarized as below:

Ø Overall ∼24.7 to 42.3% savings in encoding time.
Ø Overall ∼3.5 to 7% gain in PSNR.
Ø Overall ∼1.6 to 4% increase in bitrate.

Though this research has been carried out on a specific configuration (four core architecture), it can be used on any hardware universally. This implementation works on servers and Personal Computers. Parallelization in this thesis has been done at the frame level.

Future Work

The OpenMP framework is a very simple yet easy to adapt framework that aids in a thread-level parallelism. Powerful parallel programming APIs are available which can be used in offloading the serial code to the GPU. Careful efforts need to be invested in investigating the right choice of software and functions in the software chosen to be optimized. If optimized appropriately, huge savings in the encoding time can be achieved.

Intel® vTune^{TM} amplifier is a very powerful tool which makes it possible for analysis of different types to be carried at the code level as well as at the hardware level. The analysis that has been made use of in this thesis is Basic Hotspot analysis. There are other options available in the tool, one of which helps us to identify the regions of the code which cause the maximum number of locks and waits and also the number of cache misses that occur. Microprocessor and assembly level optimization of the code base can be achieved by diving deep into this powerful tool.

See [E351], parallel programming of CPU threads is achieved using OpenMP parallel threads. More efficient times can be obtained (reduced encoding time) using GPU (Graphics Processing Unit) programming with OpenCL framework. Explore this.

Intel vTune Amplifier basic hotspot analysis has been used to identify the top hotspots (functions and loops) in the code using Basic Hotspot Analysis which is used for code optimization. There are several different analysis options available in Intel vTune Amplifier, which can be used to optimize the code at the assembly level. Assembly level optimizations can further increase the efficiency of the codec and reduce the encoding time.

2.10 Projects   101

Investigate the suggested optimizations based on these criteria.

P.2.197 Kodpadi [E352] in her M. S. thesis (EE Dept., UTA) entitled "Evaluation of coding tools for Screen content in High Efficiency Video Coding", has suggested:
The future work can be on reducing the encoding time by parallelizing the different methods or by developing fast algorithms on the encoder side. Explore this.

P.2.198 See P.2.197 Continue to evaluate the coding performance of the newly adopted tools and their interaction with the existing HEVC tools in the Main profile and range extensions. See OP5 and OP7.

P.2.199 See P.2.197 Study latency and parallelism implications of SCC coding techniques, considering multicore and single-core architectures.

P.2.200 See P.2.197 Analyse the complexity characteristics of SCC coding methods with regard to throughput, amount of memory, memory bandwidth, parsing dependencies, parallelism, pixel processing, chroma position interpolation, and other aspects of complexity as appropriate.

P.2.201 Mundgemane [E353] in her M. S. thesis entitled "Multi-stage prediction scheme for Screen Content based on HEVC", has suggested:
The future work can be on reducing the encoding time by parallelizing the independent methods on the encoder side. Implement this.

P.2.202 See P.2.201 Implement fast algorithms for screen content coding tools that can help in decreasing the encoding time.

P.2.203 See P.2.201 The coding efficiency of HEVC on surveillance camera videos can be explored. Investigate this.

P.2.203 See [H69]. This paper describes several rate control algorithms adopted in MPEG-2, MPEG-4, H.263 and H.264/AVC. The authors propose a rate control mechanism that significantly enhances the RD performance compared to the H.264/AVC reference software (JM 18.4). Several interesting papers related to RDO of rate control are cited at the end. Investigate if the rate control mechanism proposed by the authors can be adopted in HEVC and compare its performance with the latest HM software. Consider various test sequences at different bit rates. Develop figures and tables similar to Figures 8–10 and Tables I to III shown in [H69] for the HEVC. This project is extensive and can be applied to M.S. Thesis.

102   HEVC Encoder

P.2.204 See P.2.203. In [H69], computational complexity has not been considered as a performance metric. Investigate this. It is speculated that the proposed rate control mechanism invariably incurs additional complexity compared to the HM software.

P.2.205 Hamidouche, Raulet, and Deforges [E329] have developed a software parallel decoder architecture for the HEVC and its various multilayer extensions. Advantages of this optimized software are clearly explained in the conclusions. Go through this paper in detail and develop similar software for the HEVC decoder for scalable and multiview extensions.

P.2.206 Kim et al., [E355] have implemented the cross component prediction (CCP) in HEVC (CCP scheme is adopted as a standard in range extensions) and have shown significant coding performance improvements for both natural and screen content video. Note that the chroma residual signal is predicted from the luma residual signal inside the coding loop. See Figure 2 that shows the block diagrams of encoder and decoder with CCP. This scheme is implemented in both RGB and YCbCr colour spaces.

In the conclusions, the authors state that more study can be made on how to facilitate software and hardware implementation. Investigate this.

P.2.207 See P.2.206. The authors also state "We leave it as a further study how to apply CCP to various applications including HDR image coding". Investigate this.

P.2.208 Fong, Han and Cham [E356] have developed recursive integer cosine transform (RICT) and have demonstrated from order-4 to order-32. The proposed RICT is implemented into reference software HM 13.0. By using different test sequences, they show that RICT has similar coding efficiency (see Tables VI through XI in this paper) as the core transform in HEVC. See references 18 through 22 listed in [E356]. Using the recursive structure, develop order-64 and order-128 RICTs and draw the flow graphs similar to Figure 1 and the corresponding matrices similar to Eq. 16. The higher order transforms are proposed in beyond HEVC [BH1, BH2].

P.2.209 Hsu and Shen [E357] have designed a deblocking filter for HEVC that can achieve 60 fps for the video with 4Kx@2K resolution assuming an operating frequency of 100 MHz. Go through the VLSI architecture and hardware implementation of this filter and extend

2.10 Projects 103

the design so that the deblocking filter can operate on the video with 8K × 4K resolution. This requires major hardware design tools.

P.2.210 Francois et al., [E339] present the roles of high dynamic range (HDR) and wide colour gamut (WCG) in HEVC in terms of both present status and future enhancements. They describe in detail the various groups involved in HDR/WCG standardization including the work in MPEG and JCT-VC. They conclude that the groups' efforts involve synchronization to ensure a successful and interoperable market deployment of HDR/WCG video. The proposals on HDR/WCG video coding submitted to JCT-VC can lead to several projects. The references related to these proposals are listed at the end in [E339]. In the conclusions the authors state "Finally following the conclusions of the CfE for HDR and WCG video coding, MPEG has launched in June 2015 a fast-track standardization process to enhance the performance of the HEVC main 10 profile for HDR and WCG video, that would lead to an HEVC extension for HDR around mid-2016." (CfE: Call for evidence). Figure 6 shows the application of CRI (colour remapping info.) for the HDR and WCG to SDR conversion. Implement this and show the two displays (SDR and HDR/WCG).

P.2.211 See P.2.210. Two methods for conversion from 8 bit BL resolution BT.709 to 10 bit EL resolution BT.2020 are shown in Figure 10. The authors state that the second method (Figure 10(b)) can keep the conversion precision and has better coding performance compared with the first method (Figure 10(a)). Implement both methods and confirm this comparison.

P.2.212 See P.2.210 Another candidate for HDR/WCG support is shown in Figure 11. Using a 16 bit video source, split into two 8 bit videos followed by two legacy HEVC coders and combining into 16 bit reconstructed video. Implement this scheme and evaluate this in terms of standard metrics.

P.2.213 See P.2.210 An alternate solution for HDR video coding is illustrated in Figures 12 and 13. Simulate this scheme and compare with the techniques described in P.2.210 through P.2.212.

P.2.214 See P.2.210 HDR/WCG video coding in HEVC invariably involves additional computational complexity. Evaluate the increase in complexity over the HEVC coding without these extensions. See [E84].

P.2.215 Tan et al., [SE4] have presented the subjective and objective results of a verification test in which HEVC is compared with H.264/AVC.

Test sequences and their parameters are described in Tables II and III and are limited to random access (RA) and low delay (LD). Extend these tests to all intra (AI) configuration and develop figures and tables as shown in section IV Results. Based on these results confirm the bit rate savings of HEVC over H.264/AVC.

P.2.216 Lee et al., [E358] have developed a fast RDO quantization for HEVC encoder that reduces the quantization complexity for AI, RA and LD with negligible coding loss. See Tables X through XIII. They propose to include a fast context update with no significant coding loss. Explore this. Develop tables similar to these specifically showing further reduction in complexity.

P.2.217 Georgios, Lentaris, and Reisis [E360] have developed a novel compression scheme by sandwiching the traditional video coding standards such as H.264/VAC and HEVC between down sampling at the encoder side and upsampling at the decoder side. They have shown significant encoder/decoder complexity reduction while maintaining the same PSNR specifically at low bitrates. The authors have proposed SR (super resolution) Interpolation algorithm called L-SEABI (low-complexity back-projected interpolation) and compared their algorithms with state-of-the-art algorithms such as bicubic interpolation based on various test sequences in terms of PSNR, SSIM and BRISQUE (blind/reference less image spatial quality evaluator) (See Table I). Implement all the interpolation algorithms and verify Table I.

P.2.218 See P.2.217. Please go through the conclusions in detail. The authors state that their SR compression scheme out performs the H.264/HEVC codecs for bitrates up to 10 and 3.8 Mbps respectively, consider BD bitrate and BD-PSNR [E81, E82, E96, E198] as comparison metrics. Include these in all the simulations and develop tables similar to Figure 7.

P.2.219 See P.2.218. Implement this novel SR compression scheme in VP9 and draw the conclusions.

P.2.220 See P.2.219. Replace VP9 by AVS China and implement the SR compression scheme.

P.2.221 Repeat P.2.220 for DIRAC codec developed by BBC.

P.2.222 Repeat P.2.220 for VC1 codec developed by SMPTE.

P.2.223 Kuo, Shih, and Yang [E361] have improved the rate control mechanism by adaptively adjusting the Lagrange parameter and have shown that their scheme significantly enhances the RD performance

2.10 Projects

compared to the H.264/AVC reference software. Go through this paper and all related references in detail.

Can a similar scheme be applied to HEVC reference software. If so, justify the improvement in RD performance by applying the adaptive scheme in various test sequences (See corresponding Figures and Tables in [E361]).

P.2.224 Jung and Park [E332] by using an adaptive ordering of modes have proposed a fast mode decision that significantly reduces the encoding time for RA and LD configurations. Can you extend this scheme to AI case? If so, list the results similar to Tables XI, XII and XIII based on the test sequences outlined in Table IX followed by conclusions. Use the latest HM software.

P.2.225 Zhang, Li and Li [E362] have developed a fast mode decision for inter prediction in HEVC resulting in significant savings in encoding time compared with HM16.4 anchor. See Tables VI through XII and Figure 9. The test sequences are limited to 2560×1600 resolution. Extend the proposed technique to 4K and 8K video sequences and develop similar Tables and Figure Based on the results, comment on the corresponding savings in encoder complexity.

P.2.226 Fan, Chang and Hsu [E363] have developed an efficient hardware design for implementing multiple forward and inverse transforms for various video coding standards. Can this design be extended to HEVC also? Investigate this in detail? Note that in HEVC 16×16 and 32×32 INTDCTs are also valid besides 4×4 and 8×8 INTDCTs.

P.2.227 Park et al., [E364] have proposed a 2-D 16×16 and 32×32 inverse INTDCT architecture that can process 4K@30fps video. Extend this architecture for 2-D inverse 64×64 INTDCT that can process 8K@30 fps video. Note that 64×64 INTDCT is proposed as a future development in HEVC (beyond HEVC). (See the papers in BEYOND HEVC).

P.2.228 Au-Yeung, Zhu, and Zeng [H70] have developed partial video encryption using multiple 8×8 transforms in H.264 and MPEG-4. Can this encryption approach be extended to HEVC? Note that multiple size transforms are used in H.265. Please access the paper by these authors titled "Design of new unitary transforms for perceptual video encryption" [H71].

106 HEVC Encoder

P.2.229 See [H72]. Can the technique of embedding sign-flips into integer-based transforms be applied to encryption of H.264 video be extended to H.265. Explore this in detail. See also [H73].

P.2.230 See [E365]. The authors have proposed a collaborative inter-loop Zvideo encoding framework for CPU+GPU platforms. They state "the proposed framework outperforms single-device executions for several times, while delivering the performance improvements of up to 61.2% over the state of the art solution." This is based on the H.264/AVC inter-loop encoding. They also state extending this framework to other video coding standards such as HEVC/H.265 as future work. Explore this future work in detail.

P.2.231 See [Tr11]. The authors have presented an approach for accelerating the intra CU partitioning decision of an H.264/AVC to HEVC transcoder based on Naïve-Bayes classifier that reduces the computational complexity by 57% with a slight BD-rate increase. This is demonstrated by using classes A through E test sequences. (See Table 1). Go through this paper in detail and confirm their results. Extend these simulations to 4K and 8K test sequences and verify that similar complexity reduction can be achieved.

P.2.232 See [E191] and P.2.109. Extend the comparative assessment of HEVC/H.264/VP9 encoders for random access and all intra applications using 1280×720 60 fps test sequences. Extend this to UHDTV test sequences. Consider also implementation complexity as another metric.

P.2.233 Jridi and Meher [E367] have derived an approximate kernel for the DCT of length 4 from that defined in HEVC and used this for computation of DCT and its inverse for power-of-2 lengths. They discuss the advantages of this DCT in terms of complexity, energy efficiency, compression performance etc. Confirm the results as shown in Figures 8 and 9 and develop the approximate DCT for length 64.

P.2.234 See [E368]. Section VII Conclusion is reproduced here.

"In this paper, fast intra mode decision and CU size decision are proposed to reduce the complexity of HEVC intra coding while maintaining the RD performance. For fast intra mode decision, a gradient-based method by using average gradients in the horizontal (AGH) and vertical directions (AGV) is proposed to reduce the candidate modes for RMD and RDO. For fast CU size decision, the homogenous CUs are early terminated first by employing

2.10 Projects 107

AGH and AGV. Then two linear SVMs which employ the depth difference, HAD cost ratio (and RD cost ratio) as features are proposed to perform early CU split decision and early CU termination decision for the rest of CUs. Experimental results show that the proposed method achieves a significant encoding time saving which is about 54% on average with only 0.7% BD-rate increase. In the future, average gradients along more directions can be exploited to obtain the candidate list with fewer modes for RMD and RDO. More effective features such as the intra mode and the variance of the coding block can be also considered in SVM for CU size decision to further improve the prediction accuracy".

Explore the future work suggested here and comment on further improvements in fast intra mode and CU size decision for HEVC. Extend the performance comparison similar to Figures 12, 14, and 15 and Tables II through IV.

P.2.235 See [E369]. In this paper Liu et al., have proposed an adaptive and efficient mode decision algorithm based on texture complexity and direction for intra HEVC prediction. They have also presented a detailed review of previous work in this area. By developing the concepts of CU size selection based on texture complexity and prediction mode decision based on texture direction they have shown significant reduction in encoder complexity with negligible reduction in BD-rate using standard test sequences. Go through the texture complexity analysis and texture direction analysis described in this paper and confirm the comparison results shown in Tables V through X. Extend this comparison to 4K and 8K test sequences.

P.2.236 See [E270, E271]. Using the early termination for TZSearch in HEVC motion estimation, evaluate the reduction in HEVC encoding time based on HDTV, UHDTV (4K and 8K) test sequences for different profiles.

P.2.237 S.K. Rao [E372] has compared the performance of HEVC intra with JPEG, JPEG 2000, JPEG XR, JPEG LS and VP9 intra using SD, HD, UHD and 4K test sequences. HEVC outperforms the other standards in terms of PSNR and SSIM at the cost of increased implementation complexity. Consider BD-PSNR and BD bit rate [E81, E82, E96, E198] as comparison metrics.

P.2.238 See P.2.237. Extend the performance comparison to 8K test sequences.

P.2.239 See [E373]. The impact of SAO in-loop filter in HEVC is shown clearly by Jagadeesh [E373] as part of a project. Extend this project to HDTV, UHDTV, 4K and 8K video sequences.

P.2.240 See [E374]. Following the technique proposed by Jiang and Jeong (see reference 34 cited in [E374]) for fast intra coding, Thakur has shown that computational complexity can be reduced by 14%–30% compared to HM 16.9 with negligible loss in PSNR (only slight increase in bit rate) using Kimono (1920 × 1080), BQ Mall (832 × 488) and Kristen and Sara (1280 × 1080) test sequences. Extend this technique to 4K and 8K test sequences and evaluate the complexity reduction.

P.2.241 See [E375]. Following reference 2 cited in this paper, Sheelvant analysed the performance and computational complexity of high efficiency video encoders under various configuration tools (see Table 2 for description of these tools) and has recommended the configuration settings (see conclusions). Go through this project in detail and see if you agree with these settings. Otherwise develop your own settings.

P.2.242 See [E357]. Hsu and Shen have developed VLSI architecture of a highly efficient deblocking filter for HEVC. This filter can achieve 60 fps for the 4K × @k video under an operating frequency of 100 MHz. This design can achieve very high processing throughput with reduced or comparable area complexity. Extend this deblocking filter HEVC architecture for 8K × 4K video sequences.

P.2.243 See [E377]. Bae, Kim, and Kim have developed a DCT-based LDDP distortion model and then proposed a HEVC-complaint PVC scheme. Using various test sequences (see V through VII), they showed the superiority of their PVC approach compared with the state-of-the-art PVC scheme for LD and RA main profiles. Extend this technique to AI main profile.

P.2.244 See P.2.243. Extend this technique to 4K and 8K test sequences and develop Tables similar to those shown in [E377]. Based on these simulations, summarize the conclusions.

P.2.245 See [E379]. Pagala's thesis focusses on multiplex/demultiplex of HEVC Main Profile video with HE-AAC v2 audio with lip synch. Extend this to 8K video test sequences. Give a demo.

P.2.246 See [E379]. In the multiplex/demultiplex/lip synch scheme, replace HEVC Main Profile with VP10 video. Make sure that the time lag

2.10 Projects 109

between video and audio is below 40ms. See [VP5, VP15]. Give a demo.

P.2.247 See [E379], P.2.245 and P.2.246. Implement multiplex/demultiplex/ lip sync scheme using AV1 video and HE AAC v2 audio.
AV1 codec is developed by alliance for open media (AOM). http://tinyurl.com/zgwdo59

P.2.248 See [E379]. Replace AV1 codec in P.2.247 by DAALA codec. See the references on DAALA.

P.2.249 See [E379]. Can this scheme be extended to SCC?

P.2.250 See [E382]. This paper explains clearly the fully pipelined architecture for intra prediction in HEVC and achieves a high throughput of 4 pels per clock cycle. It can decode 3840 × 2160 videos at 30 fps. In the conclusions the authors state "in the future work, we plan to implement the proposed architecture on ASIC platform to increase the system frequency, aiming at achieving real-time video decoding of higher resolution(higher than 4K videos)". Explore this.

P.2.251 See P.2.250. The authors also state "Meanwhile the proposed architecture will be integrated with inter prediction engine, inverse transforming engine and other engines to construct the entire encoding/decoding system". This is a major project. Implement this HEVC codec.

P.2.252 See [E281]. In the conclusions, it is stated "Then, two flexible and HEVC compliant architectures, able to support the DCT of size 4, 8, 16, 32 have been proposed". In beyond HEVC, DCT of size 64 also being considered. Extend these HEVC compliant architectures to support the DCT of size 64. See the sections BEYOND HEVC and PROJECTS ON BEYOND HEVC towards the end.

P.2.253 Based on a set of Pareto-efficient encoding configurations identified through a rate-distortion-complexity analysis Correa et al., [E383] have developed a scheme that accurately limits the HEVC encoding time below a predefined target for each GOP. The results also indicate negligible BD-rate loss at significant complexity reduction. Go through this paper in detail and confirm the results shown in Tables II–IV and Figures 1 and 4–9 using the test sequences listed in Table I.

P.2.254 See [E384]. After describing the mosaic videos with arbitrarycolour filter arrays (Figure 1), the authors propose a novel chroma subsampling strategy (4:2:0 format) for compressing mosaic videos in

H.264/AVC and HEVC. They claim that this strategy has the best quality and bitrate compared with previous schemes.

For the seven typical RGB-CFA structures shown in Figure 1 (captured mosaic videos) apply the demosaicking designed for each structure and as well the universal demosaicking and obtain the demosaicked full-colour RGB images.

P.2.255 See P.2.254. For the demosaicked videos convert from RGB to YUV and then to 4:2:0 six subsampling strategies (Figure 6). Convert from 4:2:0 to 4:4:4 YUV format and then to RGB reconstructed video. Compare these strategies in terms of CPSNR and CMSE for these demosaicked videos.

P.2.256 Implement the chroma subsampling strategy described in [E384] along with other strategies (Figures 2–4) in both H.264/AVC and HEVC and confirm the results shown in Figures 5–8 and Tables I–VI.

P.2.257 Apply the universal demosaicking algorithm based on statistical MAP estimation developed in [E387] to the seven typical RGB-CFA structures (see P.2.284) and obtain the demosaicked full colour RGB images. Compare the effectiveness of this universal algorithm with those described in [E384] in terms of the PSNR. See the conclusions section in [E387].

P.2.258 See [E388]. Also the abstract. Implement the fast prediction mode decision in HEVC developed in [E388] and confirm the simulation results shown in Tables 1–3.

P.2.259 See P.2.258. Apply this technique to 8K test sequences and develop tables similar to Tables 1–3 described in [E388].

P.2.260 Review [E389] in detail. The last sentence in Conclusion section states, "The future work will focus on extending algorithm to the remaining coding structures (i.e., PUs and TUs) and other configurations in order to further expedite the encoding process with minimal impact on the coding efficiency". Explore this and develop tables similar to Tables V–VII and draw the conclusions.

P.2.261 See [AVS12]. The authors proposed a fast intra coding platform for AVS2 (called iAVS2) leading to higher speeds and better balance between speed and compression efficiency loss especially for large size videos. The authors state "Owing to their (AVS2 and HEVC) similar frameworks, the proposed systematic solution and the fast algorithms can also be applied in HEVC intra coding design". Go through this paper in detail and explore how the speed up methods

P.2.262 See [E392]. SDCT and its integer approximation have been proposed and applied in image coding. The authors suggest the possibility to implement efficiently an integer SDCT in the HEVC standard.

Integrating inside HEVC may require a significant amount of work, as the transform has to be inserted in the rate-distortion optimization loop, and auxiliary information may have to be signalled. RD optimization is feasible but, the way at least the HEVC software is written, it is not necessarily easy. What can be easily done is to take the HEVC integer transform and rotate the transform using the technique developed in [E392] for obtaining a rotated transform. Implement this in all profiles and compare with the HEVC HM software. Consider both options.

P.2.263 See P.2.262 Using SDCT and its integer approximation in HEVC invariably results in increased implementation complexity. Investigate this thoroughly.

P.2.264 Performance comparison of SDCT with DCT is shown in Tables I, II and IV [E392]. However the block sizes are limited to 32×32. For super high resolution videos it is suggested that even larger block sizes such as 64×64 are suggested. Extend these Tables for 64×64 block size.

P.2.265 See P.2.265. Extend these Tables to 4K and 8K video sequences.

P.2.266 Performance comparison of INTSDCT with DCT is shown in Table V [E392]. Extend this comparison to 4K and 8K video sequences and also larger block size such as 64×64.

P.2.267 In P.2.264 through P.2.266, consider implementation complexity as another comparison metric.

P.2.268 Huang et al., [E393] have developed false contour detection and removal (FCDR) method and applied it to HEVC and H.264 videos as post processing operation. They also state "It will be interesting to adopt it as part of the in loop decoding process (Figure 2.5). This idea demands further investigation and verification". Investigate this in detail and see how FCDR can be embedded as an in loop operation besides the deblocking and SAO filters. Assuming FCDR embedding is successful, compare this with FCDR as a post processing operation in terms of PSNR, SSIM and subjective quality (See the corresponding figures and tables in [E393].) using various test sequences.

can be applied to intra HEVC using various test sequences based on the standard performance metrics.

112 HEVC Encoder

P.2.269 See P.2.268. Consider implementation complexity as another comparison metric in the FCDR process (in loop vs. post processing) in both HEVC and H.264.

P.2.270 Chen et al., [394] proposed a novel block-composed background reference (BCBR) scheme and is implemented in HEVC. They claim that the new BCBR algorithm can achieve better performance than baseline HEVC. This technique is however limited to sequences captured by static cameras (surveillance and conference sequences). They also suggest, "For moving camera cases, the long-term temporal correlation due to background is also worthy of investigation and the block-composed sprite coding would be a good choice". Investigate this in detail. Consider also encoding and decoding complexity as another performance metric.

P.2.271 See P.2.270. The authors in [394] conclude "We would like to extend our BCBR to more generic cases in our future work". Explore this.

P.2.272 Min, Xu, and Cheung [E397] proposed a fully pipeline architecture, which achieves higher throughput, smaller area and less memory, for intra prediction of HEVC. They conclude "In the future work, we plan to implement the proposed architecture on ASIC platform to increase the system frequency, aiming at achieving real time video decoding of higher resolution". Implement this.

P.2.273 See P.2.272. Min, Xu, and Cheung [E397] further state "Meanwhile, the proposed architecture will be integrated with inter prediction engine, inverse transforming engine, and the other engines to construct the entire full scale encoding/decoding system." Explore this in detail and implement the same.

P.2.274 In [H.78], the authors state "We would like to extend this sketch attack framework to handle different video coding standards such as High Efficiency Video Coding (HEVC), Audio Video Standard (AVS) and Google VP9 as our future work." Extend this attack to HEVC Main Profile.

P.2.275 In implementing the image codec based on SDCT, Francastoro, Fosson and Magli [E392] fixed the 8 quantization levels for the angles distributed uniformly between 0 and π. They state "In order to improve the compression performance, as future work, we may consider a non-uniform angle quantization." Investigate this thoroughly. Consider all possible non-uniform angle quantizations, the main objective being improved compression performance.

2.10 Projects   113

P.2.276 Wang et al., [E399] have described the MPEG internet video coding (IVC) standard and compared its performance (both objective and subjective – see Figures 10–13 for RA and LD) with web video coding (WVC), video coding for browsers (VCB) and AVC high profile. They have also compared the coding tools of IVC with those of AVC HP (Table IV). Using the test sequences and their configurations (Tables V–VII) confirm the results shown in Table VIII and Figures 10–13. This project requires detailed and extensive simulations and intensive viewing by nave and expert subjects using the double stimulus impairment scale (DSIS).

P.2.277 See P.2.276. Simulations are based on classes A, B and D test sequences (See Table V in [E399]). Extend these simulations to classes C, E, and F test sequences and draw the conclusions based on the IVC, WVC, VCB and AVC HP. As in P.2.276 this project also requires detailed and extensive simulations and intensive viewing by nave and expert subjects using the double stimulus impairment scale (DSIS).

P.2.278 In the conclusion section Wang et al., [E399] state "Our team has set up an open source project xIVC, which aims to develop a real time IVC codec for HD sequences. As of now, the decoder of XIVC can decode 1080P sequences in real-time on PC platform". Develop this decoder.

P.2.279 See P.2.277. Develop the real time IVC codec (encoder/decoder) for HD sequences.

P.2.280 Xu et al., [E400] have developed video saliency detection using features in HEVC compressed domain and claim that their method superior to other state-of-the-art saliency detection techniques. They also state that their method is more practicable compared to uncompressed domain methods as both time and storage complexity on decoding videos can be saved. In the conclusions and future work the author's state

There exist three directions for the future work. (1) Our work in its present form merely concentrates on the bottom up model to predict video saliency. In fact, videos usually contain some top-down cues indicating salient regions, such as human faces. Indeed, an ideal vision system, like the one of humans, requires the information flow in both directions of bottom-up and top-down. Hence, the protocol, integrating the top-down model into our bottom-up saliency detection method, shows a promising trend in future. Investigate this in detail.

P.2.281 See P.2.280. "Many advanced tracking filters (e.g., Kalman filter and particle filter) have emerged during the past few decades. It is quite an interesting future work to incorporate our method with those filters, rather than the forward smoothing filter of this paper. In that case, the performance of our method may be further improved." Replace the forward smoothing filter adopted in [E400] by Kalman filter and particle filter and evaluate how the performance of the video saliency detection can be further improved.

P.2.282 See P.2.280. "A simple SVM learning algorithm, the C-SVC, was developed in our work for video saliency detection. Other state-of-the-art machine learning techniques may be applied to improve the accuracy of saliency detection, and it can be seen as another promising future work". Explore this future work.

P.2.283 See [E355]. In conclusion, the authors state "In the future more study can be made on how to facilitate software and hardware implementation." Implement this.

P.2.284 See P.2.283. The authors [E355] further state "We also leave it as a further study how to efficiently apply CCP to various applications including high dynamic range video coding." Explore this for HDR video coding.

P.2.285 See [E355]. Following the range extensions in HEVC (see the references at thee end), the authors have demonstrated by using the CCP, significant coding performance improvements for both natural and screen content video. Verify this by confirming the experimental results shown in Table I through V and Figures 3 and 4.

P.2.286 See [E401]. Section V Conclusion is repeated here:

In conclusion, it is evident that encoder could generate decoder resource optimized video bit streams by exploiting the diversities of the decoder complexity requirements of the HEVC coding modes. In this context, the proposed complexity model for HEVC inter-frame decoding predicts the decoding complexity with an average prediction error less than 5% for both uni- and bi-predicted frames. Furthermore, the proposed encoding algorithm is capable of generating HEVC bit streams that can achieve an average decoder complexity n reduction of 28.06% and 41.19% with a BD-PSNR

2.10 Projects 115

loss of −1.91 dB and −2.46 dB for low delay P and random access configurations, respectively, compared to the bit streams generated by the HM reference encoder. The future work will focus on extending the framework to consider both the rate and the distortion alongside the decoder's complexity, to generate more optimized HEVC video bit streams.

Explore this future work and see how you can optimize this.

P.2.287 See [E402]. Section V Conclusion is repeated here:

In this work, an unthreaded version of the OpenHEVC decoder supporting parallel decoding at slice level is presented. Based on this decoder, implementations for several multicore chips, with and without threads support, have been tested with good results in both, performance and speedup. In the future, parallel decoding at frame level will be integrated, which certainly will improve the speedups with more than 4 cores

Integrate the parallel decoding Cabaratat the frame level and evaluate how the HEVC decoder can be speeded up.

P.2.288 A novel perceptual adaptive quantization method (AVQ) based on an adaptive perceptual CU splitting that improves the subjective quality of HEVC encoder along with less encoding time is proposed in [E403]. Confirm the simulation results shown in Table I and Figures 3 and 4.

P.2.289 See P.2.288. In Section 5 Conclusion, the authors [E403] state, "In the future, we will take the both the spatial and temporal characteristics into consideration to improve coding performance further". Explore this.

P.2.290 In Section V, the authors [E404] state, "The performance of proposed algorithm is more noticeable at high packet error rate and low bit rate conditions. Therefore, the proposed algorithm is suitable for video transmission applications in error-prone wireless channels which have bandwidth constraints. The perceived video quality can be improved by applying more sophisticated error concealment technique at the decoder." Apply error concealment techniques at the HEVC decoder and investigate how the subjective video quality can be improved.

P.2.291 Boyadjis et al., [E405] have proposed a novel method for selective encryption of H.264/AVC (CABAC) and HEVC compressed bit streams. And compared with other encryption schemes (see the extensive list of references at the end). See the last para in the conclusion and future work of his paper. Explore these topics (studies).

P.2.292 Cabarat, Hamidouche, and Deforges [E410] have implemented a software parallel hybrid SHVC decoder (AVC decoder for BL and HEVC decoder for EL) (see Figure 2). The software design enables a real time decoding of the HEVC EL at 2160p60 while the AVC base layer is decoded at 1080p60 for $x2$ spatial scalability. Implement this SHVC software and confirm the results listed in Tables 1–4. The software sites listed in the references will be very helpful.

P.2.293 See P.2.292 The authors [E410] state that only RA configuration is used for hybrid SHVC decoder. Extend their research to AI and LD configurations. Based on the simulation results, what conclusions can be made?

P.2.294 See [E405] and Section V Conclusion.

Implement the proposed extended selective encryption and reconstruct the encrypted images shown in Figures 6.8. Repeat this scheme using other test sequences.

P.2.295 See P.2.294

In section V Conclusion, the authors state

"We did not provide in this paper an exhaustive analysis of CABAC regular mode cipherable syntax elements. Some of them are easily identifiable (for example, the process proposed in this paper could almost identically apply to the intra-prediction modes for chroma components), but a large majority of these cipherable elements would require a very specific monitoring, which we did not address in our study." Carry out this exhaustive analysis.

P.2.296 See P.2.295 and P.2.296.

In section V Conclusion, the authors further state

"An analysis of the *maximum* ES for both H.264/AVC and HEVC, based on the proposed *regular* mode encryption, is thus one major track for future research. Nevertheless, such a study would have to pay particular attention to a critical tradeoff we highlighted in our study and the underlying correlation between the improvement in the scrambling performance and its consequences on the compression efficiency."

Carry out this further research.

P.2.297 Oliveria et al., [DCT34] have implemented video coding based on low-complexity approximation of DTT and compared with

2.10 Projects 117

H.264/AVC. Replace the integer DCT in HEVC [E61] by the integer DTT and compare its performance with the standard HEVC in video coding. Develop graphs similar to Figures 1 and 2 in [E62]. See DCT-P28 through DCT-P34.

P.2.298 See P.2.297. The rate distortion graphs shown in Figures 1 and 2 in [E62] are for 1280×720 60 HZ and 1920×1080, 24 Hz video test sequences and class E and class B (see Table VIII in [E62]). Extend the performance comparison (integer DTT and integer DCT) based on video test sequences–classes A, C, D and E' listed in Table VIII.

P.2.299 Lainema et al., [E78] provided an overview of intra coding techniques in HEVC and compared its performance with H.264/AVC intra coding. Replace INTDCT by INTDTT [DCT34] in HEVC intra coding and compare its performance with H.264/AVC intra coding. Develop tables similar to those in the work done by Lainema et al., [E78] using the test sequences (Table IX).

P.2.300 The encoder block diagram for screen content coding is shown in Figure 2.3. Replace the blocks "Transform & Quant" and "De-quant & Inv. Transform" by the blocks "DTT & Quant" and "De-quant and Inverse DTT" respectively. Implement the SCC coding with the DTT/Inverse DTT and compare the results shown in [6] of Chapter 2 Screen Content Coding. Based on these results what conclusions can be drawn. This requires thorough understanding of screen content coding extension of HEVC as described in the review paper [6] of Chapter 2. Details on DTT are described in [E62].

P.2.301 Atta and Ghanbari [H79] developed an efficient rate control scheme for a joint temporal-quality scalability in H.264/AVC. They showed that this algorithm achieves better coding efficiency with low computational complexity compared to earlier ones. Explore how this rate control scheme can be adopted for a joint temporal-quality scalability in HEVC/H.265. See the overview paper [OP6] on scalable coding in HEVC (SHVC).

P.2.302 The performance of HEVC is compared [E62] with earlier standards such as H.262/MPEG-2 MP video, H.263 HLP, MPEG-4 ASP visual, and H.264/MPEG-4 AVC HP based on PSNR and subjective tests. The used test sequences are listed in Table VIII [E62]. Confirm the objective test results by accessing the software from the ftp site listed at the end just before the appendix in [E62].

P.2.303 See P.2.302 Compare the performance of HEVC with AVS China.

P.2.304 See P.2.302 Compare the performance of HEVC with DIRAC (BBC).
P.2.305 See P.2.302 Compare the performance of HEVC with DIRAC (BBC).
P.2.306 See P.2.302 Compare the performance of HEVC with THOR video codec.
P.2.307 See P.2.302 Compare the performance of HEVC with WMV9 (VC1).
P.2.308 See P.2.302 Compare the performance of HEVC with VP9.
P.2.309 See P.2.302 Compare the performance of HEVC with DAALA.
P.2.310 See P.2.302 Compare the performance of HEVC with AV1 codec (AOM).
P.2.311 See [E413] and the conclusions. By incorporating separately three Parts 1) AIP with SDT-SVD, 2) TMP with SDT-SVD and 3) AIP with SDT-SVD+TMP (see Figure 5) the authors have showed some improvements in HEVC intra coding performance. However the encoding/decoding complexities are much higher (See Section C: computational complexity).

They state "In the future, fast algorithms, such as early skip for SDT-SVD transform and fast template searching method, can be exploited to reduce the complexity of the proposed method." Explore this.
P.2.312 See P.2.311. The authors further state "To further improve the efficiency of the SDT-SVD transform, more work can be done to design an efficient block with more similar structures as the residual block for deriving SVD transforms." Explore this.
P.2.313 See [E414]. The authors have integrated several techniques including AC energy of the CU that has resulted in significant reduction in encoder complexity for all intra mode (up to 71% reduction) with negligible increase in bit rate at nearly the same rate distortion. Review this paper thoroughly and confirm the simulation results shown in Figure 3.
P.2.314 In the conclusions, the authors [E417] state

"The proposed RDO-QPC can also be incorporated into rate control for the random-access configuration in H.265/HEVC to achieve better R-D performance, which is currently under our investigation." Explore this.
P.2.315 See P.2.314. The authors [E414] further state "So far the issue of scene changes has not been considered in this paper. But, the

proposed algorithm can be well adapted to the scenario with scene changes, where detection of scene changes is the key (while out of the scope of this paper). Once the scene change is detected, the first two frames can be pre encoded to obtain the optimized QP for the subsequent frames in the same scene". Explore this.

P.2.316 See [E418]. In the conclusions the authors state "The proposed method can be seamlessly incorporated into the SHVC encoding with little increase in computational complexity". Please incorporate this in SHVC and evaluate the performance.

P.2.316 W.-S. Kim et al., [E355] have demonstrated the improvement in compression efficiency by incorporating the cross component prediction (CCP) in HEVC (both for RGB and YCbCr coding). The CCP has been adopted in HEVC RExt 4:4:4 profiles. In the conclusions the authors state, "In the future, more study can be made on how to facilitate software and hardware implementation." Develop the software and hardware implementation of CCP in HEVC.

P.2.317 See P.2.316 The authors [E355] further state, "We also leave it as a further study how to efficiently apply CCP to various applications including high dynamic range video coding". Explore this.

P.2.318 Min, Xu and Cheung [E397] have developed a fully pipelined architecture for intra prediction of HEVC that can support 3840×2160 videos at 30 frames/sec. In future work the authors state, "we plan to implement the proposed architecture on ASIC platform to increase the system frequency, aiming at achieving real-time video decoding of higher resolution." Explore this. Note that besides hardware capability familiarity with ASIC is required.

P.2.319 See P.2.318. The authors [E397] further state "Meanwhile, the proposed architecture will be integrated with inter prediction engine, inverse transforming engine, and the other engines to construct the entire full scale encoding/decoding system". Implement this. This requires team effort and dedicated facilities.

P.2.320 See [E419]. The authors have developed a quality (SNR) scalability in HSVC [E325] that is superior to the benchmark and at reduced encoder complexity. At the end the authors state, "As future work, the proposed DSVC-HEVC framework can be extended to support spatial scalability and the SI (side information) creation mechanism can be further improved by including a more powerful machine learning technique to perform the SI selection". Explore this.

120 HEVC Encoder

P.2.321 See [E76]. Shu, Sun and Wu have proposed SSVC scheme for HEVC that is much more effective than the prior schemes. In section VII Conclusions, they state "advanced learning and mapping methods should be especially at high bit rates investigated to enhance the quality of the refined prediction schemes.' Explore this.

P.2.322 See P.2.321. Shu, Sun and Wu [E76] further state "The quadtree structure and the correlation between two layers should be further exploited to reduce the overhead bits of Enhancement layer." Investigate this.

P.2.323 See P.2.321 and P.2.322. The authors [E76] conclude "Joint optimization in terms of determining both motion vectors as well as quad-tree structures should be further studied with regard to both the base layer and enhancement layer coding." Explore this.

P.2.324 See the book by M. Wien [Book2] in books on HEVC.

In Section 4.1.3 (pp. 112–113) splicing of video sequences is described. Please access any two different video sequences (See test sequences) and see how they can be spliced together (See Figure 4.7 for illustration). This implies that both the video sequences need to be coded in the same HEVC profile and the coded video sequences are to be spliced into a joint NAL unit stream. To verify the successful splicing, decode the joint NAL unit stream and display the spliced video sequences. This is a major project.

P.2.325 See [E426]. Bilal et al., have proposed detection framework that is computationally less expensive as well as more accurate than HOG-linear SVM (hog: Histogram of oriented gradients). This paper has not only an excellent literature search on pedestrian detection schemes including survey papers and also web sites related to pedestrian databases (see Figure 11) to train and test the performance of detectors on video sequences shot in urban environments. Also MATLAB evaluation code for standardized pedestrian detection datasets [65] and FPGA-GPU architecture (also other hardware oriented papers [67–76]) are provided. Go through this valuable paper including all related references and implement this efficient pedestrian detection scheme. Confirm the performance results shown in the Figures and Tables [E426]. Note that this a major project and involves group effort (with different skills). Both hardware/software capabilities are required. In conclusion section, the authors [E426] state "The proposed HSG-HIK framework is proved to be a better candidate than HOG-linear SVM for incorporation in advanced

2.10 Projects   121

pedestrian detection schemes employing hybrid features and cascaded detectors". (HSG: Histogram of significant gradients, HIK: Histogram intersection kernel).

P.2.326 See [E427]. Conclusion section is repeated here for ready reference.

In this paper, we have proposed an NALF based on image low-rank prior for video compression. Every video picture is divided into different image patch groups based on their similarity. Soft-thresholding operation is utilized to solve the proposed compression noise reduction problem. We propose an efficient method to derive content-dependent thresholds for every image patch group according to image content characteristics, QPs, and coding modes. We also explore the appropriate position to integrate our proposed NALF in the current HEVC framework and the filter ON/OFF control mechanism at different levels to achieve better coding performance. Experimental results show that the proposed NALF can further improve the compression performance for HEVC, especially for video sequences with lots of similar structure patterns. As one of the first attempts on the coding performance improvement utilizing image nonlocal correlation, the proposed method also has the computational complexity issue that should be improved in the future. This proposes new challenges to the loop filter research with image nonlocal correlations and opens up new space for future exploration.

Go through this paper in detail. Explore how the computational complexity of the proposed method can be reduced, we would like to extend this sketch attack framework to handle different video standards such as High Efficiency Video Coding (HEVC), Audio Video Standard (AVS) and Google VP9 as our future work.

P.2.327 See [H78]. In the conclusions the authors state "we would like to extend this sketch attack framework to handle different video standards such as High Efficiency Video Coding (HEVC), Audio Video Standard (AVS) and Google VP9 as our future work". Extend this research to HEVC.

P.2.328 See [E355]. In the conclusions, the authors state, "In the future, more study can be made on how to facilitate software and hardware implementation. We also leave it as a further study how to efficiently apply CCP to various applications including high dynamic range video coding." Explore these further studies. These are major projects.

P.2.329 E.M.-Enriquez et al., [E429] have applied directional transforms for video coding based on lifting on graphs. These transforms include curvelets, contourlets, bandlets, directionlets, directional DCTs etc.

Conclusions and Future Work are Repeated Here

A broad class of graph-based lifting transforms and their optimization for video coding have been proposed. These transforms follow three-dimensional (spatio-temporal) high-correlation filtering paths through the video signal, and can be considered a generalization of classical separable wavelet-based encoders. To obtain an efficient graph-based lifting transform for video coding, some optimization problems have been discussed, namely: (i) the construction of suitable graph representations of the original video signal, including the graph weighting, that aims to capture the correlation between samples; (ii) the design of prediction filters for a given arbitrary weighted graph and update filters that are orthogonal to the prediction filters of its neighbours; (iii) U/P assignment techniques that aim to find a bipartition of the graph that leads to an efficient transform, and (iv) the extension of the transform to multiple levels of decomposition. Besides, we have proposed new reordering and entropy coding methods to obtain a complete coding scheme. The proposed method shows improved performance over a MCTF-based encoder and a simple encoder derived from H.264/AVC (JM15.1 configured to use similar tools as our proposed encoder), outperforming the JM15.1 configured to use 5 reference frames and all modes enabled, but subpixel ME disabled, at medium-high qualities for some sequences.

There are some interesting directions for future work. The flexibility of the transform and the good results obtained in a video coding application provides confidence that it may be successfully applied in a broad kind of signals and applications. For example, it may be used for multichannel-audio coding, image and video denoising, or biomedical signals compact representation, where one usually has multiple signals that present correlation in different domains (e.g., data extracted from the temporal evolution of different brain sensors resent spatial and temporal correlation). Our current work is focused in the rate-distortion optimization of the encoder and the incorporation of subpixel ME.

Explore all these Directions for Future Work

P.2.330 See P.2.329. The proposed method is compared with JM 15.1 (H.264). Extend this comparison to HEVC/H.265.

2.10 Projects   123

P.2.331 See P.2.329 The complete encoder/decoder block diagram for directional transforms video coding is described in Figure 6 [E429]. Implement this codec and confirm the results shown in Figures 7 and 8 and Table III. This implies implementing H.264 and MCTF encoders.

P.2.332 See [E429]. In the conclusions the authors state, "Our current work is focused in the rate-distortion optimization of the encoder and the incorporation of subpixel ME". Explore this.

P.2.333 Apply the encoder/decoder block diagram described in Figure 6 [E429] and compare its performance with AVS China (IEEE 1857.4) codec.

P.2.334 See P.2.333. Extend this performance comparison to VP9 codec.

P.2.335 See P.2.333. Extend the performance comparison to AV1 (alliance for open media) codec.

P.2.336 See [E429]. Performance comparison of video coding by directional transforms with H.264 (See Figure 6) is based on PSNR vs. bit rate for the test sequences. PSNR, however, is not the best representation of perceptual video quality. A better reflection of perceptual video quality is structural similarity metric (SSIM). Several references related to SSIM are listed. Implement this comparison based on SSIM for all the test sequences.

P.2.337 See [E430] and the conclusion section (last paragraph is repeated here)

A possible future work of this paper can be extension of the bottom-up motion vector prediction to FME search and merge mode decision, which simply exploits correlation among motion vectors of blocks in different hierarchical depths of the block. However, the proposed bottom-up MV prediction may not be effective if it is applied to FME as it is. Because the MV correlation between different depths is not large enough to be able to predict the fractional MV. Therefore, additional study on bottom-up encoding order is required for FME.

Explore this Future Work

P.2.338 See [E412]. The section Conclusion is repeated here,

There are great challenges to implementing IME under the HEVC standard. This paper has presented a hardware-oriented IME algorithm that achieved a decrease in BD rate as compared to HM 15.0.

Our corresponding hardware solution succeeded in supporting 4K × 2K@ 30 fps videos with a working frequency of no more than 500 MHz. These results demonstrate that our proposed solution offered desirable improvement in both coding speed and coding performance. For future study, we will investigate other possible algorithms further and test them on the proposed hardware platform.

Investigate these algorithms ad test them on the proposed hardware platform. This is a major project that requires the expertize of research groups and excellent lab (hardware/software) facilities.

P.2.339 See [E379] and P.2.245 through P.2.249. Replace AV1 codec in P.2.247 by DIRAC.

P.2.340 See [E433]. Section 7.5 Conclusions and future work is repeated here:

This thesis consists of implementation of serial and parallel algorithms to improve the computation efficiency of the SAO in HEVC. The algorithm is implemented in parallel using NVIDIA GPU platform programmed with CUDA language. Compared with the original serial algorithm implemented on CPU, the experimental results show that parallel algorithms can provide a significant improvement on the encoding time with an overall more than 25x speedup for high definition video sequences.

Post processing methods can be used to further improve the quality of the video. Implementing this post processing filter in parallel would be a much efficient way of computation. A significant amount of time is spent in transferring data from host to device and vice versa. In the future there is scope to decrease the overhead of pushing data to GPU and CPU back and forth. Research on increasing parallel data throughputs for further improving the encoding time can be a part of the future work.

Implement this future work. Note that there are many parts in this future work.

P.2.341 See P.2.340. Harsha's thesis [E433] involves implementation and comparison of serial and parallel algorithms of SAO in HEVC. Implement similar algorithms for deblocking filter in HEVC and show the results.

P.2.342 See P.2.340 and P.2.341. Combine the two i.e., parallel implementation of both SAO and deblocking filter in HEVC and show the results.

2.10 Projects 125

P.2.343 See [E432] thesis by Swaroop Krishna Rao, Section 10 Future work is repeated here.
This thesis mainly focusses on encoding the 360-degree video. One can research on decoding side of 360-degree video.
Explore about the decoding side of 360-degree video.

P.2.344 See P.2.343 "Also, future research on projection scheme may lead to a new projection scheme, which may be efficient than present schemes." Explore this.

P.2.345 "One can use CUDA to parallel program the proposed scheme, so that computational time can be reduced with greater speedups and without any loss in quality." Explore this.

P.2.346 See P.2.343 "The proposed scheme can be implemented on other video codecs such as VP10, DAALA and AV1 for comparison and for 8K videos in future." Investigate and implement these projects.

P.2.347 See [E438] and go through all the references listed at the end. The future work on reference clip may proceed in several directions. First, integrating reference clip into random-access setting is necessary. Investigate this.

P.2.348 See P.2.347. Second, how to allocate memory between reference frames and reference clips for different videos is important to improve coding efficiency. Explore this.

P.2.349 See P.2.347. Third, using external data like a prebuilt dictionary of clips in video coding is very interesting to study. This is another challenging project.

P.2.350 See [E438]. The section F. Computational Complexity is repeated here. Table VII presents the encoding and decoding time results of our scheme compared to the anchor. Under the same memory constraint, using clips in addition to frames for reference will increase the computational complexity of both encoder and decoder, but the increases are limited. For encoder, the increased computations are cost mainly on retrieving clips. The computations of both histogram matching and fast motion estimation are dependent on the size of clips, and thus increase (TABLE VII ENCODING AND DECODING TIME OF OUR SCHEME ANCHORED ON HEVC Class LDB LDP Enc. Time Dec. Time Enc. Time Dec. Time Class B 165% 135% 199% 137% Class C 117% 119% 124% 122% Class D 111% 110% 117% 110% Class E 127% 122% 137% 123% Class F 120% 137% 129% 135% Average 128% 125% 141% 126%) more for higher resolution videos. More advanced retrieval methods can be developed to accelerate. Explore this.

P.2.351 See P.2.350. "For decoder, the complexity increases come from generating and managing clips, which is not significant in encoder. Note that it is possible to record the positions of the generated clips in the bitstream to avoid decoder's computation. Investigate this and show how this can reduce the decoder complexity for the LDB and LDP test sequences.

P.2.352 See [E440]. The authors state that based on the proposed CCML algorithm they are able to significantly reduce the implementation complexity for the HEVC intra encoder. Go through this paper in detail and verify the test results described in the tables and graphs.

P.2.353 See [E441]. Part of the conclusion and discussions section is repeated here.
"In proposed TL-λ-PRC, Firstly, a given sequence will be classified into slow motion or fast motion category based on the motion of the sequence. Then sequences in different category will be treated differently. The tested sequences in this paper are the specified tested sequences for H.265/HEVC. The classification is processed on the basis of sequences. However, a motion classification at picture level or GOP level or consider λ-RC and TL-λ-PRC as selectable modes may further bring performance improvement compared with TL-λ-PRC. This is also a valuable following research. Explore this follow on research.

P.2.354 See [E445]. Extensive and detailed performance comparison of AV1, JEM, VP9 and HEVC encoders is implemented. Implement this and verify the results shown in Tables 5–12 and in Figures 1 and 2. Use improved and updated software. Note that this is a major project and may require group effort. Performance comparison of other video coding standards are cited in references 13–15 of [E445].

P.2.355 See [E438] and the conclusion section. The HEVC encoder block diagram is modified to include also reference clips for inter prediction. Draw the corresponding decoder block diagram and identify the additional blocks.

P.2.356 See P.2.355. In [E438], Ma et al., have implemented the reference clip inter prediction for low delay B and low delay P configurations only. Extend this process to random access configuration also.

2.10 Projects   127

P.2.357 See P.2.355 and P.2.356. Reference clip inter prediction is implemented at the CTU level only. Extend this to CU and PU levels also. Evaluate the implementation complexity at these levels.

P.2.358 In [E438], reference clip inter prediction is implemented for test sequences (classes B through F–resolutions 1920×1080 through 416×240). Extend this to 4K and 8K video test sequences and compare with anchor HEVC coder.

P.2.359 Extend the reference clip inter prediction proposed by Ma et al., [E438] for HEVC to VP9 video codec.
Access https://www.webmproject.org/vp9 for details on VP9.

P.2.360 See P.2.359. Extend this to JEM proposed by JVET. ACCESS JVET JEM software (JEM Joint Exploration Test Model) https://jvet.hhi.fraunhofer.de

P.2.361 See P.2.361. Extend this AV1 codec finalized by AOM.
AV1 codec (Alliance for open media – AOM)
http://tinyurl.com/zgwdo59
Website: http://www.streamingmedia.com/Articles/Editorial/What-Is-.../What-is-AV1-111497.aspx

P.2.362 See [E446]. The section Conclusion and future work is repeated here.
"In this paper, we proposed a novel video coding framework that uses a self-adaptive redundant dictionary consisting of inter and intra-prediction candidates as dictionary atoms, to directly find a sparse representation of video frame blocks. To solve the sparse coding problem in the context of video coding, a rate-distortion aware orthogonal least squares algorithm is proposed, which iteratively finds the orthonormalized atom with largest bits-normalized distortion reduction. We further propose a two-stage coding framework with a RD aware adaptive switching algorithm. The proposed coder is extended to incorporate variable block sizes and a fast partition mode decision algorithm is proposed. We show the proposed coder can outperform the latest video coding standard HEVC with P-frame coding in the lower bit rate range and is competitive with that at the higher rate range.

Although our proposed coder has only shown limited gain over HEVC P-frame, and falls short to HEVC with B-frame coding, additional gains are likely with further optimization of the various components. First of all, the rate-distortion behaviours of both the first stage for coding the SeAD atoms and the second stage for coding the DCT atoms exhibit a very smooth trend, indicating that some analytical models for the rate-distortion curves of these two stages may be established, which may be utilized to yield a better rate distortion optimized decision on the adaptive switching between the two stages and on the block partition decision." Explore along these lines.

P.2.363 See P.2.362 "Secondly, how to design a better SeAD dictionary and a better second stage dictionary can be further explored. Specifically, the proposed framework only uses atoms from the previous frame in SeAD, while one could investigate using multiple frames, with both previous and future frames. For the second stage dictionary, although our piloting study on designing a special fixed dictionary through dictionary learning returned similar result as with using the DCT, other new development along this line might be borrowed to design a better dictionary or to adapt the dictionary."

Investigate these research directions.

P.2.364 See P.2.362 and P.2.363 "Thirdly, we believe significant gains are possible with more careful design of the entropy coder for coding the first stage atom locations. Lastly, although our preliminary attempt at extending the two stage coding frame work to intra-frame coding has not yielded promising results, further optimization of the entropy coder, especially for coding the intra-prediction candidate locations, is likely to make it more competitive than HEVC." Explore these research areas.

Note that P.2. 362 through P.2. 364 are aimed at improving the efficiency of the proposed encoder and at the same time reducing the implementation complexity.

P.2.365 See [E447] The section Conclusion is repeated here:

This paper presents a fast IME algorithm designed for HEVC that supports a highly flexible block partitioning structure. This new algorithm reduces its computational complexity significantly by searching narrow regions around multiple accurate MVPs. The accurate MVPs are derived from the MVs of the PUs in the CUs in the lower layer in the coding block hierarchy. Thanks to the accuracy of the bottom-up MV prediction, the reduced search range does not significantly affect the degradation of the coding efficiency. Experimental results show that the proposed IME reduces computational complexity up to 81.95% with 0.39% BD-BR increase. Furthermore,

2.10 Projects 129

the proposed IME is designed to maintain compatibility with other fast algorithms that are based on top-down encoding order [37–50], which is a crucial issue when designing a new scheme because many fast algorithms can conflict with each other, thereby not being able to achieve performance improvement when they are used together. In order to keep compatibility with HEVC standard, the bottom-up MVPs are utilized for only IME and bitstream is generated by AMVP. A possible future work of this paper can be extension of the bottom-up motion vector prediction to FME search and merge mode decision, which simply exploits correlation among motion vectors of blocks in different hierarchical depths of the block. However, the proposed bottom-up MV prediction may not be effective if it is applied to FME as it is. Because the MV correlation between different depths is not large enough to be able to predict the fractional MV. Therefore, additional study on bottom-up encoding order is required for FME.

Explore thoroughly this future work.

P.2.366 See [E449]. The conclusion section is repeated here.

"There are great challenges to implementing IME under the HEVC standard. This paper has presented a hardware-oriented IME algorithm that achieved a decrease in BD rate as compared to HM 15.0. Our corresponding hardware solution succeeded in supporting 4K × 2K@ 30 fps videos with a working frequency of no more than 500 MHz. These results demonstrate that our proposed solution offered desirable improvement in both coding speed and coding performance. For future study, we will investigate other possible algorithms further and test them on the proposed hardware platform.

Implement this future study.

P.2.367 See [E448]. Can this hardware implementation be extended to AV1/VP9 encoders. http://aomedia.org

P.2.368 See [E355], P.2.206 and P.2.207. Can this cross component prediction be applied to AV1/VP9 encoders. Investigate thoroughly. http://aomedia.org

P.2.369 As AV1 coder relates to video only, please multiplex the coded bit stream with HE-AAC version 2 audio bit stream followed by demultiplex and decode of the two bit streams and maintain the lip synch. See[E379]. See also (arib std-b32 version 3.9-E1, "Video coding, Audio coding, and Multiplexing Specifications for Digital Broadcasting", 2016.)

P.2.370 See [E449]. "The conclusion section is repeated here. There are great challenges to implementing IME under the HEVC standard. This paper has presented a hardware-oriented IME algorithm that achieved a decrease in BD rate as compared to HM 15.0. Our corresponding hardware solution succeeded in supporting 4K × 2K@ 30 fps videos with a working frequency of no more than 500 MHz. These results demonstrate that our proposed solution offered desirable improvement in both coding speed and coding performance. For future study, we will investigate other possible algorithms further and test them on the proposed hardware platform."

Explore these algorithms and see how you can implement them in the proposed hardware platform.

P.2.371 See P.2.370. In [E449] hardware implementation of integer ME algorithm in HEVC is carried out. As HEVC includes fractional ME [E202] see how you can develop a hardware oriented FME algorithm followed by hardware implementation.

P.2.372 See [E451]. Abstract is repeated here.

"The increased compression ratios achieved by the High Efficiency Video Coding (HEVC) standard lead to reduced robustness of coded streams, with increased susceptibility to network errors and consequent video quality degradation. This paper proposes a method based on a two-stage approach to improve the error robustness of HEVC streaming, by reducing temporal error propagation in case of frame loss. The prediction mismatch that occurs at the decoder after frame loss is reduced through the following two stages: (i) at the encoding stage, the reference pictures are dynamically selected based on constraining conditions and Lagrangian optimization, which distributes the use of reference pictures, by reducing the number of prediction units (PUs) that depend on a single reference; (ii) at the streaming stage, a motion vector (MV) prioritization algorithm, based on spatial dependencies, selects an optimal sub-set of MVs to be transmitted, redundantly, as side information to reduce mismatched MV predictions at the decoder. The simulation results show that the proposed method significantly reduces the effect of temporal error propagation. Compared to the reference HEVC, the proposed reference picture selection method is able to improve the video quality at low packet loss rates (e.g., 1%) using the same bitrate, achieving quality gains up to 2.3 dB for 10% of packet loss ratio. It is shown, for instance, that the redundant MVs are able to boost the performance achieving quality gains of 3 dB when compared to the reference HEVC, at the cost using 4% increase in total bitrate."

Implement this proposed robust (error resilient) HEVC coding and streaming under various packet loss ratios and evaluate its resilience and recovery wen the coded bit stream is transmitted over noisy channels.

P.2.373 See [E452]. A section of this paper related to HEVC is reproduced here.

"Over the past two decades, a great deal of research using the compressed domain has been conducted, with most studies focusing on existing compression standards, including MPEG2, MPEG-4 and H.264/AVC. The latest compression standard, H.265/HEVC, is now replacing the others as it contains many advanced features. Although some researchers have embedded the watermark in the H.265/HEVC domain, as very few papers have been published, this field is relatively young. Therefore, it seems that there is a great deal of scope to explore this area for developing a watermarking technique. As the robustness of existing techniques to image or video compression are assessed using any of the existing compression methods, there is potential for further research into watermarking techniques that can deal with H.265/HEVC compression which is the next-generation compression standard."

Explore this and develop a robust imperceptible watermarking scheme for HEVC encoded video.

P.2.374 See [E452] and P.2.373. Repeat this for AV1 encoded video.

P.2.375 See [E453]. The section conclusion is repeated here.

"HEVC has achieved high coding performance in comparison with former coding standards. It is an open problem on how to design more efficient transforms to improve coding efficiency. In this paper, we propose the variable block-sized Signal Dependent Transforms (SDTs) and implement them in HEVC and JEM to explore a possible way to enhance performance. Being capable of exploring non-local correlations and adaptive to image/video contents, the proposed SDTs have high coding efficiency. Experimental results demonstrate that the proposed transform brings the BD-rate savings of 2.2%, 2.4%, 3.3%, and 7.1% on average for AI-Main10, RA-Main10, RA-Main10, and LP-Main10 configurations, compared to the test model HM-12.0 of HEVC. Thanks to the introduction of the large block size SDTs, the BD-rate savings is up to 23.3%."

132   HEVC Encoder

The last sentence before the CONCLUSION is repeated here. "But the further exploration of fast algorithms for fast searching and derivation of bases are still expected and will be left as future work". Explore this in detail.

P.2.376 See [E454]. J. He et al., "Adaptive quantization parameter selection for H.265/HEVC by employing inter-frame dependency", IEEE Trans. CSVT, Early access.

Part of the conclusion and future work is repeated here. "In the future we will continue to work on the following issues. First, this paper only focuses on the QP selection problem for the RA coding structure. In the RA structure, the frames in the highest coding level (L4) are never used as references, thus the distortion propagation ends at this level. However, in the LB and LP coding structures, all frames are used as references, which results in a more complicated distortion propagation. Therefore, it would be desirable to further investigate how to appropriately utilize similar approaches on LB and LP coding structures." LB and LP are low delay B frames and low delay P frames. RA is random access. Investigate along these lines.

P.2.377 See P.2.376. "Second, in the proposed algorithm, the distortion dependencies are formulated at the frame level by using the energy of the average prediction residuals of a frame. However, due to the fact that different areas of a frame may have different impacts on the future frames, further gains may be achieved by extending the proposed algorithm and distortion dependency model to the CTU level." Explore tis and see if additional gains may be achieved.

P.2.378 See P.2.376 and P.2.377. "Third, the proposed algorithm offers a guidance for rate control, and it can be easily combined with the current R-λ based rate control scheme in HM [36]. Thus potential gains can be expected by utilizing the similar approach in the rate control scheme." Investigate this and see if any potential gains can be achieved.

P.2.379 See [E456]. The section conclusion Is repeated here.

"In this paper, we proposed a very deep residual highway convolutional neural networks (RHCNN) in-loop filtering in HEVC. It is made up of several cascaded highway units, convolutional layers and shortcuts. Each highway unit contains two small convolutional layers, followed by ReLU activation layers and one identity skip connection. It is noteworthy that the shortcut helps on recovering clean image details and tackling the gradient vanishing problem. The RHCNN is employed to approximate the inverse processing of the quantization components, which are complicated nonlinear ingredients in HEVC. To improve the accuracy and robustness, the entire quantization parameter (QP) range is divided into several QP bands, where a dedicated RHCNN is trained for each QP band. Furthermore, we adopt a progressive training scheme for the RHCNN where the QP band with lower value is used for early training and their weights are used as initial weights for QP band of higher values in a progressive manner. Extensive experimental results validate the excellent performance of our RHCNN. In addition, except QP and frame type, the compression distortion is also affected by the texture complexity, motion types of moving objects and cameras, motion amounts, etc. It is worthy to employ these information during the training of the proposed RHCNN, and it will be our future work." Explore this.

P.2.380 See [E457] There are several interesting references related to error concealment techniques in video. The conclusions section is repeated here.

"This paper proposed a novel video error–concealment approach that combines the block reliability with the boundary–difference minimization along the block boundary. More specifically, a downhill simplex approach was adopted to minimize the computational complexity of the boundary matching and fine-tune the motion vectors. Intra-coded blocks or unreliable blocks were chosen for motion-vector refinements. The initial motion vectors for intra-coded blocks were considered as zeros, and the motion vectors of the collocated blocks were used for the initial motion vectors of unreliable blocks. According to the simulation results, the proposed approach shows good performance based on both subjective and quantitative evaluations. However, the best boundary matching does not always guarantee the best motion vector of a lost block. The proposed downhill simplex approach may be combined with other error–concealment approaches to reduce mismatches along the block boundary. For future research, the selection of the initial simplex can be investigated to avoid detecting a local minimum."

134   HEVC Encoder

Combine the downhill simplex approach with other error-concealment techniques with the goal of reducing the mismatches along the boundary.

P.2.381 See P.2.380 and [E457]. It says "For future research, the selection of the initial simplex can be investigated to avoid detecting a local minimum". Explore this.

P.2.382 MPEG by ISO/IEC and VCEG by ITU-T are jointly exploring video coding beyond HEVC. While the compression format is focused on traditional block based motion compensated prediction followed by transform/entropy coding with in-loop filters, the experts are looking into multiple size transforms (DCT/DST/Hadamard etc), additional intra directional predictions (See Figure 2.4). Explore this and compare its performance with the traditional HEVC in terms of complexity, BD rates, and other metrics.

P.2.383 See [E459] and the abstract. Explore this fast motion estimation approach and confirm the results described in this Paper.

P.2.384 See [E460]. Last sentence of the conclusion says "Furthermore, the methodology we applied can also be used for other HEVC encoding decisions, such as the PU and the TU partitioning." Explore this.

P.2.385 HEVC [E202] specifies video coding only. Multiplex HEVC video coded bit streams (various profiles and levels) with HE-AAC verson2 audio coded bit streams, transmit the multiplexed bit streams, demultiplex and decode the video and audio and maintain the lip synch. See [E379] and also (ARIB STD-B32 version 3.9-E1, "Video coding, Audio coding and Multiplexing Specifications for Digital Broadcasting", 2016.) and http://www.3gpp.org/DynaReport/26410.htm – HE ACC V2 software
High efficiency advanced audio codec
https://www.ffmpeg.org/download.html – ffmpeg software

P.2.386 See [E461]. Explore the future work suggested in the section Conclusions.

P.2.387 See [E464]. Explore the future work and research directions suggested in the section Conclusion.

P.2.388 See [E465]. In the Conclusion section, three promising directions as future work are described. Explore these.

P.2.389 See [E466]. Part of the abstract is repeated here:

"Transform and quantization account for a considerable amount of computation time in video encoding process. However, there are a large number of discrete cosine transform (DCT) coefficients which are finally quantized into zeros. In essence, blocks with all zero quantized coefficients do not transmit any information, but still occupy substantial unnecessary computational resources. As such, detecting All Zero Block (AZB) before transform and quantization has been recognized to be an efficient approach to speed up the encoding process." Several techniques (See the references listed in [E469].) have been developed for detecting all zero blocks before prediction/transform/quantization for different block sizes in both H.264/AVC and HEVC. Review all these techniques in detail (also implement them) and evaluate their capabilities in terms of encoder complexity reduction and suggest future directions.

P.2.390 See [E467] and the conclusion section. Explore the ongoing work.

P.2.391 Several papers have been published in the literature that focus on reducing the HEVC encoder complexity. Review these in detail and develop a meaningful and practical unified approach for reducing this complexity. Implement this approach and evaluate is effectiveness based on standard metrics, (See the Section 2.9 summary). As an example see [E468] (also references listed at the end) that focus on reducing the encoder complexity based on intra mode decision.

P.2.392 See [E470]. The last sentence of the conclusion section says "In an upcoming paper, we will show that when extending the proposed techniques to the context of TZ-Search even more significant speedups can be reached." Explore this.

P.2.393 See [E469] and the abstract. Extend this scheme to all intra (AI) configuration.

P.2.394 See [E472] and the conclusion section. The authors state, "There are several related research issues to be resolved in the future." Explore these research issues in depth.

P.2.395 See [E475]. Explore the subject of future research suggested in the conclusion section. While there has been an extensive research related to error control strategies for H.264 coded bit streams (see the references listed at the end), not much has been done on H.265 (to our knowledge). This opens up a number of projects on error resilience, error detection/correction, error concealment etc applied to H.265 coded bit streams (various profiles/levels).

P.2.396 See [JPEG29] and the conclusion section. A joint image compression and encryption scheme is proposed and applied to JPEG. The authors suggest that their scheme can be extended to other image/video compression standards. Extend this to HEVC (all profiles).

P.2.397 See [E476] and the conclusion section. Explore the suggested future research.

P.2.398 See [E477] and the conclusion section. The authors state there are lot of open issues for further study. Explore these. Please note that fractional pel ME/MC has been adopted in many video coding standards including HEVC.

P.2.399 See [E479]. Explore the future work suggested in the conclusion section.

P.2.400 See [E480]. The ongoing work is suggested in the conclusion section. Explore this.

P.2.401 See [E355]. In the conclusion section it says

"In the future, more study can be made on how to facilitate software and hardware implementation." Explore this software and hardware implementation of cross-component prediction in HEVC.

P.2.402 See [E355] and P.2.401 In the conclusion section it further states "We also leave it as a further study how to efficiently apply CCP to various applications including high dynamic range video coding." Explore this further study,

P.2.403 See [E482] and the conclusion section. It says

"Additionally, it is possible to design a hardware architecture specifically designed to implement the proposed fast algorithm." Design this hardware architecture.

P.2.404 See [E483]. Explore the future work suggested in the conclusion section.

P.2.405 See [E446]. Explore the various comments/suggestions cited in the conclusion section for improving the performance of this video coding framework.

P.2.406 See [E484]. Explore the future work suggested in the conclusions and future directions section (just before the references).

P.2.407 See [E485]. Explore how the quality enhancement CNN based and time constrained quality enhancement optimization techniques can be embedded in HEVC decoders using ASIC.

P.2.408 See [E486]. The conclusion and future works section says "In the future, it is desirable to investigate the parallelization of HEVC encoding on various parallel processing architectures and to reduce the overhead of parallel task scheduling. In the current work, the proposed CSPS focuses on parallelizing HEVC encoding within frames." Explore this parallelization approach in detail.

P.2.409 See [E487]. Conclusion section is repeated here.

Conclusion

In this paper, we proposed an optimization-based method to reconstruct the 3-D human body shape with a single depth camera. In our method, we only need two depth images from which we estimate the body shape separately. We merge the two half body shapes to a complete one by solving a quadratic problem. We also propose a novel mid-points registration of the same human body with different poses. The left is the front side in the "A" pose, and the right is the back side in the "T" pose. Both of the depth images have been converted into 3-D point clouds for visualization in this figure. At this stage, in the section of pose estimation we find rigid rotation and translation by optimization method. In the further work, we will focus on how to speed up pose estimation. Human body part classifier based on deep learning can be used to initialize the body pose and pose and further reduce the number of optimization iterations."

Explore this further work.

P.2.410 See [E488]. Explore the future work suggested in the conclusion and perspective section at the end.

3
Screen Content Coding for HEVC

Abstract

Screen content (SC) video is generated by computer programs and displayed on screen without any signal noise. The SC picture usually contains a lot of discontinuous textures and edges. Patch of SC picture is often identical in many regions in the picture. Based on these differences from camera-captured video, it has been developed as an extension of HEVC [6] We discuss, here, the four screen content coding tools, lossless, and lossy coding performance, fast algorithms, quality assessment, and other algorithms developed recently.

Keywords

Adaptive color transform, Adaptive motion vector resolution, Intra block copy, Palette coding, Residual DPCM, Sample-based prediction, Screen contents, Screen image quality assessment, String matching, Template matching, Transform skip

3.1 Introduction to SCC

High efficiency video coding (HEVC) discussed in Chapter 1 is the latest video compression standard of the Joint Collaborative Team on Video Coding (JCT-VC), which was established by the ITU-T Video Coding Experts Group (VCEG) and the ISO/IEC Moving Picture Experts Group (MPEG) [1]. It demonstrates a substantial bit-rate reduction over the existing H.264/AVC standard [2]. Several extensions and profiles of HEVC have been developed according to application areas and objects to be coded. However, both the HEVC and the H.264/AVC focused on compressing camera-captured video sequences, mainly consisting of human objects and complex textures (Figure 3.1). Although several different types of test sequences were used during the development of these standards, the camera-captured sequences exhibited common characteristics such as the presence of sensor noise and an abundance of translational motion. Thus, video compression exploits both

Figure 3.1 Image example from camera captured video content [4].

temporal and spatial redundancies. A frame that is compressed by exploiting the spatial redundancies is termed as intra frame and the frames which are compressed by exploiting the temporal redundancies are termed as inter frames. The compression of an inter frame requires a reference frame which will be used to exploit the temporal redundancies. The inter frame is usually of two types namely a P frame and a B frame. The P frame make use of one already encoded/decoded frame which may appear before or after the current picture in the display order i.e. a past or a future frame as its reference, whereas the B frame makes use of two already encoded/decoded frames, one of which is a past and the other being the future frame as its reference frames, thus, providing higher compression but also higher encoding time as it has to use a future frame for encoding [3]. Furthermore, conventional video coders, in general, remove high-frequency components for compression purposes, since the human visual sensitivity is not so high in high frequencies.

Recently, however, there has been a proliferation of applications that display more than just camera-captured content. These applications include displays that combine camera-captured and computer graphics, wireless displays, tablets as second display, control rooms with high resolution display wall, digital operating room (DiOR), virtual desktop infrastructure (VDI), screen/desktop sharing and collaboration, cloud computing and gaming, factory automation display, supervisory control and data acquisition (SCADA) display, automotive/navigation display, PC over IP (PCoIP), ultra-thin client, remote sensing, etc. [5, 6]. The type of video content used in these applications can contain a significant amount of stationary or moving computer graphics and text, along with camera-captured content, as shown in Figure 3.2. However, unlike camera-captured content, screen content frequently contains no sensor noise, and such content may have large uniformly flat areas, repeated patterns, highly saturated or a limited number of different

Figure 3.2 Images of screen content: (a) slide editing, (b) animation, (c) video with text overlay, (d) mobile display [4].

colours, and numerically identical blocks or regions among a sequence of pictures. These characteristics, if properly managed, can offer opportunities for significant improvements in compression efficiency over a coding system designed primarily for camera-captured natural content. Unlike natural images/video, screen contents may not be very smooth. They usually have totally different statistics. For text or graphics contents, it is much sharper and with high contrast [7]. Because of the high contrast, any little artefact caused by removing high frequency components in conventional video coders may be perceived by users. Thus, all coding techniques supported by HEVC RExt and additional coding tools such as intra block coping, palette coding, and adaptive colour space transform are required to compress screen content. Features of screen contents are summarized as:

- Sharp content: Screen content usually includes sharp edges, such as in graphic or animation content. To help encoding sharp content, transform skip has been designed for screen content [8].
- Large motion: For example, when browsing a web page, a large motion exists when scrolling the page. Thus, new motion estimation algorithms to handle the large motions for screen content may be required.
- Artificial motion: For example, when fading in or fading out, the conventional motion model may not be easy to handle it.
- Repeating patterns: For example, the compound images may contain the same letter or objects many times. To utilize the correlation among repeating patterns, Intra Block Copy (IBC) [6, 20, 43, 69, 72] has been developed.

Joint Call for Proposals (CfP) was released in Jan. 2014 with the target of developing extensions of the HEVC standard including specific tools for screen content coding [9]. The use cases and requirements of the CfP are described in [5] and common conditions for the proposals are found in [10]. These documents identified three types of screen content: mixed content, text and graphics with motion, and animation. Up to visually lossless coding performance was requested for RGB and YC_bC_r 4:4:4 format having 8 or 10 bits per colour component. After seven responses to the CfP were evaluated at the JCT-VC meeting [11], several core experiments (CEs) were defined including intra block copying extensions, line-based intra copy, palette mode, string matching for sample coding, and cross-component prediction and adaptive colour transforms. As a result of evaluating the outcome of the CEs and related proposals, the HEVC Screen Content Coding Test Model 6 [12] and Draft Text 5 [13] were published in Oct. 2015. All the documents are available in [14]. Test sequences [15], reference software [16] and manual [17] are also available.

3.2 Screen Content Coding Tools

HEVC-SCC is based on the HEVC framework while several new modules/tools are added as shown in Figure 3.3, including intra block copy (IBC) [72], palette coding, adaptive colour transform, and adaptive motion vector resolution.

Figure 3.3 Encoding block diagram for screen content coding [6].

3.2.1 Intra Block Copy

HEVC-SCC introduces a new CU mode in addition to the conventional intra and inter modes, referred to as intra block copy (IBC). The IBC mode performs like an inter mode prediction but the prediction units (PU) of IBC coded coding units (CU) predict reconstructed blocks in the same picture, taking the advantage of exploiting the repeated patterns that may appear in screen content. Similar to inter mode, IBC uses block vectors to locate the predictor block [12].

Since screen content is likely to have similar or repeated patterns on a screen, such spatial redundancy can be removed by quite different way from the conventional intra prediction schemes. The most significant difference is the distance and shapes from neighbouring objects [18]. Removable spatial redundancy in the conventional intra prediction schemes refers to the similarity between the boundary pixels of the block to be coded and the adjacent pixels located spatially within one pixel. However, the removable spatial redundancy in IBC mode refers to the similarity between the area in the reconstructed picture and the block to be coded. A target 2D block/object is predicted from a reconstructed 2D block/object that is more than one pixel distant from it using the motion or location information (motion vector) from the reference block/object.

Figure 3.4 shows the concept of IBC and the block vector (BV), which is conceptually similar to the motion vector (MV) of inter prediction. In terms of the accuracy of the vectors, MV has in usual quarter-pel accuracy to ensure

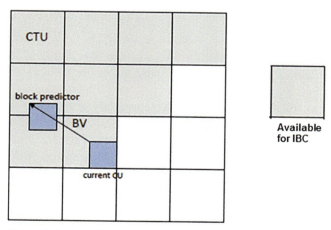

Figure 3.4 Intra block copy prediction in the current picture [6].

improved prediction accuracy, whereas BV is enough to be in integer-pel accuracy. This is because of the characteristics of the screen content in IBC mode. For example, objects in computer graphics are generated pixel by pixel and repeated patterns are found in integer-pel accuracy. Block compensation, which is similar to motion compensation, of IBC is conducted on the reconstructed area of the current frame not previously coded, or decoded frames. In addition, the BV should be sent to the decoder side, but it is derived by prediction to reduce the amount of data in a manner similar to the motion vector. Prediction may be independent from the MV prediction method or in the same way as the MV prediction method.

The global BV search is performed for 8×8 and 16×16 blocks. Search area is a portion of the reconstructed current picture before loop filtering, as depicted in Figure 3.4. Additionally, when slices/tiles are used, the search area is further restricted to be within the current slice/tile. For 16×16 blocks, only a one-dimensional search is conducted over the entire picture. This means that the search is performed horizontally or vertically. For 8×8 blocks, a hash-based search is used to speed up the full picture search. The bit-length of the hash table entry is 16. Each node in the hash table records the position of each block vector candidate in the picture. With the hash table, only the block vector candidates having the same hash entry value as that of the current block are examined [19].

The hash entries for the current and reference blocks are calculated using the original pixel values. The 16-bit hash entry H is calculated as

$$H = \mathrm{MSB}(DC0, 3) \ll 13 + \mathrm{MSB}(DC1, 3) \ll 10 + \mathrm{MSB}(DC2, 3)$$
$$\ll 7 + \mathrm{MSB}(DC3, 3) \ll 4 + \mathrm{MSB}(Grad, 4) \quad (3.1)$$

where $\mathrm{MSB}(X, n)$ represents the n MSB of X, DC0, DC1, DC2, DC3 denote the DC values of the four 4×4 sub-blocks of the 8×8 block, and Grad denotes the gradient of the 8×8 block. Operator '$\ll$' represents arithmetic left shift.

In addition to full block vector search, some fast search and early termination methods are employed in the HEVC-SCC. The fast IBC search is performed after evaluating the RD cost of inter mode, if the residual of inter prediction is not zero. The SAD-based RD costs of using a set of block vector predictors are calculated. The set includes the five spatial neighbouring block vectors as utilized in inter merge mode (as shown in Figure 3.5) and the last two coded block vectors. In addition, the derived block vectors of the blocks pointed to by each of the aforementioned block vector predictors are also included. This fast search is performed before the evaluation of intra prediction mode. It is applied only to $2N \times 2N$ partition of various CU sizes.

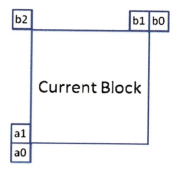

Figure 3.5 Candidates of block vector predictor [20].

3.2.2 Palette Mode

For screen content, it is observed that for many blocks, a limited number of distinct colour values may exist. In this case, the set of colour values is referred to as the palette. Thus, palette mode enumerates those colour values and then for each sample, sends an index to indicate to which colour it belongs. In special cases it is also possible to indicate a sample that is outside the palette by signalling an escape symbol followed by component values as illustrated in Figure 3.6. Palette mode can improve coding efficiency when the prediction does not work due to low redundancy and when the number of pixel values for the block is small [18, 21]. According to the results by Xiu, et al., [22], coding gain of the palette-based coding increases up to 9.0% in the average BD-rate for lossy coding and up to 6.1% for lossless coding mode.

3.2.2.1 Palette Derivation

In the SCM-7.0 software [16], for the derivation of the palette for lossy coding, a modified k-means clustering algorithm is used. The first sample of the block is added to the palette. Then, for each subsequent sample from the block, the SAD from each of the current palette entries is calculated. If the distortion for each of the components is less than a threshold value for the palette entry corresponding to the minimum SAD, the sample is added to the cluster belonging to the palette entry. Otherwise, the sample is added as a new palette entry. When the number of samples mapped to a cluster exceeds a threshold, a centroid for that cluster is calculated and becomes the palette entry corresponding to that cluster.

In the next step, the clusters are sorted in a decreasing order of frequency. Then, the palette entry corresponding to each entry is updated. Normally, the cluster centroid is used as the palette entry. But a rate-distortion analysis is

Figure 3.6 Example of indexed representation in the palette mode [6].

performed to analyse whether any entry from the palette predictor may be more suitable to be used as the updated palette entry instead of the centroid when the cost of coding the palette entries is taken into account. This process is continued till all the clusters are processed or the maximum palette size is reached. Finally, if a cluster has only a single sample and the corresponding palette entry is not in the palette predictor, the sample is converted to an escape symbol. Additionally, duplicate palette entries are removed and their clusters are merged.

For lossless coding, a different derivation process is used. A histogram of the samples in the CU is calculated. The histogram is sorted in a decreasing order of frequency. Then, starting with the most frequent histogram entry, each entry is added to the palette. Histogram entries that occur only once are converted to escape symbols if they are not a part of the palette predictor.

After palette derivation, each sample in the block is assigned the index of the nearest (in SAD) palette entry. Then, the samples are assigned to 'INDEX' or 'COPY_ABOVE' mode. For each sample for which either 'INDEX' or 'COPY_ABOVE' mode is possible, the run for each mode is determined. Then, the cost (in terms of average bits per sample position) of coding the mode, the run and possibly the index value (for 'INDEX' mode) is calculated. The mode for which the cost is lower is selected. The decision is greedy in the sense that future runs and their costs are not taken into account.

3.2.2.2 Coding the Palette Entries

For coding of the palette entries, a palette predictor is maintained. The maximum size of the palette as well as the palette predictor is signalled in the sequence parameter set (SPS). In SCM 4, a palette_predictor_initializer_present_flag is introduced in the PPS. When this flag is 1, entries for initializing the palette predictor are signalled in the bitstream. The palette predictor is initialized at the beginning of each CTU row, each slice and each tile. Depending on the value of the palette_predictor_initializer_present_flag, the palette predictor is reset to 0 or initialized using the palette predictor initializer entries signalled in the picture parameter set (PPS). In SCM 5, palette predictor initialization at the SPS level was introduced to save PPS bits when a number of PPS palette predictor initializers shared common entries. In SCM 6, a palette predictor initializer of size 0 was enabled to allow explicit disabling of the palette predictor initialization at the PPS level.

For each entry in the palette predictor, a reuse flag is signalled to indicate whether it is part of the current palette. This is illustrated in Figure 3.7. The reuse flags are sent using run-length coding of zeros. After this, the number of new palette entries is signalled using exponential Golomb code of order 0. Finally, the component values for the new palette entries are signalled.

previous palette

Index	G/Y	B/Cb	R/Cr
0	G0	B0	R0
1	G1	B1	R1
2	G2	B2	R2
3	G3	B3	R3
4	G4	B4	R4
5	G5	B5	R5

current palette

Pred flag	Index	G/Y	B/Cb	R/Cr
1	0	G0	B0	R0
0				
1	1	G2	B2	R2
1	2	G3	B3	R3
0				
0				
	3	G3N	B3N	R3N
	4	G4N	B4N	R4N

Re-used palette entries (3)

New palette entries (2), signalled

Figure 3.7 Use of palette predictor to signal palette entries [6].

3.2.2.3 Coding the Palette Indices

The palette indices are coded using three main palette sample modes: INDEX mode, COPY_ABOVE mode, and ESCAPE mode as illustrated in Figure 3.8. In the INDEX mode, run-length coding is conducted to explicitly signal the colour index value, and the mode index, colour index, and run-length are coded. In the COPY_ABOVE mode, which copies the colour index of the row above, the mode index and run-length are coded. Finally, in the ESCAPE mode, which uses the pixel value as it is, the mode index and the quantized pixel value are coded. When escape symbol is part of the run in 'INDEX' or 'COPY_ABOVE' mode, the escape component values are signalled for each escape symbol.

3.2.3 Adaptive Colour Transform (ACT)

Conventional natural content is usually captured in RGB colour format. Since there is strong correlation among different colour components, a colour space conversion is required to remove inter-component redundancy. However, for screen content, there may exist many image blocks containing different features having very saturated colours, which leads to less correlation among colour components. For those blocks, coding directly in the RGB colour space may be more effective. ACT enables the adaptive selection of colour-space conversion for each block. To keep the complexity as low as possible, the

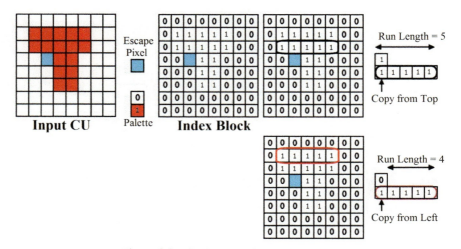

Figure 3.8 Coding the palette indices [21].

colour-space conversion process is applied to the residual signal as shown in Figure 3.9 and after the intra- or inter-prediction process, the prediction residuals are selected to perform forward colour-space transform as shown in Figure 3.10.

3.2.3.1 Colour Space Conversion

To handle different characteristics of image blocks in screen content, a RGB-to-YC_oC_g conversion [24] was investigated to use it for forward and backward lossy and lossless coding.

Forward transform for lossy coding (non-normative):

$$\begin{bmatrix} Y \\ Co \\ Cg \end{bmatrix} = \begin{bmatrix} 1 & 2 & 1 \\ 2 & 0 & -2 \\ -1 & 2 & -1 \end{bmatrix} \begin{bmatrix} R \\ G \\ B \end{bmatrix} /4 \qquad (3.2)$$

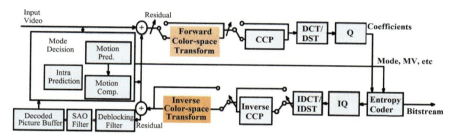

Figure 3.9 Location of the ACT in the encoder [23].

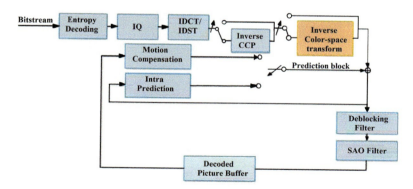

Figure 3.10 Location of the ACT in the decoder [23].

Forward transform for lossless coding (non-normative):

$$\begin{aligned} Co &= R - B \\ t &= B + (Co \gg 1) \\ Cg &= (G - t) \\ Y &= t + (Cg \gg 1) \end{aligned} \qquad (3.3)$$

Backward transform (normative):

$$\begin{aligned} &if(lossy)\{ \\ &\quad Co = Co \ll 1 \\ &\quad Cg = Cg \ll 1 \\ &\} \\ t &= Y - (Cg \gg 1) \\ G &= Cg + t \\ B &= t - (Co \gg 1) \\ R &= Co + b \end{aligned} \qquad (3.4)$$

3.2.3.2 Encoder Optimization

In the ACT mode, encoder complexity increases double because the mode searching is performed in both the original colour space and the converted colour space. To avoid this, following fast methods are applied:

- For intra coding mode, the best luma and chroma modes are decided once and shared between the two colour spaces.
- For IBC and inter modes, block vector search or motion estimation is performed only once. The block vectors and motion vectors are shared between the two colour spaces.

3.2.4 Adaptive Motion Vector Resolution

For natural video content, the motion vector of an object is not necessarily exactly aligned to the integer sample positions. Motion compensation is, therefore, not limited to using integer sample positions, i.e. fractional motion compensation is more efficient to increase compression ratio. Computer-generated screen content video, however, is often generated with knowledge of the sample positions, resulting in motion that is discrete or precisely aligned with sample positions in the picture. For this kind of video, integer motion vectors may be sufficient for representing the motion. Savings in bit-rate can be achieved by not signalling the fractional portion of the motion vectors.

Adaptive MV resolution allows the MVs of an entire picture to be signalled in either quarter-pel precision (same as HEVC version 1) or integer-pel precision. Hash based motion statistics are kept and checked in order to properly decide the appropriate MV resolution for the current picture without relying on multi-pass encoding. To decide the MV precision of one picture, blocks are classified into the following categories:

- C: number of blocks matching with collocated block
- S: number of blocks not matching with collocated block but belong to smooth region. For smooth region, it means every column has a single pixel value or every row has a single pixel value.
- M: number of blocks not belonging to C or S but can find a matching block by hash value.

The MV resolution is determined as:

- If CSMRate < 0.8, use quarter-pel MV.
- Otherwise, if C == T, use integer-pel MV.
- Otherwise, if AverageCSMRate < 0.95, use quarter-pel MV.
- Otherwise, if M > (T−C−S)/3, use integer-pel MV.
- Otherwise, if CSMRate > 0.99 and MRate > 0.01, use integer-pel MV.
- Otherwise, if AverageCSMRate + AverageMRate > 1.01, use integer-pel MV.
- Otherwise, use quarter-pel MV.

T is the total number of blocks in one picture. CSMRate = (C+S+M)/T, MRate = M/T. AverageCSMRate is the average CSMRate of current picture and the previous 31 pictures. AverageMRate is the average MRage of the current picture and the previous 31 pictures.

4

Lossless and Visually Lossless Coding Algorithms

4.1 Residual DPCM

Differential pulse code modulation (DPCM) has been widely used to reduce spatial and temporal redundancy in the video content. The subtraction of the prediction signal from the current block signal generates the residual signal or the prediction error, containing the part of the original signal which could not be predicted by the selected predictor [25]. The residual signal can be further compressed by any method. In the HEVC, compression can be achieved by the application of a transformation, which is applied to represent the correlated parts of the residual signal in the residual block by a potentially small number of transform coefficients. These coefficients are then quantized and coded into the bitstream.

Sample-by-sample residual DPCM (RDPCM) of intra-predicted residuals was proposed [26] in the context of H.264/AVC lossless coding. When using this technique, instead of performing conventional intra-prediction each residual sample is predicted from neighbouring residuals in the vertical or horizontal direction when the intra prediction is equal to one of these two directions. Average bitrate reductions of 12% were reported using this technique compared with conventional H.264/AVC lossless coding. This technique was later extended and adapted to the HEVC standard [27] achieving on average 8.4% bitrate reductions on screen content sequences.

The residuals of each sample are calculated by sample-by sample DPCM in vertical and/or horizontal mode. When the intra prediction mode is vertical, the RDPCM elements $\tilde{r}_{i,j}$ is given by

$$\tilde{r}_{i,j} = \begin{cases} r_{i,j}, & i=0, 0 \leq j \leq (N-1) \\ r_{i,j} - r_{(i-1),j}, & 1 \leq i \leq (M-1), 0 \leq j \leq (N-1) \end{cases} \quad (4.1)$$

154 *Lossless and Visually Lossless Coding Algorithms*

or when the intra prediction mode is horizontal mode

$$\tilde{r}_{i,j} = \begin{cases} r_{i,j}, & 0 \leq i \leq (M-1), j = 0 \\ r_{i,j} - r_{i,(j-1)}, & 0 \leq i \leq (M-1), 1 \leq j \leq (N-1) \end{cases}. \quad (4.2)$$

In the vertical mode, the samples in the first row in the block are left unchanged. All other samples are predicted from the sample immediately above in the same column. The horizontal intra RDPCM is given in a similar way.

The RDPCM elements are signalled to the decoder so that the original residual samples are reconstructed by

$$r_{i,j} = \sum_{k=0}^{i} \tilde{r}_{k,j}, \quad 0 \leq i \leq (M-1), 0 \leq j \leq (N-1), \text{ in vertical mode}$$

$$r_{i,j} = \sum_{k=0}^{j} \tilde{r}_{i,k}, \quad 0 \leq i \leq (M-1), 0 \leq j \leq (N-1), \text{ in horizontal mode.}$$

$$(4.3)$$

Thus, the RDPCM is implemented in one-dimensional direction in the HEVC-SCC. However, two dimensional RDPCM was suggested as [28]

$$\tilde{r}_{i,j} = \alpha_1 r_{i,j-1} + \alpha_2 r_{i-1,j} + \alpha_3 r_{i-1,j-1} + \alpha_4 r_{i-2,j} + \cdots \quad (4.4)$$

where α_i denotes weighting factor for neighbouring pixels.

To find the best mode for the current residual block, the SAD distortion metric can be used for each mode (i.e. horizontal, vertical or no RDPCM mode). The mode with minimum SAD is selected as the best.

When performing the RDPCM on a block, samples in the first column and the first row for horizontal and vertical RDPCM, respectively, are not predicted. Therefore it is beneficial to exploit redundancy by performing prediction on these samples in the direction orthogonal to the main RDPCM direction, as shown in Figure 4.1.

4.1.1 Sample-Based Weighted Prediction with Directional Template Matching

The sample-based weighted prediction (SWP) algorithm is proposed in [30] to introduce a weighted averaging of neighbouring pixels for intra-prediction

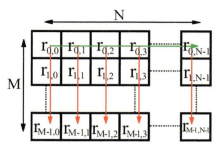

Figure 4.1 Secondary prediction (green line) after vertical RDPCM (red lines) [29].

of the current pixel. The predicted pixel $p_{SWP}[i]$ is calculated as follows:

$$p_{SWP}[i] = \text{round}\left(\sum_{j \in S} w_{int}[i,j]g[j] / \sum_{j \in S} w_{int}[i,j]\right), \quad (4.5)$$

where $g[j]$ is the pixels around the reconstructed current pixel, S is the set of supporting pixels, and $w_{int}[i,j]$ are the integer weighting values, which are calculated as follows:

$$w_{int}[i,j] = \text{round}\left(a_{SWP} \cdot b_{SWP}^{-SAD(P_g^k[i], P_g^k[j])/h_{dist}}\right), \quad (4.6)$$

where the factor a_{SWP} is chosen to be 2^{11}, if the internal bit depth is 8, the basis factor b_{SWP} is 2 for exponential decaying weights, and the parameter h_{dist} is empirically chosen to be 4.75 for luma and for 4:4:4 chroma. $SAD(\cdot)$ is the operator for the sum of differences between two patches, $P_g^k[i]$ and $P_g^k[j]$, which are formed by causally neighbouring pixels (typically four pixels as shown in Figure 4.2). Thus, the $SAD(\cdot)$ can be defined as a similarity measure in the supporting area

Figure 4.2 Causally neighbouring pixels for prediction and patches. Pixel X is to be predicted from the patch with pixels a, b, c, and d (left, centre); Patch around pixel b (right) [30].

and given by

$$SAD\left(P_g^k[i], P_g^k[j]\right) = \sum_{v \in N_0} |g[i+v] - g[j+v]|. \qquad (4.7)$$

Pixel X is predicted by a weighted average from the candidate pixels a, b, c, and d. For each candidate pixel, the SAD of the corresponding patches is calculated, e.g., for calculation of the SAD between X and b, the patch for X (pixels a, b, c, and d in the center of the figure) is compared to the patch for b (shaded blue area on the right hand side of the figure).

The prediction performance of the SWP is well for natural images that maintain high correlation among neighbouring pixels, while degrades for such as text images that contain sharp edges among letters. With the weighted/averaged prediction, sharpness may be lost by averaging effect. Thus, another compromising prediction technique, directional template matching (DTM) has to be introduced. The main idea is to reduce the averaging effect by selecting the minimum SAD patch within the support area. The sharp edges could be kept without smoothing.

4.1.2 Sample-Based Angular Intra-Prediction

HEVC has adopted block-based angular intra-prediction which is useful for lossless coding to exploit spatial sample redundancy in intra coded CUs. A total of 33 angles (Figures 2.8 and 2.9) are defined for the angular prediction that can be categorized into three classes: 1 diagonal, 16 horizontal, and 16 vertical predictions. In HEVC, the total number of intra prediction modes is 35, including Mode 0 for INTRA_PLANAR, Mode 1 for INTRA_DC, and Mode 2 to 34 for INTRA_ANGULAR [31]. Given an $N \times N$ prediction unit (PU), the number of reference samples is $4N + 1$, i.e., 2N upper, 2N left, and 1 diagonal, belonged to neighbouring PUs. All the samples inside the PU share the same prediction angle in the block-based angular intra-prediction. The value of prediction angle should be informed to the decoder. However, the sample-based angular prediction is performed sample by sample. Since four effective intra prediction block sizes ranging from 4×4 to 32×32 samples, each of which supports 33 distinct directions, there are 132 combinations of block sizes and prediction directions. The prediction accuracy is 1/32 in the horizontal or vertical direction via linear interpolation.

The coding gain of the SAP provides a 1.8% to 11.8% additional bitrate reduction on average [32] in the lossless coding mode. In addition the SAP provides more gain in 10-bit configurations due to the fact that there are more blocks with large prediction residuals, the difference between the original pixel value and its prediction, in the 10-bit video than in 8-bit video. The SAP also improves coding efficiency by increasing the usage of angular intra prediction and the number of intra coded CUs, while decreasing the usage of the planar and DC modes.

4.1.3 Sample-Based Angular Intra-Prediction with Edge Prediction

Another type of compound contents includes whole slide images (WSIs), which is the digitized version of microscope glass slides. Pathologists can send WSIs to others for sharing, collaborating, consulting and making diagnosis. WSIs usually feature a high number of edges and multifunctional patterns due to great variety of cellular structures and tissues. They should be scanned at high resolutions, resulting in huge file sizes. Therefore, designing efficient and accurate lossless or visually lossless compression algorithms are an important challenge. Edge prediction is proposed in [33] as a post-processing step on the residual signal computed by the original intra coding process. This method adds an extra coding step to the pipeline and alters the block-wise coding structure of HEVC as in [30].

In order to maintain the inherent block-wise coding and decoding structure of HEVC, an alternative intra coding modes are suggested in [34], while HEVC-RExt includes the optional use of the SAP, which is limited to the horizontal and vertical directions. For the case of the DC mode, a sample prediction is computed as an average of neighbouring samples at positions $\{a, c\}$; $P_{x,y} = (a+c) \gg 1$ in Figure 4.2. For the case of the PLANAR mode, an edge predictor is proposed and calculated as follows:

$$P_{x,y} = \begin{cases} \min(a,c) & \text{if } c \geq \max(a,c) \\ \max(a,c) & \text{if } c \leq \min(a,c) \\ a+c-b & \text{otherwise} \end{cases} \quad (4.8)$$

The edge predictor mode and the SAP modes require that samples be decoded sequentially and be readily available for the prediction and reconstruction of subsequent samples. This inevitably breaks the block-wise decoding structure of HEVC. However, the reconstruction can be regarded as a spatial residual transform that only depends on the residual samples and the reference samples. Such spatial residual transform can be expressed in matrix form and applied during the decoding process in order to maintain the block-wise decoding structure. This technique improves coding efficiency by an average of 6.64% compared to SAP-1, which applies the SAP in all angular modes with a constant displacement among any two adjacent modes, and by an average of 7.67% compared to SAP-HV, which applies the SAP only in the pure horizontal and vertical directions [34]. This is mainly due to the introduction of a DC mode based on the SAP and an edge predictor in lieu of the PLANAR mode.

Although the edge predictor is capable of detecting horizontal and vertical edges accurately by selecting the best predictor for each pixel, more efficiency needs to be achieved in textual smooth regions by taking median values of adjacent five pixels as

$$P_{x,y} = \text{median}\{a, b, c, d, e\} \tag{4.9}$$

By adding the median prediction, coding efficiency increases by 16.13% on average compared to HEVC intra prediction coding [35].

4.2 Fast Coding Algorithms

As discussed in Chapter 1, the design of HEVC was based on two main objectives: achieving higher coding efficiency compared to previous standards and attaining low enough complexity to enable video applications on mobile devices. The second objective, however, is not easily fulfilled due to highly complex algorithms such as quadtree-based coding structure, large block transforms, an advanced motion prediction, additional filtering operations, and a high number of intra-prediction modes [36]. Among those properties, motion prediction and intra-prediction can be taken into account to design the screen content codec, responding to different nature from camera captured sequences.

4.2.1 Adaptive Motion Compensation Precision

Fractional precision motion compensation usually improves the video coding efficiency, especially for natural video. But considering the fact that slow moving or text-based screen content may be motion compensated in integer precision, using fractional precision is a waste of bits. Thus, adaptive precision methods are developed to improve coding efficiency of screen content, saving bits by using integer motion vectors in some cases. Encoder is designed to use integer precision when encoding the content captured from a screen and to use fractional precision when encoding the content captured from a normal camera.

To adopt the adaptive precision method in the HEVC structure, encoder should signal a flag at the slice header level to indicate whether integer precision or fractional precision is used for the current slice. Moreover, two-pass encoding is required to decide the precision of motion vectors, taking approximately double the encoding time, which is undesirable in practical use. Thus, fast algorithm has to be developed to minimize the encoding time while preserving most of the benefits brought by adaptive motion compensation precision.

In [37], fast algorithm is developed to efficiently design hash-based block matching scheme, using cyclic redundancy check (CRC) as the hash function to generate the hash value for every block. The complexity of CRC operation is about $O(m \cdot n \cdot w \cdot h)$, by considering the block size and picture size. More than 8G operations are required to calculate all the hash values of 64×64 blocks in 1080p video [38], for example. To reduce the complexity of hash table generation, reuse of the intermediate results is suggested. In fact, there exists $(m-1)$ overlapping rows between two block operations. Thus, 63 of the intermediate data could be reused in the next 64×64 block, resulting in reduced complexity $O(m \cdot w \cdot h + n \cdot w \cdot h)$, i.e., about 256M operations for the same example. Next step is the block matching using the hash values and another reduction is possible. This can be done by checking all the blocks having the same hash value and selecting a block to predict the current block. The complexity of block matching is about $O(l \cdot (m \cdot n))$, where l denotes the number of blocks in a picture, resulting in about 2M operations for example above. The overall complexity of the proposed hash-based block matching method is about $O(m \cdot w \cdot h + n \cdot w \cdot h + l \cdot m \cdot n)$, resulting in reduced complexity from 4T (full picture search) to 258M, which is more than $15,000 \times$ speedup.

Although we can reduce the complexity in the block matching operations, further reduction is possible using the adaptive precision decision for screen content. For example, in [37], the blocks are classified into four categories: collocated matched blocks (C) that the optimal motion vector should be (0, 0), smooth blocks (S) that every row or column has a single pixel value, matched blocks (M) that an exact match by hash values can be found, and other extra blocks (O). Some logical analysis can be given as follows:

- If the percentage of O is too large, fractional precision MV is used.
- If the percentage of M is not large enough, fractional precision MV is used.
- If the percentage of C, S, and M is larger than a threshold, integer precision MV is used.

Using the adaptive precision MV, the maximum bit saving was 7.7% for the YUV Desktop sequence under Low Delay coding structure without a significant impact on encoding time.

4.2.2 Fast Intra Coding

HEVC SCC has adopted several enhanced coding methods to improve compression efficiency by considering the properties of computer generated contents. Due to lot of redundancy in the spatial domain, intra prediction and processing are mainly dealt with in the standard. Slight changes can be made to generate coding tree unit (CTU) and other intra-based tools such as intra block copy, palette mode, adaptive colour transform are newly introduced. Many researches are focused on these intra coding techniques that a large amount of computation is required.

CTU is a quad-tree structure that can be a CU or can be split into four smaller units, if necessary. Since the SCC often includes a lot of redundancy in the spatial domain, the CTU structure may be different from that of normal HEVC contents. Fast CTU partition algorithm is suggested based on entropy of the individual CU [39]. The entropy is quite low in the screen content, because most of area is smooth and the pixel values are mostly equal. Several rules are obtained to terminate CTU partition earlier. For example, if the entropy of a 64×64 CU is 0, if the entropy of its four 32×32 sub-blocks are equal, or if there are two 32×32 sub-blocks that have the same entropy as the other two 32×32 sub-blocks, then partition may be terminated. Using this method, the encoding time attained 32% reduction on average, BD rate exhibited 0.8% loss, and PSNR obtained a 0.09% loss, compared to the test mode HM-12.1+RExt-5.1. Fast CU partition decision using machine learning with features that describe CU statistics and sub-CU homogeneity is suggested in [40], achieving 36.8% complexity reduction on average with 3.0% BD-rate increase.

The IBC is a prediction method that finds a matched block within a current frame and sends a block vector (BV) information to the decoder. Since the amount of BV data is significant, the block vector prediction (BVP) is used to reduce the BV data. A block vector difference (BVD), which is calculated by the difference between a BV and a BVP, is coded by the $3rd$ order exponential Golomb coding. The IBC using these techniques provides a significant coding gain up to 69.39% BD-rate [41] at the cost of computational complexity increased by more than 30% [42]. Therefore, fast algorithm for block vector search in the IBC is considered. One way is to terminate the block vector search early as possible using a threshold when computing SAD between the block and predicted block. The threshold value is defined based on statistical experimental results. The average time savings was 29.23% with BD-rate 0.41% compared to SCM-2.0 [42]. Other algorithms [43] using thresholds and block activities were reported to reduce the block matching time.

HEVC SCC has also adopted the transform skip mode (TSM). Since screen content usually has more regular and sharper edges, prediction method might work well with no transform that may be inefficient and even worsen the coding performance. For these TUs, DCT or DST is skipped and transform_skip_flag (TSF) is signalled to the decoder. Thus, it is based on the similar principle of IBC that sample values are predicted from other samples in the same picture. These two techniques are highly related in terms of statistical occurrence. For example, up to 94.5% of TSM occurs in IBC-coded CUs [44]. Note that coded_block_flag (CBF) is sent for every TU, indicating all transform coefficients are zero, if CBF is set to 0 and any of transform coefficients are non-zero, if it is set to 1. Thus, we can save bits for two flags by careful modification of signalling, such that, for a 4×4 TB of IBC-coded CU, TSF is not signalled and CBF plays the same role as TSF. That is, when CBF is 1, it indicates TSM is selected in encoder. By the method and clever adjustment, [44] obtained reduction of 4×4 transform block encoding time by 28.2% with a slight increase of BD-rate by 0.22%.

The most probable mode (MPM) is used to reduce bits in the 4×4 intra-prediction instead of nine prediction modes. The encoder estimates the MPM for the current block based on the availability of neighbouring blocks. If the MPM is the same as the prediction mode, we need to send only one bit instead of four bits. In screen content, if all boundary samples are exactly having the same value, all samples can be filled by the boundary samples with the MPM index and any rate-distortion optimization (RDO) can be skipped, named simple intra prediction (SIP) in [45]. Direct prediction from boundary samples is possible by introducing single colour mode [46] and independent uniform prediction mode [47, 48].

4.3 Visual Quality Assessment

The main objective of image/video coding is to compress data for storage and transmission, while retaining visual quality reasonable to human eye. The simplest way to assess visual quality is peak signal-to-noise ratio (PSNR) that is computed by difference between the original data and reconstructed data.

If it is infinite, that means there is no difference and the quality loss does not exist. However, the lossy coding usually results in degradation due to quantization after prediction and transform. Therefore, it is necessary to take out the irrelevant data to human sensitivity that should be defined by certain dedicated models.

4.3.1 Screen Image Quality Assessment

Numerous researches have been performed to develop perceptual quality assessment for images (IQA) that can be classified into three categories depending on the type of contents: natural image quality assessment (NIQA), document image quality assessment (DIQA), and screen image quality assessment (SIQA). It can be further classified into two categories depending on the measurement method: objective and subjective quality assessment. Objective assessment is preferred with its advantage: firstly, they are usually low complexity and secondly, we can classify distortions into several known components such as blocking, ringing, and blurring. Widely used objective assessment metrics are PSNR, SSIM (Structural Similarity) [49], gradient-based [50], image feature-based, and machine learning-based algorithms. There, of course, have been some limitations that they may not be exactly suitable for human observers. Subjective quality assessment is a human judgment-based method. Several test procedures have been defined in ITU-R Rec. BT.500-11: namely, SS (Single Stimulus), SC (Stimulus Comparison), SSCQE (Single Stimulus Continuous Quality Evaluation), DSIS (Double Stimulus Impairment Scale), SDSCE (Simultaneous Double Stimulus Continuous Evaluation), and DSCQS (Double Stimulus Continuous Quality Scale) [SE2]. Another classification is possible depending on the existence of reference images: full-reference, reduced-reference, and no-reference IQA algorithms [51]. The result of quality assessment is reported as either a scalar value or a spatial map denoting the local quality of each image region. Some of best-performing algorithms have been shown to generate quality estimates that correlate with human ratings, typically yielding Spearman rank-order and Pearsonlinear correlation coefficients in excess of 0.9. These IQA algorithms are summarized in Figure 4.3 [51].

Firstly, NIQA has been studied tremendously during the last several decades. Natural images are usually obtained by visual camera that produces pictorial data. Recently, DIQA has attracted attention in the research community due to the necessity of digitization of old documents or imaged documents that their original features should be maintained. Most DIQA

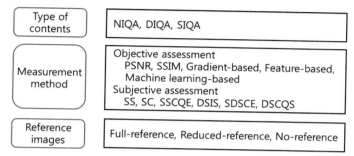

Figure 4.3 Classification of IQA depending on type of images, measurement method, and reference images.

algorithms are designed in no-reference manner, since the original documents may not exist. The effectiveness of DIQA methods can be expressed by accuracy of character recognition. Since SCIs include pictorial regions beside textual regions without environmental degradations, features quite differ from those of the document images and DIQA methods cannot be directly adopted to evaluate the visual quality. The NIQA methods cannot be applied to evaluate the quality of SCIs either. Thus, new screen image database and quality assess metrics have to be developed. In [52], 20 reference and 980 distorted SCIs are included in database that can be downloaded in [53]. Distorted images are generated by applying the typical seven distortions: Gaussian noise, Gaussian blur, motion blur, contrast change, JPEG, JPEG2000 (see Chapter 1), and layer segmentation based coding (LSC) [54] that firstly separates SCIs into textual and pictorial blocks with a segmentation method and applies different encoding method.

4.3.2 Objective Quality Assessment

It has been observed that natural images and textual images have different properties in terms of energy in the spatial frequency domain. To examine this, we decompose images using Fourier transform and then compute energy of the frequency coefficients. Energy of natural images linearly falls off from low to high frequency, while that of textual images has a peak at high frequency, since there are lot of small characters and sharp edges. Thus, SCIs consisting of two or more different contents need to be evaluated by relevant IQA metrics for each content. Since the final decision for quality assessment is to be made for the compound whole images rather than regional images, we still have to develop how to aggregate them.

There are various ways to classify textual and pictorial content, such as gradient-based [55], text detection [56], and segmentation-based, etc. In [57], a block classification approach is suggested by making use of the information content map computed based on the local variance in the 4×4 block. Since textual regions contain high contrast edges, the local information is higher than in pictorial regions. By applying an empirical threshold on the mean of the block information, the textual and pictorial regions can be separated. The quality of each content can be assessed by any methods, although the most popular one would be the SSIM that combines local luminance, contrast and structural similarities. The three types of similarity between the reference and distorted images are pooled into an aggregated quality index. In SCIs, however, some incorrect quality scores happen from the average pooling. Therefore, [58] suggests a structure induced quality metric (SIQM) based on structural degradation model (SDM) defined by

$$\text{SIQM}(\mathbf{r},\mathbf{d}) = \frac{\sum_{i=1}^{M} \text{SSIM\_MAP}(r_i, d_i) \cdot \text{SDM}(r_i)}{\sum_{i=1}^{M} \text{SDM}(r_i)} \qquad (4.10)$$

where $\mathbf{r}$ and $\mathbf{d}$ denotes reference and distorted image signal and $\text{SDM}(r_i)$ is defined by

$$\text{SDM}(\mathbf{r}) = 1 - \text{SSIM}(\mathbf{r},\mathbf{r}_f) \qquad (4.11)$$

where $\mathbf{r}_f$ is generated by applying a simple circular-symmetric Gaussian low-pass filter. Distortion maps generated by this method show more highlighted around the texts than in the pictorial regions. Performance of the SIQM is 0.852 on average Spearman and Pearson correlation coefficient, while the SSIM produces 0.750.

Another pooling method is suggested in [57], based on weighted average of textual quality Q_T and pictorial quality Q_P, defined by

$$Q_S = \frac{Q_T \cdot E(\omega_T) + Q_P \cdot E(\omega_P)}{E(\omega_T) + E(\omega_P)} \qquad (4.12)$$

where $E(\omega_T)$ and $E(\omega_P)$ denote the expectation of the local energy for the textual and pictorial regions, respectively. These quantities take a role of weighting factor for each content. The higher local energy, the more importance in the region. Q_T and Q_P are computed by another weighted SSIM metric. Performance of this method is 0.851 on average Spearman and Pearson correlation coefficient, while the SSIM produces 0.744, which are similar to those in [58].

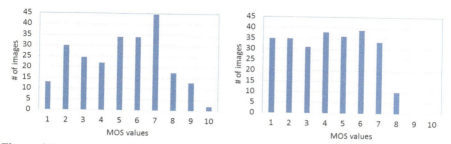

Figure 4.4 Histogram of the MOS values for (left) SCC and (right) HEVC [59]. Higher MOS values are achieved by SCC.

4.3.3 Subjective Quality Assessment

Subjective testing methodologies can be roughly categorized into two types: the single stimulus and double stimulus. The former asks the viewers to rate the quality of one distorted image, while the later asks the viewers to rate the quality between reference and distorted images. After testing, mean opinion score (MOS) of ten levels is computed. The higher MOS value is, the more correlation with human eye is. It reveals that the subjective quality scores for SCC is better than that for HEVC [59]. That is, SCC provides better performance than HEVC at the same distortion level as shown in Figure 4.4. However, there are many factors affecting human vision when viewing SCIs, including area ratio and region distribution of textual regions, size of characters, and content of pictorial regions, etc. [60]. When testing by subjects, the consistency of all judgments for each image should be examined. It can be measured by the confidence interval derived from the value and standard deviation of scores. Generally, with a 95% confidence level, the testing scores is regarded as confident.

4.4 Other SCC Algorithms

4.4.1 Segmentation

Since the SCIs are mixed with text, graphics, and natural pictures, researchers are interested in segmentation of these into several regions and applied different compression algorithms to different image types. In [61], two-step segmentation was developed: block classification and refinement. The first step is to classify 16 × 16 non overlapping blocks into text/graphics blocks and picture blocks by counting the number of colours in each block. If the number of colours is larger than a certain threshold, the block is classified as picture block. The underlying reason is that natural pictures generally have a large number of colours, while text has a limited number of colours. If the number of colours is more than a threshold, it will be classified into pictorial block, otherwise as text/graphics. In this step, it produces a coarse segmentation, because it may contain different image types. Therefore, a refinement segmentation is followed to extract textual pixels from pictorial pixels. Shape primitives such as horizontal/vertical line or rectangle with the same colour are extracted and compared the size and colour with some threshold. Thus, it is called shape primitive extraction and coding (SPEC) [61].

In [62], foreground and background separation algorithm is developed. They use the smoothness property of the background and the deviation property of the foreground. The overall segmentation algorithm is summarized as: firstly, if all pixels in the block have the same colour, it can be background or foreground by considering neighbouring blocks. Second, if all pixels can be predicted with small enough error using least square fitting [63] method, it can be background. Third, they run the segmentation algorithm after decomposing the block size into four smaller blocks until the size 8×8. This algorithm outperforms SPEC in terms of precision and recall.

4.4.2 Rate Control

The rate control is always an important factor to define codec's performance even in SCC. It is required to decide importance of different types of images. The more bit rate for textual regions, the finer quality we obtain for them, while obtaining worse quality for pictorial regions. In video coding it is related to the frame rate. It helps to utilize the bandwidth more efficiently. Rate control can be performed in two procedures: bit allocation and bit control. Once available bits are allocated in GOP level, picture level, and CU level, the next step is to adjust the coding parameters so that the actual amount of bits consumed is close to the pre-allocated target bits. It is desired for a video codec to minimize the bit rate as well as to minimize the distortion which is caused by data compression coding. Thus, the rate control problem is formulated to minimize the distortion D, subject to a rate constraint R to derive the optimal coding parameter P_{opt}:

$$P_{opt} = \arg\min_{P} D \quad s.t. \quad R \leq B \quad (4.13)$$

where B is the given bit budget. The coding parameter is a set including coding mode, motion estimation, and quantization parameter (QP). Such constrained problem which is too complicated to be solved in real video codec is converted into unconstrained optimization problem, called rate-distortion optimization (RDO) by using Lagrange multiplier λ as

$$P_{opt} = \arg\min_{P} (D + \lambda R). \qquad (4.14)$$

λ serves as a weighting factor for the rate constraint and also indicates the slope of the R-D curve [64]. In the practical applications, however, 4.14 is formulated in a simple form such as the quadratic R-D function [65], which has been adopted by most of video coding standards, defined by

$$R = aQ^{-1} + bQ^{-2} \qquad (4.15)$$

where Q is quantization scale instead of distortion due to its simplicity.

The work in [64] proposes λ domain rate control for HEVC. The benefits of $R - \lambda$ rate control are: it is more equivalent to finding the distortion on the R-D curve and it can be more precise than adjusting integer QP since λ can take any continuous positive values. It outperforms the R-Q model in 4.15 by 0.55 dB on average.

However, since the screen content has different characteristics, e.g., abrupt changes, a more appropriate rate control scheme is required. The work in [66] proposes an enhanced algorithm based on the $R - \lambda$ model. First, they analyse the complexity of each picture using a sliding window to handle the discontinuities. Then, bits are allocated and the parameters of the model are adjusted. As a result, it decreases the distortion by 2.25% on average and improves the coding efficiency by 5.6%. Another important aspect in screen content is that there are many repeating patterns among pictures and in the same picture as stated earlier. This feature is utilized to introduce the IBC in the screen content coder. Problem is that bit rate at picture level should be maintained at all coding process. The work in [67] proposes weighted rate distortion optimization (WRDO) solution for screen content coding. A weighting factor ω is now applied to 4.14 as

$$P_{opt} = \arg\min_{P} (\omega D + \lambda R). \qquad (4.16)$$

This algorithm was already implemented in HEVC test model, HM-16.2 with uniform distortion weight for all blocks in the picture. ω is kept as 1 in the normal picture which has less influence, while $\omega > 1$ in the more important picture. In case of hierarchical coding structure, pictures in the highest temporal level are never used as reference pictures and $\omega = 1$.

As ω is only determined by the coding structure, it is not related to the content. To solve the problem, the block distortion weight should be determined by each block's influence instead of the fixed uniform weight. In [67], block distortion weight is calculated in two ways: inter weight among pictures considering temporal correlations and intra weight within one picture considering the correlations in IBC process. Then, the overall distortion weight is obtained be taking both aspects together. It results in 14.5% of coding gain for the IBBB coding structure at the cost of 2.9% of coding complexity increase.

4.5 Summary

In the context of screen content coding (SCC), various aspects have to be taken into account, due to different characteristics in the natural camera captured content. The SCC is an extension of HEVC standard with several new tools, including IntraBC [72], palette coding, adaptive colour transform, and adaptive motion vector resolution, which were discussed in Section 3.2. IntraBC is a kind of motion estimation/compensation used in the natural video coding but it is performed in intra picture, since there are many redundancies in the spatial domain. By using palette mode, we can increase coding gain by sending an index instead of real colour value. For screen content, there may exist blocks containing different features of colours, which lead to less correlation. For these blocks, direct coding in the RGB space may be more effective. This gives motivation for adopting adaptive colour transform. Motion vector resolution should also be adaptively decided in multi-featured screen content. In some cases, it can be in fractional resolution, while in other cases it should be in integer resolution.

Since there is strong correlation in the spatial domain of screen content, intra prediction is a key factor to be developed. Though it has been developed and standardized in modern video coding standards including H.264 and HEVC, further developments are required for screen content video. For example, sample-based angular intra-prediction with edge prediction is useful for the SCC discussed in Section 4.1.

Development of fast coding algorithms is necessary, since the SCC often requires high computational complexity. Intra coding and motion compensation are two main parts that need to be speeded up as discussed in Section 4.2.

The main objective of image/video coding is to compress data for storage and transmission, while retaining visual quality reasonable to human eye. Thus, screen image quality assessment has to be introduced in the research arena. We discussed and compared quality assessment methods depending on the type of contents: natural images, document images, and screen content images in Section 4.3.

Due to different nature in the screen content, many other algorithms are still being developed such as segmentation and rate control as discussed in Section 4.4. A lot of research results are still being introduced in the literature. Some of them appeared in Section 4.6, giving readers updated projects.

4.6 Projects

P.4.1 Tao, et al., [68] proposed a re-sampling technique for template matching that can increase prediction performance at the cost of overhead information: position, index, and value. Among these three, position will consume more number of bits than the other two. As a solution, they applied the similarity of non-zero prediction error of pixel positions. Implement the encoder and find the compression performance.

P.4.2 Zhang, et al., [69] proposed a symmetric intra block copy (SIBC) algorithm in which utilizes symmetric redundancy. They conducted a simple flipping operation either vertically or horizontally on the reference block before it is used to predict the current block. The filpping operation is easy to implement with low cost. They achieved up to 2.3% BD-rate reduction in lossy coding on some sequences that have a lot of symmetric patterns. Please investigate to find the amount of symmetricity in all test sequences that can be downloaded from [15]. Compare the performance to normal IBC using the reference software [16].

P.4.3 Tsang Chan and Siu [70] have developed a fast local search method that can be used for hash based intra block copy mode. Due to the high computational complexity for the IBC [19], they proposed fast local search by checking the hash values of both current block and block candidates. The encoding time is reduced by up to 25% with only negligible bitrate increase. Implement this algorithm and evaluate the performance using the latest reference software.

P.4.4 The Intra Block Copy (IntraBC) tool efficiently encodes repeating patterns in the screen contents as discussed in Section 3.2.1. It is also applicable for coding of natural content video, achieving about 1.0% bit-rate reduction on average. Chen, et al., [71] have worked for further improvements on IntraBC with a template matching block vector and a fractional search method. The gain on natural content video coding increased up to 2.0%, which is not so big, of course, comparing to the efficiency on screen content video. For example, Pang, et al., [72] achieves 43.2% BD rate savings. However, it reveals that there exists some sort of redundancy in the spatial domain of natural video. Carefully design and implement intra prediction with block copy method for natural video content. Evaluate the coding performance in terms of the BD rate.

P.4.5 Fan, et al., [73] have developed quantization parameter offset scheme based on inter-frame correlation, since the inter-frame correlation among adjacent frames in screen content videos is very high. Firstly, they define a measurement of inter-frame correlation and then, quantization parameter offset for successive frames is appropriately adjusted. Number of correlated sub-blocks is counted based on SAD and thresholding between two frames. The more correlation, the larger quantization parameter may be required. The maximum BD rate gain was over 3.8% and the average performance gain is over 2% compared with the reference software. Implement this kind of optimization problem using different correlation measures and block sizes.

P.4.6 Zhao, et al., [74] proposed Pseudo 2D String Matching (P2SM) for screen content coding. Redundancy of both local and non-local repeated patterns is exploited. Different sizes and shapes of the patterns are also considered. They achieved up to 37.7% Y BD rate reduction for a screen snapshot of a spreadsheet. Implement the P2SM and confirm their results.

P.4.7 Sample-based weighted prediction was discussed in Section 4.1.1. Sanchez [75] also proposed sample-based edge prediction based on gradients for lossless screen content coding in HEVC, since a high number of sharp edges causes inefficient coding performance for screen content video using current video coding tools. It is a DPCM-based intra-prediction method that combines angular prediction with gradient-based edge predictor and a DPCM-based DC predictor. Average bit-rate reduction was 15.44% over current HEVC intra-prediction. Implement the edge predictor and apply it to the latest reference software. Describe the advantage obtained by the edge predictor.

P.4.8 The intra coding in the HEVC main profile incorporates several filters for reference samples, including a bi-linear interpolation filter and a smoothing filter [31]. Kang [76] developed adaptive turn on/off filters for inra-prediction method. The decision is based on two criteria: statistical properties of reference samples and information in the compressed domain. For the former, they used Mahalanobis distance for measuring distances between samples and their estimated distributions. For the latter, they used R-D optimization model in the compressed domain. Implement this method using different distance measures and different transforms. Analyze benefits from turning filters on and off. Derive the most efficient transform for coding screen content by taking into account structural information [77].

P.4.9 Natural video has been commonly coded in 4:2:0 sampling format, since the human visual system is less sensitive to chroma. Screen content video, however, is coded in full chroma format, since the downsampling of chroma introduces blur and colour shifting. This is because of the anisotropic features in compound contents [78]. Nevertheless, downsampling is required to increase compression ratio as much as possible. S. Wang, et al., [79] proposed an adaptive downsampling for chroma based on local variance and luma guided chroma filtering [80]. Coding performance is measured in terms of PSNR and SSIM. Implement their adaptive downsampling scheme using screen content reference software. Compare the performance to that using full chroma format in terms of different performance criteria including BD rate.

P.4.10 Three transform skip modes (TSM) have been proposed during the development of HEVC, including vertical transform skipping, horizontal transform skipping, and all 2D transform skipping. Vertical transform skipping means applying only the vertical 1D transform and skipping the horizontal 1D transform. This is called 1D TSM. If the prediction errors have minimal correlation in one or both directions, transform doesn't work well. Thus, it can be skipped. The 1D TSM is efficient in the screen content which has strong correlations in one direction but not the other. J.-Y. Kao, et al., [81] developed the 1D TSM based on dynamic range control, modification of the scan order, and coefficient flipping. D. Flynn, et al., [82] discusses more on TSM. Evaluate their effectiveness using the latest SCM software.

P.4.11 Implement and compare the three intra coding techniques: intra string copy [83], intra block copy [20, 84], and intra line copy [85].

P.4.12 Just Noticeable Difference (JND) has been used for measuring image/video distortions. Wang, et al., [86] developed the JND modeling to be used for compressing screen content images. Each edge profile is decomposed into luminance, contrast, and structure, and then evaluate the visibility threshold in different ways. The edge luminance adaptation, contrast masking, and structural distortion sensitivity are also studied. Develop the JND model for lossless and lossy coding of screen content video and evaluate it in terms of human visual sensitivity.

P.4.13 Some fast algorithms for coding screen content are discussed in Section 4.2, mainly dealing with intra coding and motion compensation. Another approach is proposed by Zhang, et al., [87] to speed up intra mode decision and block matching for IntraBC. If we could take proper background detection, encoding time can be saved by skipping mode decision process in the region. Background region can be detected by some sort of segmentation technique discussed in Section 4.4.1. Derive the background detection algorithms by your own means and apply it for coding screen content.

P.4.14 F. Duanmu, et al., [88] developed a transcoding framework to efficiently bridge the HEVC standard and it's SCC extension. It can achieve an average of 48% re-encoding complexity reduction with less than 2.14% BD-rate increase. It is designed as a pre-analysis module before intra frame mode selection. Coding modes are exchanged based on statistical study and machine learning. Implement this type of transcoding tool and confirm their results.

P.4.15 C. Chen, et al., [89] proposed a staircase transform coding scheme for screen content video coding, that can be integrated into a hybrid coding scheme in conjunction with conventional DCT. Candidates for the staircase transform include Walsh-Hadamard transform and Haar transform [90]. The proposed approach provides an average of 2.9% compression performance gains in terms of BD-rate reduction. Implement this hybrid coder and evaluate the performance.

P.4.16 See [99]. Explore the future research suggested in the conclusion section.

References

[1] D. Flynn, J. Sole and T. Suzuki, High efficiency video coding (HEVC) range extensions text specification, Draft 4, JCT-VC, Retrieved 2013-08-07.

[2] I. E. G. Richardson, H.264 and MPEG-4 video compression: Video coding for Next Generation Multimedia, New York: Wiley, 2003.

[3] S. Kodpadi, "Evaluation of coding tools for screen content in High Efficiency Video Coding," Dec. 2015. http://www.uta.edu/faculty/krrao/dip. Click on courses and then click on EE5359. Scroll down and go to Thesis/Project Title and click on S. Kodpadi.

[4] N. Mundgemane, "Multi-stage prediction scheme for screen content based on HEVC," *Master's thesis at University of Texas at Alrington*, 2015. http://www.uta.edu/faculty/krrao/dip. Click on courses and then click on EE5359. Scroll down and go to Thesis/Project Title and click on N. Mundgemane.

[5] H. Yu, et al., "Requirements for an extension of HEVC for coding on screen content," *ISO/IEC JTC 1/SC 29/WG 11 Requirements subgroup*, Jan. 2014.

[6] J. Xu, R. Joshi, and R.A. Cohen, "Overview of the Emerging HEVC Screen Content Coding Extension," *IEEE Trans. CSVT,* vol. 26, no. 1, pp. 50–62, Jan. 2016.

[7] C. Lan, et al., "Intra and inter coding tools for screen contents," *JCTVC-E145*, Mar. 2011.

[8] D. Milovanovic and Z. Bojkovic, "Development of mixed screen content coding technology : SCC-HEVC Extensions Framework," *Proc. of the 13th Int. Conf. on Data Networks, Communications, Computers (DNCOCO'15)*, pp. 44–50, Budapest, Hungary, Dec. 2015.

[9] ISO/IEC JTC1/SC29/WG11 and ITU-T SG16 Q6/16, "Joint Call for Proposals for Coding of Screen Content," 17 Jan. 2014.

[10] H. Yu, et al., "Common conditions for screen content coding tests," *JCTVC – U1015-r2,* June 2015.

[11] ISO/IEC JTC1/SC29/WG11 N14399, "Results of CfP on Screen Content Coding Tools for HEVC," Apr. 2014.

[12] ISO/IEC JTC1/SC29/WG11 N15779, "HEVC Screen Content Coding Test Model 6 (SCM 6)," Oct. 2015.

[13] R. Joshi, et al., "High Efficiency Video Coding (HEVC) Screen Content Coding: Draft 5," *JCTVC-V1005,* Oct. 2015.

[14] Access to JCT-VC documents, "http://phenix.int-evry.fr/jct/," [Online].

[15] Test sequences for SCC, "http://pan.baidu.com/share/link?shareid=3128894651&uk=889443731," [Online].

[16] HEVC SCC Extension Reference Software, "https://hevc.hhi.fraunhofer.de/svn/svn_HEVCSoftware/tags/HM-16.8+SCM-7.0/," [Online].

[17] HEVC SCC Software reference manual, "https://hevc.hhi.fraunhofer. de/svn/svn_HEVCSoftware/branches/HM-SCC-extensions/doc/ software-manual.pdf," [Online].
[18] Y.-J. Ahn, et al., "Analysis of Screen Content Coding Based on HEVC," *IEIE Trans. on Smart Processing and Computing*, vol. 4, no. 4, pp. 231–236, Aug. 2015.
[19] W. Zhu, et al., "Hash-Based Block Matching for Screen Content Coding," *IEEE Trans. on Multimedia*, vol. 17, no. 7, pp. 935–944, July 2015.
[20] X. Xu, et al., "Block Vector Prediction for Intra Block Copying in HEVC Screen Content Coding," *IEEE Data Compression Conference (DCC)*, pp. 273–282, 2015.
[21] L. Guo, et al., "Color palette for screen content coding," *IEEE ICIP*, pp. 5556–5560, Oct. 2014.
[22] X. Xiu, et al., "Palette-based Coding in the Screen Content Coding Extension of the HEVC Standard," *IEEE Data Compression Conference (DCC 2015)*, pp. 253–262, April 2015.
[23] L. Zhang, et al., "Adaptive Color-Space Transform for HEVC Screen Content Coding," *IEEE Data Compression Conference (DCC)*, pp. 233–242, 2015.
[24] H. S. Malvar, G. J. Sullivan, and S. Srinivasan, "Lifting-based reversible color transformations for image compression," *SPIE Applications of Digital Image Processing*, vol. 7073, Aug. 2008.
[25] M. Wien, High efficiency video coding: Coding tools and specification, Springer, 2015.
[26] Y.-L. Lee, K.-H. Han, and G.J. Sullivan, "Improved lossless intra coding for H.264/MPEG-4 AVC," *IEEE Trans. on Image Processing*, vol. 15, no. 9, pp. 2610–2615, Sept. 2006.
[27] S. Lee, I.-K. Kim and C. Kim, "Residual DPCM for HEVC lossless coding," *JCTVC-M0079*, 13th JCT-VC meeting, Incheon, KR, Apr. 2013.
[28] K. Kim, J. Jeong, and G, Jeon, "Improved residual DPCM for HEVC lossless coding," *27th SIBGRAPI Conf. on Graphics, Patterns and Images*, pp. 95–102, Aug. 2014.
[29] M. Naccari, et al., "Improving inter prediction in HEVC with residual DPCM for lossless screen content coding," *IEEE PCS*, pp. 361–364, San Jose, CA, Dec. 2013.

[30] E. Wige, et al., "Sample-based weighted prediction with directional template matching for HEVC lossless coding," *IEEE PCS*, pp. 305–308, Dec. 2013.
[31] V. Sze, M. Budagavi, and G.J. Sullivan, High efficiency video coding (HEVC): Algorithms and architectures, Springer, 2014.
[32] M. Zhou, et al., "HEVC lossless coding and improvements," *IEEE Trans. CSVT,* vol. 22, no. 12, pp. 1839–1843, Dec. 2012.
[33] Y. H. Tan, C. Yeo and Z. Li, "Residual DPCM for lossless coding in HEVC," *Proc. IEEE Int. Conf. Acoustics, Speech Signal Processing (ICASSP),* pp. 2021–2025, May 2013.
[34] V. Sanchez, et al., "HEVC-based Lossless Compression of Whole Slide Pathology Images," *Proc. 2014 IEEE Global Conference on Signal and Information Processing (GlobalSIP),* pp. 452–456, Dec. 2014.
[35] V. Sanchez, "Lossless screen content coding in HEVC based on sample-wise median and edge prediction," *IEEE International Conference on Image Processing (ICIP),* pp. 4604–4608, 2015.
[36] V. Sanchez, "Fast intra-prediction for lossless coding of screen content in HEVC," *2015 IEEE Global Conference on Signal and Information Processing (GlobalSIP),* pp. 1367–1371, 2015.
[37] B. Li and J. Xu, "A Fast Algorithm for Adaptive Motion Compensation Precision in Screen Content Coding," IEEE *Data Compression Conference (DCC),* pp. 243–252, 2015.
[38] B. Li, J. Xu, and F. Wu, "A Unified Framework of Hash-based Matching for Screen Content Coding," *IEEE VCIP'14,* pp. 530–533, Dec. 7-10, 2014.
[39] M. Zhang, Y. Guo, and H. Bai, "Fast intra partition algorithm for HEVC screen content coding," *IEEE Visual Communications and Image Processing Conf.,* pp. 390–393, 2014.
[40] F. Duanmu, Z. Ma, and Y. Wang, "Fast CU partition decision using machine learning for screen content compression," *IEEE Int. Conf. on Image Processing (ICIP),* pp. 4972–4976, 2015.
[41] G. Bjontegaard, "Calculation of average PSNR differences between RD-curves," *ITU-T SG 16 Q 6, VCEG-M33,* Mar. 2001.
[42] J. Ma and D. Sim, "Early Termination of Block Vector Search for Fast Encoding of HEVC Screen Content Coding," *IEIE Trans. on Smart Processing and Computing,* vol. 3, no. 6, pp. 388–392, Dec. 2014.

[43] D.-K. Kwon and M. Budagavi, "Fast intra block copy (intra BC) search for HEVC Screen Content Coding," *IEEE ISCAS,* pp. 9–12, June 2014.

[44] D. Lee, et al., "Fast transform skip mode decision for HEVC screen content coding," *IEEE Int. Symp. on Broadband Multimedia Systems and Broadcasting (BMSB),* pp. 1–4, 2015.

[45] S.–H. Tsang, Y.–L. Chan and W.–C. Siu, "Fast and Efficient Intra Coding Techniques for Smooth Regions in Screen Content Coding Based on Boundary Prediction Samples," *IEEE ICASSP,* pp. 1409–1413, Brisbane, Australia, Apr. 2015.

[46] Y.-W. Chen, et al., "SCCE3 Test D.1: Single color intra mode for screen content coding," *JCTVC of ITU-T SG 16 WP 3 and ISO/IEC JTC1/SC29/WG11, JCTVCR0058, Sapporo, Japan,* June-July 2014.

[47] X. Zhang, R. A. Cohen and A. Vetro, "Independent Uniform Prediction mode for screen content video coding," *IEEE Visual Communications and Image Processing Conference,* pp. 129–132, 7–10 Dec. 2014.

[48] X. Zhang, and R. Cohen, "SCCE3 Test D.2: Independent Uniform Prediction (IUP) Mode," *JCT-VC of ITU-T SG 16 WP 3 and ISO/IEC JTC1/SC29/WG11, JCTVC-R0200, Sapporo, Japan,* June-July 2014.

[49] Z. Wang, et al., "Image quality assessment: From error visibility to structural similarity," *IEEE Trans. on Image Processing,* vol. 13, no. 4, pp. 600–612, Apr. 2004.

[50] J. Zhu and N. Wang, "Image quality assessment by visual gradient similarity," *in Proceedings of the IEEE Trans. on Image Processing,* vol. 21, no. 3, pp. 919–933, Mar. 2012.

[51] D. M. Chandler, "Seven challenges in image quality assessment: Past, present, and future research," *ISRN Signal Process.,* vol. 2013, pp. 1–53, 2013.

[52] H. Yang, Y. Fang and W. Lin, "Perceptual Quality Assessment of Screen Content Images," *IEEE Trans. on Image processing,* vol. 24, no. 11, pp. 4408–4421, Nov. 2015.

[53] SIQAD (Perceptual Quality Assessment of Screen Content Images), "https://sites.google.com/site/subjectiveqa/," [Online].

[54] Z. Pan, et al., "A low-complexity screen compression scheme for interactive screen sharing," *IEEE Trans, CSVT,* vol. 23, no. 6, pp. 949–960, June 2013.

[55] S. Wang, et al., "Content-aware layered compound video compression," *IEEE ISCAS,* pp. 145–148, 2012.

[56] A. Risnumawan, et al., "A robust arbitrary text detection system for natural scene images," *Expert Systems with Applications*, vol. 41, no. 18, pp. 8027–8048, Dec. 2014.

[57] S. Wang, et al., "Perceptual screen content image quality assessment and compression," *IEEE Int. Conf. on Image Processing (ICIP)*, pp. 1434–1438, 2015.

[58] K. Gu, et al., "Screen image quality assessment incorporating structural degradation measurement," *IEEE Int. Symp. on Circuits and Systems (ISCAS)*, pp. 125–128, 2015.

[59] S. Shi, et al., "Study on subjective quality assessment of Screen Content Images," *IEEE Picture Coding Symposium (PCS)*, pp. 75–79, 2015.

[60] H. Yang, et al., "Subjective quality assessment of screen content content images," *International Workshop on Quality of Multimedia Experience (QoMEX)*, pp. 257–262, 2014.

[61] T. Lin and P. Hao, "Compound Image Compression for Real-Time Computer Screen Image Transmission," *IEEE Trans. on Image Processing*, vol. 14, no. 8, pp. 993–1005, Aug. 2005.

[62] S. Minaee and Y. Wang, "Screen content image segmentation using least absolute deviation fitting," *IEEE Int. Conf. on Image Processing (ICIP)*, pp. 3295–3299, 2015.

[63] Y. Li and G.R. Arce, "A maximum likelihood approach to least absolute deviation regression," *EURASIP Journal on Applied Signal Processing*, pp. 1762–1769, Sept. 2004.

[64] B. Li, et al., "Lambda Domain Rate Control Algorithm for High Efficiency Video Coding," *IEEE Trans. on Image Processing*, vol. 23, no. 9, p. 3841–3854, Sept. 2014.

[65] T. Chiang and Y.-Q. Zhang, "A new rate control scheme using quadratic rate distortion model," *IEEE Trans. on CSVT*, vol. 7, no. 1, pp. 246–250, Jan. 1997.

[66] Y. Guo, et al., "Rate control for screen content coding in HEVC," *IEEE Int. Symp. on Circuits and Systems (ISCAS)*, pp. 1118–1121, 2015.

[67] W. Xiao, et al., "Weighted rate-distortion optimization for screen content coding," *IEEE Trans. on CSVT*, 2016 (Early Access).

[68] P. Tao, et al., "Improvement of re-sample template matching for lossless screen content video," *IEEE International Conference on Multimedia and Expo (ICME)*, pp. 1–6, 2015.

[69] K. Zhang, et al., "Symmetric intra block copy in video coding," *IEEE International Symposium on Circuits and Systems (ISCAS)*, pp. 521–524, 2015.

[70] S.-H. Tsang, Y.-L. Chan, and W.-C. Siu, "Hash based fast local search for Intra Block Copy (IntraBC) mode in HEVC screen content coding," *Asia-Pacific Signal and Information Processing Association Annual Summit and Conference (APSIPA)*, pp. 396–400, 2015.

[71] H. Chen, et al., "Improvements on Intra Block Copy in natural content video coding," *IEEE International Symposium on Circuits and Systems (ISCAS)*, pp. 2772–2775, 2015.

[72] C. Pang, et al., "Intra Block Copy for HEVC Screen Content Coding," *IEEE Data Compression Conference (DCC)*, p. 465, 2015.

[73] J. Fan, et al., "Inter-frame Correlation Based Quantization Parameter Offset Optimization for Screen Content Video Coding," *IEEE International Conference on Multimedia Big Data (BigMM)*, pp. 420–423, 2015.

[74] L. Zhao, et al., "Pseudo 2D String Matching Technique for High Efficiency Screen Content Coding," *IEEE Transactions on Multimedia*, vol. 18, no. 3, pp. 339–350, 2016.

[75] V. Sanchez, "Sample-based edge prediction based on gradients for lossless screen content coding in HEVC," *IEEE Picture Coding Symposium (PCS)*, pp. 134–138, 2015.

[76] J.-W. Kang, "Sample selective filter in HEVC intra-prediction for screen content video coding," *The Institute of Engineering and Technology (IET) Journals & Magazines*, vol. 51, no. 3, p. 236–237, Feb. 2015.

[77] J.-W Kang, "Structured sparse representation of residue in screen content video coding," *Electronics Letters*, vol. 51, no. 23, pp. 1871–1873, 2015.

[78] C. Lan, G. Shi, and F. Wu, "Compress compound images in H.264/MPGE-4 AVC by exploiting spatial correlation," *IEEE Trans. on Image Processing*, vol. 19, no. 4, pp. 946–957, 2010.

[79] S. Wang, et al., "Joint Chroma Downsampling and Upsampling for screen Content Image," *IEEE Trans. on CSVT*, vol. 26, no. 9, pp. 1595–1609, Sept. 2016.

[80] T. Vermeir, et al., "Guided Chroma Reconstruction for Screen Content Coding," *IEEE Trans. on CSVT*, vol. 26, no. 10, pp. 1884–1892, Oct. 2016.

[81] J.-Y. Kao, et al., "Improved transform skip mode for HEVC screen content coding," *Int. Conf. on Image Processing Theory, Tools and Applications (IPTA)*, pp. 504–509, 2015.

[82] D. Flynn, et al., "Overview of the Range Extensions for the HEVC Standard: Tools, Profiles, and Performance," *IEEE Trans. on CSVT*, vol. 26, no. 1, pp. 4–19, Jan 2016.

[83] F. Zou, et al., "Hash Based Intra String Copy for HEVC Based Screen Content Coding," *IEEE ICME,* Torino, Italy, June 2015.

[84] X. Xu, et al., "Intra Block Copy in HEVC Screen Content Coding Extensions," *IEEE J. of Emerging Selected Topics in Circuits and Systems (Early Access).*

[85] C. C. Chen and W. H. Peng, "Intra Line Copy for HEVC Screen Content Coding," *IEEE Trans. on CSVT,* vol. 27, pp. 1568–1579, July 2017.

[86] S. Wang, et al., "Just Noticeable Difference Estimation for Screen Content Images," *IEEE Trans. on Image Processing,* vol. 25, no. 8, pp. 3838–3851, 2016.

[87] H. Zhang, et al., "Fast intra mode decision and block matching for HEVC screen content compression," *IEEE ICASSP,* pp. 1377–1381, 2016.

[88] F. Duanmu, et al., "A novel screen content fast transcoding framework based on statistical study and machine learning," *IEEE ICIP,* pp. 4205–4209, 2016.

[89] C. Chen, et al., "A staircase transform coding scheme for screen content video coding," *IEEE ICIP,* pp. 2365–2369, 2016.

[90] K. R. Rao and P. C. Yip, The transform and data compression handbook, CRC Press, 2001.

[91] J. Li, et al., "Diversity-Based Reference Picture Management for Low Delay Screen Content Coding, *IEEE Trans.* CSVT, (Early Access).

[92] W. Xiao, et al., "Fast Hash-based Inter Block Matching for Screen Content Coding, "*IEEE Trans.* CSVT, (Early Access).

[93] S. Wang, et al., "Reduced-Reference Quality Assessment of Screen Content Images," *IEEE Trans. CSVT,* (Early Access)

[94] K. Zhang, et al., "Segmental Prediction for Video Coding," *IEEE Trans. CSVT,* (Early Access).

[95] W. S. Kim, et al., "Cross-Component Prediction in HEVC," IEEE Trans. CSVT, (Early Access).

[96] K. Gu et al., "No-reference quality assessment of screen content pictures" IEEE Trans. IP, (early access) 4 research directions possible projects. Several papers related to QA of SC pictures.

[97] W. Zhu et al., "Inter-palette coding in screen content coding," IEEE Trans. Broadcasting, vol. 63, pp. 673–679, Dec. 2017.

[98] G.J. Sullivan et al., "Verification testing of the compression performance of the HEVC screen content coding extensions," SPIE Applications of the digital image processing LX, vol. 10396, San Diego, CA, Oct. 2017.

[99] J. Li et al., "Diversity-based reference picture management for low delay screen content coding," IEEE Trans. CSVT, vol. 28, pp. 1369–1378 , June 2018. Conclusion section is repeated here,

[100] S. Kuanar, C. Conly and K. R. Rao, "Deep Learning based HEVC in-loop Filtering for Decoder Quality Enhancement," IEEE Picture Coding Symposium (PCS), June 2018, San Francisco, California.

Appendix

This paper proposes a diversity-based RPM scheme to improve the coding efficiency for screen content coding. To avoid high complexity, the diversity is calculated based on hash values. The experimental results show that the proposed algorithm improves coding efficiency by 4.9% on average and up to 13.7%, with saving about 5% encoding time. The visual quality can also be enhanced. In addition, the proposed algorithm is nonnormative and compatible with existing video coding standards that support flexible RPM, such as HEVC and H.264/AVC. In this paper, the diversity is modelled by the number of hash values considering the distinct characteristics of screen content, and it indeed works well for screen content. But for natural content, this method has no coding efficiency change. Applying the concept of diversity to natural content coding will be investigated in future research.

documents

http://phenix.int-evry.fr/jct

The reader is referred to Chapter 1 where references related to SCC are added.

H.264 Advance Video Coding (AVC)/MPEG-4 Part 10 References

H1. R. Schäfer, T. Wiegand, and H. Schwarz, "The emerging H.264/AVC standard," *EBU Technical Review*, pp. 1–12, Jan. 2003.

H2. JVT Draft ITU-T recommendation and final draft international standard of joint video specification (ITU-T Rec. H.264–ISO/IEC 14496-10 AVC), March 2003, JVT-G050 available on http://ip.hhi.de/imagecom_G1/assets/pdfs/JVT-G050.pdf.

H3. T. Wiegand et al., "Overview of the H.264/AVC video coding standard," *IEEE Trans. CSVT*, Special Issue on H.264/AVC, vol. 13, pp. 560–576, July 2003. Draw the JPEG baseline decoder block diagram.
[12 points] he proposed systematic solution and the fast algorithm can also be applied to

H4. H.S. Malvar et al., "Low-complexity transform and quantization in H.264/AVC," IEEE Trans. CSVT, vol. 13, pp. 598–603, July 2003.

H5. T. Stockhammer, M.M. Hannuksela, and T. Wiegand, "H.264/AVC in wireless environments," IEEE Trans. CSVT, vol. 13, pp. 657–673, July 2003.

H6. S. Wenger, "H.264/AVC over IP," IEEE Trans. CSVT, vol. 13, pp. 645–656, July 2003.

H7. A. Tamhankar and K.R. Rao, "An overview of H.264/MPEG-4 part 10," Proc. 4th EURASIP-IEEE Conf. focused on Video/Image Processing and Multimedia Communications, vol. 1, pp. 1–51, Zagreb, Croatia,
July 2003.

H8. J. Ostermann et al., "Video coding with H.264/AVC: Tools, performance, and complexity," IEEE Circuits and Systems Magazine, vol. 4, pp. 7–28, First Quarter 2004.

H9. M. Fieldler, "Implementation of basic H.264/AVC decoder," Seminar Paper at Chemnitz University of Technology, June 2004.

H10. Y. Zhang et al., "Fast 4×4 intra-prediction mode selection for H.264," in Proc. Int. Conf. Multimedia Expo, pp. 1151–1154, Taipei, Taiwan, Jun. 2004.

H11. G.J. Sullivan, P. Topiwala and A. Luthra, "The H.264/AVC advanced video coding standard: Overview and introduction to the fidelity range extensions," *SPIE Conf. on Applications of Digital Image Processing XXVII*, vol. 5558, pp. 53–74, Aug. 2004.

H12. F. Fu et al., "Fast intra prediction algorithm in H.264/AVC," Proc. 7th Int. Conf. Signal Process., pp. 1191–1194, Beijing, China, Sep. 2004.

H13. A. Puri, X. Chen, and A. Luthra, "Video coding using the H.264/MPEG-4 AVC compression standard," *Signal Processing: Image Communication*, vol. 19, pp. 793–849, Oct. 2004.

H14. F. Pan et al., "Fast intra mode decision algorithm for H.264/AVC video coding," Proc. IEEE ICIP., pp. 781–784, Singapore, Oct. 2004.

H15. Proc. IEEE, Special Issue on Advances in Video Coding and Delivery, vol. 93, pp. 3–193, Jan. 2005. This has several overview papers.

H16. T. Sikora, "Trends and perspectives in image and video coding," *Proc. IEEE*, vol. 93, pp. 6–17, Jan.2005.

H17. G.J. Sullivan and T. Wiegand, "Video compression – From concepts to H.264/AVC standard," (Tutorial), *Proc. IEEE*, vol. 93, pp. 18–31, Jan. 2005.

H18. D. Marpe and T. Wiegand, "H.264/MPEG4-AVC Fidelity Range Extensions: Tools, Profiles, Performance, and Application Areas," IEEE ICIP, vol. 1, pp. I– 596, 11-14 Sept. 2005.

H19. Y-L. Lee and K-H. Han, "Complexity of the proposed lossless intra for 4:4:4," (ISO/IEC JTC1/SC29/WG11 and ITU-T SG 16 Q.6) document JVT-Q035, 17-21 Oct. 2005.

H20. D. Kumar, P. Shastry, and A. Basu, "Overview of the H.264/AVC," 8th Texas Instruments Developer Conference India, 30 Nov.–1 Dec. 2005, Bangalore, India.

H21. Proc. IEEE, Special Issue on Global Digital Television: Technology and Emerging Services, vol. 94, Jan. 2006.

H22. G. Raja and M. Mirza, "In-loop de-blocking filter for H.264/AVC Video," Proc. the 2nd IEEE-EURASIP Int. Symp. Commun., Control and Signal Processing 2006, ISCCSP, Marrakech, Morocco, March 2006.

H23. S.-K. Kwon, A. Tamhankar, and K.R. Rao, "Overview of H.264/MPEG-4 Part 10," *J. Vis. Commun. Image Representation (JVCIR)*, vol. 17, pp. 186–216, April 2006. Special Issue on "Emerging H.264/AVC Video Coding Standard."

H24. S. Kumar et al., "Error resiliency schemes in H.264/AVC standard," *J. Visual Commun. Image Representation. (JVCIR)*, Special Issue on H.264/AVC, vol. 17, pp. 425–450, April 2006.

H25. D. Marpe, T. Wiegand, and G. J. Sullivan, "The H.264/MPEG-4 AVC standard and its applications," *IEEE Communications Magazine*, vol. 44, pp. 134–143, Aug. 2006.

H26. S. Naito and A. Koike, "Efficient coding scheme for super high definition video based on extending H.264 high profile," in Proc. SPIE Visual Commun. Image Process., vol. 6077, pp. 607727-1–607727-8, Jan. 2006.

H27. J. Kim et al., "Complexity reduction algorithm for intra mode selection in H.264/AVC video coding" J. Blanc-Talon et al., (Eds.): pp. 454–465, ACIVS 2006, LNCS 4179, Springer, 2006.

H28. S. Ma and C.C.J. Kuo, "High-definition video coding with super macroblocks," in Proc. SPIE VCIP, vol. 6508, pp. 650816-1–650816-12, Jan. 2007.

H29. T. Wiegand and G.J. Sullivan, "The H.264/AVC video coding standard," *IEEE SP Magazine*, vol. 24, pp. 148–153, March 2007. (Info on H.264/AVC resources)

H30. H.264/AVC JM 18.4 reference software: http://iphome.hhi.de/suehring/tml/download/jm18.4zip

H31. "H.264 video compression standard," White paper, Axis communications.

H32. W. Lee et al., "High speed intra prediction scheme for H.264/AVC," IEEE Trans. Consumer Electronics, vol. 53, no. 4, pp. 1577–1582, Nov. 2007.

H33. Y.Q. Shi and H. Sun, "Image and video compression for multimedia engineering," Boca Raton: CRC Press, II Edition (Chapter on H.264), 2008.

H34. B. Furht and S.A. Ahson, "Handbook of mobile broadcasting, DVB-H, DMB, ISDB-T and MEDIAFLO," Boca Raton, FL: CRC Press, 2008. (H.264 related chapters)

H35. M. Jafari and S. Kasaei, "Fast intra- and inter-prediction mode decision in H.264 advanced video coding," International Journal of Computer Science and Network Security, vol. 8, pp. 130–140, May 2008.

H36. A.M. Tourapis (January 2009), "H.264/14496-10 AVC reference software manual" [Online.]. Available: http://iphome.hhi.de/suehring/tml/JM%20Reference%20Software%20Manual%20%28JVT-AE010%29.pdf

H37. P. Carrillo, H. Kalva, and T. Pin, "Low complexity H.264 video encoding," SPIE, vol. 7443, paper # 74430A, Aug. 2009.

H38. B.M.K. Aswathappa and K.R. Rao, "Rate-distortion optimization using structural information in H.264 strictly intra-frame encoder," IEEE Southeastern Symp. on System Theory (SSST), Tyler, TX, March 2010.

H39. D. Han et al., "Low complexity H.264 encoder using machine learning," IEEE SPA 2010, pp. 40–43, Poznan, Poland, Sept. 2010.

H40. T. Sathe, "Complexity reduction of H.264/AVC motion estimation using OpenMP," M.S. Thesis, EE Dept., University of Texas at Arlington, Arlington, Texas, May 2011. http://www.uta.edu/faculty/krrao/dip click on courses and then click on EE5359 Scroll down and go to Thesis/Project Title and click on T. Sathe.

H41. D. Han, A. Kulkarni and K.R. Rao, "Fast inter-prediction mode decision algorithm for H.264 video encoder," IEEE ECTICON 2012, Cha Am, Thailand, May 2012.

H42. H.264/MPEG-4 AVC: http://en.wikipedia.org/wiki/H.264

H43. S. Subbarayappa, "Implementation and analysis of directional discrete cosine transform in H.264 for baseline profile," M.S. Thesis, EE Dept., UTA, May 2012.

H44. I.E. Richardson, "H.264/MPEG-4 Part 10 White Paper," www.vcodex.com

H45. I.E. Richardson, "Overview: What is H.264?" 2011. www.vcodex.com

H46. I.E. Richardson, "A Technical Introduction to H.264/AVC," 2011. [Online]. Available: http://www.vcodex.com/files/H.264_technical_introduction.pdf

H47. I.E. Richardson, "White Paper: H.264/AVC intra prediction," 2011. www.vcodex.com

H48. I.E. Richardson, "White Paper: H.264/AVC inter prediction," 2011. www.vcodex.com

H49. I.E. Richardson, "White Paper: 4 × 4 transform and quantization in H.264/AVC," Nov. 2010. www.vcodex.com

H50. I.E. Richardson, "White Paper: H.264/AVC loop filter," 2011. www.vcodex.com

H51. Joint Video Team (JVT), ITU-T website, http://www.itu.int/ITU-T/studygroups/com16/jvt/

H52. "RGB Spectrum's H.264 codecs hit the mark in the latest U.S. missile test program," May 2012. www.rgb.com. See [B18].

H53. ISO website: http://www.iso.org/iso/home.htm

H54. IEC website: http://www.iec.ch/

H55. ITU-T website: http://www.itu.int/ITU-T/index.html

H56. C.-L. Su, T.-M. Chen and C.-Y. Huang," Cluster-Based Motion Estimation Algorithm with Low Memory and Bandwidth Requirements for H.264/AVC Scalable Extension," IEEE Trans. CSVT, vol. 24, pp. 1016–1024, June 2014.

H.264/14496-10 Advanced video coding reference software manual 28 June–3 July, 2009
JM 16.0
http://iphome.hhi.de/suehring/tml/JM%20Reference%20Software%20Manual %20(JVT-AE010).pdf
H.264/mpeg-4 AVC reference software joint Model 18.4 available at http://iphome.hhi.de/suehring/tml/download/jml8.4.zip

186 *Appendix*

H57. FFmpeg Developers (2013): Libavcodec H.264 Decoder [Online]. Available: http://ffmpeg.org, accessed Aug 1, 2014.

H58. X264: A free H.264 Encoder [Online] Available: http://www.videolan.org/developers/x.264.html, accessed Aug. 1, 2014.

H59. D. S. Eagle, "Beyond the Full HD (UHD & 8K)," Sunday 12 April 2015, 3:30 PM to 4:45 PM, NAB Show, Las Vegas, NV, April 2015.

Sony's XAVC format employs MPEG-4 AVC/H.264 level 5.2, the highest picture resolution and frame rate video compression codec based on industry standards. XAVC enables a very wide range of operational possibilities for content production, notably: from Proxy to 4K pixel resolutions, intra frame and long GOP schemes, and 1080 50P/60P infrastructure capability.

Built with the principles of workflow efficiency, evolution and optimized image quality at its heart, Sony's XAVC can support the following content formats:

- 4K (4096 × 2160 and 3840 × 2160), HD and proxy resolution
- MPEG-4 AVC/H.264 video compression
- 12, 10 and 8 bit colour depth
- Up to 60 fps
- MXF wrapping format can be used
- 4:4:4, 4:2:2, 4:2:0 colour sampling

XAVC has been developed as an open format, providing a license program for other manufacturers in the broadcast and production industry to develop their own high quality and high frame rate products.
Website : http://www.sony.co.uk/pro/article/broadcast-products-sonys-new-xavc-recording-format-accelerates

H60. I. E. G. Richardson, "The H.264 Advanced Video Compression Standard," II Edition, Wiley, 2010.

H61. T. Wiegand et al., "Introduction to the Special Issue on Scalable Video Coding—Standardization and Beyond," IEEE Trans. CSVT, vol. 17, pp. 1099–1269, Sept. 2007.

H62. K. Muller et al., "Multi–view video coding based on H.264/AVC using hierarchical B-frames," in Proc. PCS 2006, Picture Coding Symp., Beijing, China, April 2006.

H63. T. Wiegand et al., "Joint draft 5: scalable video coding," Joint Video Team of ISO/IEC MPEG and ITU-T VCEG, Doc. JVT-R201, Bangkok, Jan. 2006.

H64. G. J. Sullivan, "The H.264/MPEG4-AVC video coding standard and its deployment status," in proc. SPIE conf. visual communications and image processing (VCIP), Beijing, China, July 2005.

H65. T. Wiegand et al., "Rate-constrained coder control and comparison of video coding standards," IEEE Trans. CSVT, vol. 13, pp. 688–703, July 2003.

H66. C. Fenimore et al., "Subjective testing methodology in MPEG video verification," in Proc. SPIE Applications of Digital Image Processing, vol. 5558, pp. 503–511, Denver, CO, Aug. 2004.

H67. MPEG industry forum resources [online] Available: http://www.mpegif.org/reswources.php

H68. H. Schwarz, D. Marpe and T. Wiegand, "Overview of the scalable H.264/AVC extension," in Proc. IEEE ICIP, Atlanta, GA, pp. 161–164, Oct. 2006.

H69. C.-H. Kuo, L.-L. Shih and S.-C. Shih, "Rate control via adjustment of Lagrange multiplier for video coding," IEEE Trans. CSVT, (EARLY ACCESS)

H70. A.K. Au-Yeung, S. Zhu and B. Zeng, 'Partial video encryption using multiple 8×8 transforms in H.264 and MPEG-4,' Proceedings of ICASSP 2011, pp. 2436–2439, Prague, Czech Republic., May 2011.

H71. A.K. Au-Yeung, S. Zhu and B. Zeng, 'Design of new unitary transforms for perceptual video encryption,' IEEE Trans. CSVT, vol. 21, pp. 1341–1345, Sept. 2011.

H72. B. Zeng et al., 'Perceptual encryption of H.264 videos: embedding sign-flips into the integer-based transforms,' IEEE Trans. Information Forensics and Security, vol. 9, no.2, pp. 309–320, Feb. 2014.

H73. S.J. Li et al., "On the design of perceptual MPEG-video encryption algorithms," IEEE Trans. CSVT, vol. 17, pp. 214–223, Feb. 2007.

H74. B.Y. Lei, K.T. Lo, and J. Feng, 'Encryption Techniques for H.264 Video," The Handbook of MPEG Applications: Standards in Practice, pp. 151–174, Edited by M.C. Angelides and H. Agius, Wiley, Feb. 2011.

H75. Overview of H.264/AVC: http://www.csee.wvu.edu/~xinl/courses/ee569/H264_tutorial.pdf

Access to HM Software Manual: http://iphome.hhi.de/marpe/download /Performance_HEVC_VP9_X264_PCS_2013_preprint.pdf

H76. M.P. Krishnan, "Implementation and performance analysis of 2D order 16 integer transform in H.264 and AVS - China for HD video coding" M.S. Thesis, EE Dept., UT-Arlington, TX, Dec. 2010.

http://www.uta.edu/faculty/krrao/dip click on courses and then click on EE5359 Scroll down and go to Thesis/Project Title and click on M. P. Krishnan.

H77. B. Dey and M.K. Kundu, "Enhanced microblock features for dynamic background modelling in H.264/AVC video encoded at low bitrate," IEEE Trans. CSVT, vol. 28, pp. 616–625, March 2018.

H78. K. Minemura et al., "A Novel Sketch Attack for H.264/AVC Format-Compliant Encrypted Video," *IEEE Trans. CSVT,* vol. 27, pp. 2309–2321, Nov. 2017.

Conclusions

In this work, we proposed a novel sketch attack for H.264/AVC format-compliant encrypted video. Specifically, the macroblock bit stream size was exploited to sketch the outline of the original video frame directly from the encrypted video. In addition, the Canny edge map was considered as the ideal outline image and an edge similar score was modified for performance evaluation purposes. Experimental results suggest that the proposed sketch attack can extract visual information directly from the format-compliant encrypted video. Although the proposed and conventional methods can sketch the outline from the encrypted INTRA-frame, only the proposed MBS sketch attack method can sketch the outline from the encrypted INTER-frame, which outnumbers INTRA-frame, by far, in compressed video. Moreover, the proposed MBS sketch attack is verified to be more robust against compression when compared to the conventional sketch attacks. In view of this proposed sketch attack framework, we suggest that this framework should be considered for format compliant video encryption security analysis. In addition to determining the encryption modules in use by analysing the encrypted video, we would like to extend this sketch attack framework to handle different video standards such as High Efficiency Video Coding (HEVC), Audio Video Standard (AVS) and Google VP9 as our future work.

H79. R. Atta and M. Ghanbari, "Low-complexity joint temporal-quality scalability rate control for H.264/SVC," IEEE Trans. CSVT, vol. 28, pp. 2331–2344, Sept. 2018.

Abstract is repeated here.

Abstract

Rate control in scalable video coding (SVC) is a very challenging problem because of the inter-layer prediction structure which makes developing an efficient rate-control algorithm complex and difficult. Little prior work is available for joint temporal-quality (T-Q) scalability considering the rate-distortion (R-D) dependency among the temporal and quality layers. However, most of the rate control algorithms in SVC suffer from high computational complexity, growing significantly with the number of layers. In this paper, a single-pass joint temporal-quality rate-control algorithm is presented for H.264/SVC. In this algorithm, by analysing the R-D dependency of joint T-Q scalability, Cauchy distribution-based rate-quantization (R–Q), and distortion-quantization (D–Q) models, a set of empirical values are first derived to estimate the initial values of the R-D model parameters for the joint temporal and quality layers. Then, a novel prediction mechanism to update these model parameters is proposed to allocate the bit budgets efficiently among the temporal and quality layers and hence to improve the performance of the proposed algorithm. Experimental results show that the proposed algorithm achieves better coding efficiency with low computational complexity compared to two other benchmark rate-control algorithms.

    H81. T. Stutz and A. Uhl, "A Survey of H.264 AVC/SVC encryption," IEEE Trans. CSVT. vol. 22, pp. 325–339, March 2012.

    H82. L. Merritt, "x264: A high performance H.264/AVC encoder," Feb. 2016. Available: http://akuvian.org/src/x264/overview_x264_v8_5.pdf

H.264-P1. See [H80]. Extend this sketch attack framework to HEVC. (See references on HEVC)

H.264-P2. See [H80]. Extend this sketch attack framework to AVS. (See references on AVS)

H.264-P3. See [H80]. Extend this sketch attack framework to VP9. (See references on VP9)

H.264-P4. See [JPEG29] and the conclusion section. A joint image compression and encryption scheme is proposed and applied to JPEG. The authors suggest that their scheme can be extended to other image/video compression standards. Extend this to H.264.

Books on H.264:

1. I. E. G. Richardson, "H.264 and MPEG-4 video compression: Video coding for Next Generation Multimedia," New York, Wiley, 2003.

2. S. Okubo et al., "H.264/AVC Textbook (in Japanese, title translated)," Tokyo, Impress Publisher, 2004.
3. S. Ono, T. Murakami, and K. Asai, Ubiquitous technology, "Hybrid Video Coding – MPEG-4 and H.264 (In Japanese, title translated)," Tokyo, Ohmsha Press, 2005.
4. L. Chiariglione (Editor), "The MPEG Representation of Digital Media," Springer, 2012.
5. K. R. Rao, D. N. Kim and J .J. Hwang, "Video coding standards: AVS China, H.264/MPEG-4 Part10, HEVC, VP6, DIRAC and VC-1," Springer, 2014. Also e-book. Translated into Spanish. Translated into Chinese by China Machine Press, 2017.

H.264 Standard, JM SOFTWARE

1. ITU–T and ISO/IEC JTC 1, "Advanced video coding for generic audiovisual services," ITU-T Rec. H.264 & ISO/IEC 14496-10, Version 1, May 2003; version 2, Jan 2004; Version 3 (with High family of profiles), sept 2004; version 4, July 2005 [online] Available: http://www.itu.int/rec/T-REC-H.264.

JM 19.0 Reference software (6/19/2015) also updates on KTA.

DCT References

DCT1. N. Ahmed, M. Natarajan and K.R. Rao, "Discrete Cosine Transform," IEEE Trans. Computers, vol. C-23, pp. 90–93, Jan. 1974.

DCT2. W. Chen, C. H. Smith, and S. C. Fralick, "A fast computational algorithm for the discrete cosine transform," IEEE Trans. Commun., vol. COM-25, pp. 1004–1009, Sept. 1977.

DCT3. M. Vetterli and A. Ligtenberg "A discrete Fourier-cosine transform chip," IEEE Journal on Selected Areas of Communications, vol. 4, pp. 49–61, Jan. 1986.

DCT4. W. K. Cham and R. J. Clarke, "Application of the principle of dyadic symmetry to the generation of orthogonal transform," IEE Proc. F: Communications, Radar and Signal Processing, vol. 133, no. 3, pp. 264–270, June 1986.

DCT5. C. Loeffler, A. Ligtenberg, and G. Moschytz, "Practical fast 1-D DCT algorithms with 11 multiplications," Proc. IEEE ICASSP, vol. 2, pp. 988–991, Feb. 1989.

DCT6. W. K. Cham, "Development of integer cosine transforms by the principle of dyadic symmetry," IEE Proc. Communications, Speech and Vision, vol. 136, issue 4, pp. 276–282, 1989.

DCT7. M. Masera, M. Martina, and G. Masera, "Adaptive approximated DCT architectures for HEVC," IEEE Trans. CSVT, vol. 27, pp. 2714–2725, Dec. 2017. This has several references related to integer DCT architectures. Same as [E381].

DCT8. S.-H. Bae, J. Kim and M. Kim, "HEVC-Based perceptually adaptive video coding using a DCT-Based local distortion detection probability," IEEE Trans. Image Processing, vol. 25, pp. 3343–3357, July 2016. Same as [E377].

DCT9. Z. Zhang et al., "Focus and blurriness measure using reorganized DCT coefficients for an autofocus application," IEEE Trans. CSVT, vol. 28, pp. 15–30, Jan. 2018.
Conclusion section is repeated here.

Conclusion

In this paper, metrics for measuring image sharpness/blurriness and an AF approach based on the proposed metrics are presented. Focus measure is first presented by utilizing optimal high-and middle-frequency coefficients extracted from a reorganized DCT representation. To quantify the focus measure, we design and use a new psychometric function-based metric. Experiments on a large number of databases show that the focus measure is highly sensitive to the best focus position and could predict stable and accurate focus values even under noisy conditions. An NR blurriness metric is also presented for measuring absolute blurriness. This metric determines image blurriness by exploiting the changes in the information in edge structures and image contrast. The relevant parameters are controlled adaptively according to the estimated noise level. An efficient MAD-based noise estimation algorithm is introduced. Comprehensive experiments were performed to evaluate the performance of this metric, and the results demonstrated that it is closely correlated well with subjective perceived scores and that it outperforms or at least is comparable with state-of-the-art methods, especially for noisy images. The focus and blurriness measures were used to develop a swift and reliable AF method in which the two metrics play a critical role. For future work, we may use the proposed sharpness/blurriness metrics for other imaging applications, such as image edge enhancement.

DCT10. S.-H. Bae and M. Kim, "DCT-QM; a dct-based quality degradation metric for image quality optimization problem," IEEE Trans. IP, vol. 25, pp. 4916–4930, Oct. 2016.

DCT11. M. Jridi and P.K. Meher, "A scalable approximate DCT architectures for efficient HEVC compliant video coding" IEEE Trans. CSVT, (early access) See also P.K. Meher et al., "Generalized architecture for integer DCT of lengths N = 8, 16 and 32," IEEE Trans. CSVT, vol. 24, pp. 168–178, Jan. 2014. See also [DCT39.]

DCT12. V. Dhandapani and S. Ramachandran, "Area and power efficient DCT architecture for image compression," EURASIP Journal on Advances in Signal Processing, vol. 2014, 2014: 23.

DCT13. M. Narroschke, "Coding efficiency of the DCT and DST in hybrid video coding," IEEE J. on selected topics in signal processing, vol. 7, pp. 1062–1071, Dec. 2013.

DCT14. G. Fracastoro, S.M. Fosson, and E. Magli, "Steerable discrete cosine transform," IEEE Trans. IP, vol. 26, pp. 303–314, Jan. 2017. Same as [E392]. G. Fracastoro and E. Magli, "Steerable discrete cosine transform," IEEE Intrnl. workshop on multimedia signal processing, pp. 1–4, Xiamen, China, Oct. 2015.

DCT15. B. Zeng and J. Fu, "Directional discrete cosine transforms-a new framework for image coding," *IEEE Trans. CSVT,* vol. 18, pp. 305–313, March 2008. Also B. Zeng and J. Fu, "Directional discrete cosine transforms for image Coding," IEEE ICME 2006.

DCT16. J.-S. Park et al., "2-D large inverse transform (16 × 16, 32 × 32) for HEVC (high efficiency video coding)," J. of Semiconductor Technology and Science, vol. 12, pp. 203–211, June 2012. Same as [E364].

DCT17. J. Dong, et al., "2-D order-16 integer transforms for HD video coding," IEEE Trans. CSVT, vol. 19, pp. 1462–1474, Oct. 2009.

DCT18. C.-K. Fong, Q. Han, and W.-K. Cham, "Recursive Integer Cosine Transform for HEVC and Future Video Coding Standards'" IEEE Trans. CSVT, vol. 27, pp. 326–336, Feb. 2017.

DCT19. C. Yeo et al., "Mode dependent transforms for coding directional intra prediction residuals," IEEE Trans. CSVT, vol. 22, pp. 545–554, April 2012.

DCT20. S.-H. Bae and M. Kim, "A DCT-based total JND profile for spatio-temporal and foveated masking effects," IEEE Trans. CSVT, vol. 20, pp. 1196–1207, June 2017.

DCT21. Z. Chen, Q. Han, and W.-K. Cham, "Low-complexity order-64 integer cosine transform design and its application in HEVC," IEEE Trans. CSVT, vol. 28, pp. 2407–2412, Sept. 2018.

DCT22. J. Lee et al., "A compressed-domain corner detection method for a DCT-based compressed image," IEEE ICCE, Las Vegas, Jan. 2017.

DCT23. Z. Qian, W. Wang, and T. Qiao, "An edge detection method in DCT domain," IEEE ICCE, Las Vegas, Jan. 2017.

DCT24. M. Wien, "High Efficiency Video Coding: Coding Tools and Specification," Springer, 2014. See Section 8.1.1 pages 206-210. This section explains how the integer DCTs adopted in HEVC are derived.

DCT25. P. Sjovall et al., "High-level synthesis implementation of HEVC 2-D DCT/DST on FPGA," IEEE ISCAS2017, pp. 1547–1551, New Orleans, Louisiana, March 2017.

DCT26. G. Pastuszak, "Hardware architectures for the H.265/HEVC discrete cosine transform," *IET Image Process.*, vol. 9, no. 6, pp. 468–477, 2015.

DCT27. R. Jeske, et al., "Low cost and high throughput multiplierless design of a 16 point 1-D DCT of the new HEVC video coding standard," *in Proc. Southern Conf.Programmable Logic*, Bento Goncalves, Spain, Mar. 2012.

DCT28. A. D. Darji and R. P. Makwana, "High-performance multiplierless DCT architecture for HEVC," *in Proc. Int. Symp.VLSI Design and Test*, Ahmedabad, India, June 2015.

DCT29. W. Zhao, T. Onoye, and T. Song, "High-performance multiplierless transform architecture for HEVC," *in Proc. IEEE ISCAS*, pp. 1668–1671, Beijing, China, May 2013,

DCT30. P. Arayacheeppreecha, S. Pumrin, and B. Supmonchai, "Flexible input transform architecture for HEVC encoder on FPGA," *in Proc. Int. Conf. Electrical Engineering/Electronics, Computer, Telecommunications and Information Tech.*, Hua Hin, Thailand, June 2015.

DCT31. G. Pastuszak and A. Abramowski, "Algorithm and architecture design of the H.265/HEVC intra encoder," *IEEE Trans. CSVT*, vol. 26, pp. 210–222, Jan. 2016.

DCT32. T. Ichita et al., "Directional discrete cosine transforms arising from discrete cosine and sine transforms for directional block-wise image representation," IEEE ICASSP2017, pp. 4536–4540, New Orleans, March 2017.

DCT33. G.C. Langelaar and R.L. Lagendik, "Optimal differential energy watermarking of DCT encoded images and video," IEEE Trans. Image Processing, vol. 10, pp. 148–158, Jan. 2001.

DCT34. P.A.M. Oliveira et al., "Low complexity image and video coding based on an approximate discrete Tchebichef transform," IEEE Trans. CSVT, vol. 27, pp. 1066–1076, May 2017.

DCT35. Y. Wang et al., "DCT regularized extreme visual recovery," IEEE Trans. IP, Early access.

DCT36. Z.M. Hafed and M.D. Levine, "Face recognition using the discrete cosine transform," International Journal of Computer Vision, vol. 48(3), pp. 167–188, 2001.

DCT37. W. Chen, M.J. Er, and S. Wu, "Illumination compensation and normalization for robust face recognition using discrete cosine transform in logarithm domain," TSMC Part B: Cybernetics, 36(2): 458–466, 2006.

DCT38. Q. Shang et al., "Single-Port SRAM-Based Transpose Memory With Diagonal Data Mapping for Large Size 2-D DCT/IDCT," IEEE Trans. Very Large Scale Integration (VLSI) Systems,, vol. 22, no.11, pp. 2422–2426, Nov. 2014.

DCT39. M. Jridi, A. Alfalou, and P.K. Meher, "A Generalized Algorithm and Reconfigurable Architecture for Efficient and Scalable Orthogonal Approximation of DCT," IEEE Trans. Circuits and Systems-1, pp. 449–457, Feb. 2015. See also [DCT11].

DCT39. T. Sugimoto, Y. Nakayama, and T. Komuru, "22.2 channel audio encoding/decoding hardware system based on MPEG-4 AAC", IEEE Trans. Broadcasting, vol. 63, pp. 426–432, June 2017. (MDCT application in MPEG-4 AAC).

DCT40. P.K. Meher et al., "Efficient integer DCT architectures for HEVC," IEEE Trans. CSVT, vol. 24, pp. 168–178, Jan. 2014.

DCT41. A. Ahmed, M. U. Shahid, and A. Rehman, "N Point DCT VLSI architecture for Emerging HEVC Standard," in *Proc. VLSI Design*, vol. 012, Article 752024, pp. 1–13, 2012.

DCT42. S. Shen et al., "A unified 4/8/16/32-point integer IDCT architecture for multiple video coding standards," in *Proc. IEEE Int. Conf. Multimedia Expo*, (ICME), pp. 788–793. July 2012.

DCT43. J.-S.-S. Park et al., "2-D large inverse transform ($16 \times 16, 32 \times 32$) for HEVC (high efficiency video coding)," *Journey of Semiconductor. Technology and Science*, vol. 12, no. 2, pp. 203–211, June. 2012.

DCT44. E.M.-Enriquez, "Directional transforms for video coding based on lifting on graphs," IEEE Trans. CSVT, vol. 28, pp. 933–946, April 2018.

DCT45. Q. Shang et al., "Single-port SRAM-based transpose memory with diagonal data mapping for large size 2-D DCT/IDCT," "IEEE Trans. on very large scale integration (VLSI) systems," vol. 22, pp. 2423–2427, Nov. 2014.
DCT46. J. R. Hernandez, M. Amado, and F. Perez-Gonzalez, "DCT-domain watermarking techniques for still images: Detector performance analysis and a new structure," IEEE Trans. Image Process., vol. 9, pp. 55–68, Jan. 2000.
DCT47. A. Briassouli and M. G. Strintzis, "Locally optimum nonlinearities for DCT watermark detection," IEEE Trans. Image Process., vol. 13, pp. 1604–1617, Dec. 2004.
DCT48. I. J. Cox, J. Kilian, F. T. Leighton, and T. Shamoon, "Secure spread spectrum watermarking for multimedia," IEEE Trans. Image Process., vol. 6, pp. 1673–1687, Dec. 1997.
DCT49. T. Biatek, V. Lorcy, and P. Philippe, "Transform competition for temporal prediction in video coding," IEEE Trans. CSVT early access. The section Conclusion is repeated here:

Conclusion

In this paper, transform competition for inter-prediction mode is thoroughly studied and evaluated. Based on a transform design that incorporates distortion and sparsity criteria, experiments have revealed the appropriateness of using separable transforms rather than non-separable transforms for inter video coding. Also, it appears that enforcing the DCT2 in a multiple transform system does not reduce its performance. Separable systems using DTT are also designed and it is revealed that this approach saturates when a certain number of transforms is reached ($N > 5$). In coding environment, an extensive evaluation of the proposed systems against HEVC is provided. It appears from experimental results that systems using one or multiple transform sets enable significant bitrate savings with both separable transforms and DTT. Considering bitrate savings, storage requirement and encoding runtime increase, systems using 9 transforms seem to be the best options for separable and DTT transforms. The recommended STS and MTS designs enable to reach bitrate savings between -1% and -1.4% depending on the configuration. As the gains observed here are complimentary to those obtained for intra-coding environments, the conclusion of this work confirms the interest of adapted transform competition for the upcoming video coding standards.

DCT50. S. Jaballah, C. Larabi and J.B. Tahar, "Asymmetric DCT-JND for luminance adaptation effects: An application to perceptual video coding in MV-HEVC," IEEE ICASSP 2018, April 2018, Calgary, Canada.

DCT51. A.C. Mert, E. Kalali and I. Hamzaoglu, "An FPGA implementation of future video coding 2D transform," IEEE ICCE-Berlin, pp. 31–36, Berlin, Germany, Sept. 2017.

Abstract is repeated here:

Abstract

Future Video Coding (FVC) is a new international video compression standard offering much better compression efficiency than previous video compression standards at the expense of much higher computational complexity. In this paper, an FPGA implementation of FVC 2D transform is proposed. The proposed FVC 2D transform hardware can perform 2D DCT-II, DCT-V, DCT-VIII, DST-I, DST-VII operations for 4×4 and 8×8 transform units. It uses two reconfigurable data paths for all 1D transforms. It implements multiplications with constants using DSP blocks in FPGA. The proposed FPGA implementation, in the worst case, can process 54 8K Ultra HD (7680 $\times$ 4320) video frames per second. The proposed FPGA implementation has up to 29% less energy consumption than the FPGA implementation of FVC 2D transform hardware in the literature.

DCT52. A.C. Mert, E. Kalali and I. Hamzaoglu, "High performance 2D transform hardware for future video coding," IEEE Trans. CE, vol. 63, pp. 117–125, May 2017. The section CONCLUSION is repeated here.

In this paper, two different high performance FVC 2D – aomedia Git repositories – Git at Google aomedia.googlesource.com

Up to 16 transform combinations with independent horizontal & vertical selection, Reduce available combinations based on block size (for efficiency)

Identity transform useful for screen content coding.

DCT54. T. Richter et al., "High-speed low-complexity video coding with EDiCTus: a DCT coding proposal for JPEG XS," SPIE Applications of the digital image processing LX, vol. 10396, San Diego, CA, Oct. 2017. Listed also as [JXS6].

DCT55. X. Zhang, F. Peng and M. Long, "Robust coverless image steganography based on DCT and LDA classification," IEEE Trans. Multimedia, EA.

Conclusion section is repeated here:

"A coverless image steganography scheme based on DCT transform is proposed in this paper. The secret information is hidden through the mapping rules, and the cover images are not modified in the steganography process. It can effectively resist the detection of the existing steganalysis algorithms. Meanwhile, LDA topic model is exploited to enhance the capability of resisting detection from subjective vision. Furthermore, the feature sequence is generated based on DC coefficients, which is helpful to improve the robustness, and the search efficiency is optimized by constructing an inverted index. Experimental results and analysis show that the proposed algorithm can achieve good performance in subjective and objective detection resistance on testing database, and it is robust against most image processing attacks, and geometric attacks. The future work will be focused on the improvement of capacity with less side information, along with robustness against rotation and the content loss attacks."

In this paper, two different high performance FVC 2D transform hardware are designed and implemented using Verilog HDL. They are the first FVC 2D transform hardware in the literature. The proposed baseline and reconfigurable hardware can process 84 and 60 8K Ultra HD (7680×4320) video frames per second, respectively. The baseline hardware with data gating technique has less energy consumption than the reconfigurable hardware. The reconfigurable hardware has smaller area than the baseline hardware. Therefore, the proposed FVC baseline 2D transform hardware can be used in high performance and low energy FVC encoders. The proposed FVC reconfigurable 2D transform hardware can be used in high performance and low cost FVC encoders.

DCT53 In AV1 codec DCT, ADCT, Flipped (reverse) ADST transforms are used. Besides square sizes, 4×8, 8×4, 16×8, 8×16, 16×32 and 32×16 are also used. Access these transforms and develop fast algorithms based on SMF.

For details on AOM, please access http://aomedia.org The code base for AOM is here https://aomedia.googlesource.com/aomedia Git repositories - Git at Google aomedia.googlesource.com

Up to 16 transform combinations with independent horizontal & vertical selection, Reduce available combinations based on block size (for efficiency). Identity transform useful for screen content coding.

DCT54. T. Richter et al., "High-speed low-complexity video coding with EDiCTus: a DCT coding proposal for JPEG XS," SPIE Applications of the digital image processing LX, vol. 10396, San Diego, CA, Oct. 2017. Listed also as [JXS6].

DCT55. X. Zhang, F. Peng and M. Long, "Robust coverless image stegonagraphy based on DCT and LDA classification," IEEE Trans. Multimedia, EA.

Conclusion section is repeated here:

"A coverless image steganography scheme based on DCT transform is proposed in this paper. The secret information is hidden through the mapping rules, and the cover images are not modified in the steganography process. It can effectively resist the detection of the existing steganalysis algorithms. Meanwhile, LDA topic model is exploited to enhance the capability of resisting detection from subjective vision. Furthermore, the feature sequence is generated based on DC coefficients, which is helpful to improve the robustness, and the search efficiency is optimized by constructing an inverted index. Experimental

DCT56. M.J. Garrido et al., "A high performance FPGA-based architecture or the future video coding adaptive multiple core transform," IEEE Trans. CE, vol. 64, no.1, pp. 53–60, Feb. 2018.

DCT57. E. Kalali, A.C. Mert and I. Hamzaoglu, "A computation and energy reduction technique for HEVC discrete cosine transform," IEEE Trans. CE, vol. 62, no.2, pp. 166–174, May 2016.

DCT58. A. Saad, A.C. Bovik, and C. Charrier, "Blind image quality assessment: A natural scene statistics approach in the DCT domain," IEEE Trans. IP, vol. 21, pp. 3339–3352, Aug. 2012.

DCT59. A. Saad, A.C. Bovik, and C. Charrier, "A DCT statistics-based blind image quality index," IEEE SP Letters, vol. 17, no.6, pp. 583–586, June 2010.

DCT60. J. Agbinya, "Interpolation using the discrete cosine transform," IEE Electronics Letters, vol. 28, no. 28, pp. 1927–1928, 1992.

DCT61. W. Dai and T.D. Tran, "Regularity-constrained pre- and post-filtering for block DCT-based systems," IEEE Trans. SP, vol. 51, pp. 2568–2581, Oct. 2003.

DCT62. S.Y. Zhu, S.-K.A. Yeung and B. Zeng, "In search of better-than-DCT unitary transforms for encoding of residual images," IEEE Trans. SP Letters, vol. 17, pp. 961–964, Nov. 2010.

DCT63. H.M. Chen and B. Zeng, "New transforms tightly bounded by DCT and KLT," IEEE Trans. SP Letters, vol. 19, pp. 344–347, June 2012.

DCT64. C. Sun and E.-H. Yang, "An efficient DCT-based image compression system based on Laplacian transparent composite model," IEEE Trans. IP, vol. 24, pp. 862–900, March 2015.

DCT65. C. Lan et al., "Variable block sized signal dependent transform for video coding," IEEE Trans. CSVT, vol. 28, pp. 1920–1933, Aug. 2018. Same as [E453]. Conclusion section is repeated here.

Conclusion

"HEVC has achieved high coding performance in comparison with former coding standards. It is an open problem on how to design more efficient transforms to improve coding efficiency. In this paper, we propose the variable block-sized Signal Dependent Transforms (SDTs) and implement them in HEVC and JEM to explore a possible way to enhance performance. Being capable of exploring non-local correlations and adaptive to image/video contents, the proposed SDTs have high coding efficiency. Experimental results demonstrate that the proposed transform brings the BD-rate savings of 2.2%, 2.4%, 3.3%, and 7.1% on average for AI-Main10, RA-Main10, RA-Main10, and LP-Main10 configurations, compared to the test model HM-12.0 of HEVC. Thanks to the introduction of the large block size SDTs, the BD-rate savings is up to 23.3%." Note that VBS SDT for video coding is adopted for evaluation of potential future video coding standard.

DCT66. M. Budagavi et al., "Core Transform Design in the High Efficiency Video Coding (HEVC) Standard," IEEE J. Sel. Topics Signal Process. vol. 7, no. 6, pp. 1029–1041, Dec. 2013. This paper explains clearly, among other things, how INTDCTs were selected in HEVC.

DCT67. B. Lee and M. Kim, "A CU-Level Rate and Distortion Estimation Scheme for RDO of Hardware-Friendly HEVC Encoders Using Low-Complexity Integer DCTs," IEEE Trans. IP, vol. 25, pp. 3787–3800, Aug. 2016. A clear (even though briefly) explanation of the HEVC flexible coding structure with hierarchical quadtree blocks is given in the introduction.

DCT68. A. Foi, V. Katkovnik, and K. Egiazarian, "Pointwise shape-adaptive DCT for high-quality denoising and deblocking of grayscale and color images," IEEE Trans. IP, vol. 16, pp. 1395–1411, May 2007.

DCT69. H.O. Dominguez and K.R. Rao, "Discrete cosine transform," II Edition, Taylor and Francis, 2019.

DCT-P1. Fong, Han and Cham [DCT18] have developed recursive integer cosine transform (RICT). This has orthogonal basis vectors and also has recursive property. Based on this, flow graph for order 32 RICT with flow graphs for orders 4, 8 and 16 embedded is shown in Figure 1. They also have implemented the RICT in HEVC HM13.0 and have demonstrated that the RICT has similar coding efficiency as the core transform in HEVC (see Tables VI through XI). Using Figure 1, write down the sparse matrix factors (SMFs) for orders 4, 8, 16 and 32 RICTs.

DCT-P2. See DCT-P1. Using Table I and the SMFs for orders 4, 8, 16 and 32 RICTs write down the corresponding transform matrices. Verify the transform matrix for order 32 with that shown in Equation 16 of [DCT18].

DCT-P3. See DCT-P1. Forward and inverse architecture of order 4 RICT is shown in Figure 4. This structure is symmetric (Hardware of order 4 inverse RICT can be shared by the order 4 forward RICT.) Develop similar symmetric hardware structures for orders 8, 16 and 32.

DCT-P4. The authors in [DCT18] state that higher order RICTs such as orders 64 and 128 can be derived using their fast algorithms. Derive these.

DCT-P5. See DCT-P4. Draw flow graph for order 64 RICT with the flow graphs for orders 4, 8, 16 and 32 embedded similar to that shown in Figure 1.

DCT-P6. Repeat DCT-P5 for order 128 RICT.

DCT-P7. See DCT-P6. Extend the suggested values for $b_{N,i}$ for orders 64 and 128 RICTs (See Table I in [DCT18]).

DCT-P8. See DCT-P3. Develop similar symmetric hardware structures for orders 64 and 128 RICT.

DCT-P9. See Table V in [DCT18]. Extend the comparison to orders 64 and 128. This requires extending Loeffler's method to orders 64 and 128. B. Lee and M. Kim, "A CU-level rate and distortion estimation scheme for RDO of hardware-friendly HEVC encoders using low-complexity integer DCTs," IEEE Trans. IP, vol. 25, pp. 3787–3800, Aug. 2016.

Books on H.264   201

DCT-P10. See [DCT11]. Jridi and Meher have developed approximate DCT architectures for efficient HEVC compliant video coding. Full-parallel and area constrained architectures for the proposed approximate DCT are described. Show that the equations given by (6a) and (6b) yield the 4-point DCT kernel given by (4). Write down the SMFs based on Figure 1 for the 4-point integer 1-D DCT.

DCT-P11. See DCT-P10. Show that the equations given by (7)–(9) yield Equation(5) for the 8-point DCT kernel.

DCT-P12. See DCT-P11. Write down the SMFs for the 8-point integer 1-D DCT based on Figure 2. Show that these SMFs yield equation (5) for the 8-point DCT kernel.

DCT-P13. See [DCT11]. Parallel architecture for approximate 8-point 1-D DCT is shown in Figure 2. Develop similar architecture for approximate 16-point 1-D DCT.

DCT-P14. Based on DCT-P13 write down the corresponding SMFs and the resulting 16×16 1-D DCT matrix.

DCT-P15. See DCT-P13. Extend this to approximate 32-point 1-D DCT.

DCT-P16. See DCT-P14 Extend this to 32-point 1-D DCT.

DCT-P17. Chen, Han and Cham [DCT21] have developed a set of low complexity integer cosine transforms (LCICTs) and used them in HEVC HM13. These have fully factorizable structures and high circuit reusability. They conclude that order-64 ICTs can significantly improve the coding performance under high QP (low bitrate) which makes them beneficial in low bitrate applications such as video conferencing and video surveillance. Using Equations 5–10, write down the LCICT matrices for order 8, 16, 32 and 64.

DCT-P18. See DCT-P17 Using Figure 3 (flow graph) write down the SMFs for LCICTs of order 8, 16, and 32.

DCT-P19. See DCT-P17. Extend the flow graph for LCICT of order 64.

DCT-P20. See DCT-P17 Write down the SMFs for LCICT of order 64.

DCT-P21. See [DCT21]. In Table IV the BD bitrates for LCICT are compared with those for another order 64 ICT (see reference 14 at the end) using various test sequences under AI, RA, and LDB. Confirm these results.

DCT-P22. The flow graph for order-32 LCICT (flow graphs for order-8 1nd 16 LCICT are embedded) is shown in Figure 1 [DCT21]. Draw the flow graph for the inverse LCICTs.

DCT-P23. Based on DCT-P22, write down the corresponding SMFs for these inverse LCICTs.

DCT-P24. See DCT-P22 and DCT-P23. Evaluate the orthoganality of these LCICTs. Hint: matrix multiply the forward LCICT with the matrix for corresponding inverse LCICT. Are these matrices truly orthogonal?

DCT-P25. M. Wien, "High Efficiency Video Coding: Coding Tools and Specification," Springer, 2014. See Section 7.3.3 determination of the interpolation filter (IF) coefficients (pages 194–196) for the theory behind how these IF coefficients are derived using DCT. Derive these IF coefficients independently and confirm that they match with the coefficients listed in Tables 5.4 and 5.5. Plot the corresponding transfer functions (See Figures 7.16 and 7.17 – pages 201–202).

DCT-P26. Sjovall et al., [DCT25] have developed a high level synthesis implementation of HEVC 2-D DCT/DST (8/16/32 point DCT units and 4-point DCT/DST unit) on FPGA. Develop this architecture and verify the results shown in Tables 1 and 2.

DCT-P27. See DCT-P26. Extend the high level synthesis to 2D- 64 point DCT.

DCT-P28. See [DCT34]. The authors have developed low complexity integer approximation of discrete Tchebichef transforms (DTT) for sizes 4 and 8. (See $\hat{T}_4$ and $\hat{T}_8$). Extend this to $\hat{T}_{16}$.

DCT-P29. See DCT-P29. Develop the sparse matrix factors (SMF) for $\hat{T}_{16}$ similar to those for $\hat{T}_8$. Draw the signal flow graph for $\hat{T}_{16}$ similar to that shown in Figure 1 [DCT34]. Compute the number of additions and bit shifting operations based on these SMF. See Table II [DCT34]. What is the diagonal matrix before $\hat{T}_{16}$ that yields $\hat{T}_{16}$?

DCT-P30. See the discussion under Section B Orthogonality and invertibility in [DCT34] for $\hat{T}_8$ extend this to $\hat{T}_{16}$.

DCT-P31. In HEVC transform sizes can go up to 32 × 32 (integer approximations of DCT) [E61]. Develop $\hat{T}_{32}$ [DCT34].

DCT-P32. See DCT-P31. Derive the SMF for $\hat{T}_{32}$ and draw the signal flow graph similar to Figure 1 in [DCT 34]. Compute the number of additions and bit shifting operations based on these SMF. See Table II [DCT34]. What is the diagonal matrix before $\hat{T}_{32}$ that yields $\hat{T}_{32}$?

DCT-P33. See [DCT34] The authors have compared the performance of original JPEG with those where the 8 × 8 DCT is replaced by the DTT (See Section 5 and Figures 1and 2 for graphs and Figure 3 for Lena image). Verify these results. The software for using the DTT can be requested from the authors. Spectral residual-based similarity [SR-SIM] is described in reference [64] cited in [DCT34].

DCT-P34. The performance of $\hat{T}_8$ in video coding (CIF resolution) is shown in Figure 4 and the first frame of the decoded Foreman sequence is shown in Figure 5 [DCT34]. Confirm these results.

DCT-P35. Wang et al., [DCT35] have applied multi-scale DCT in visual recovery that is superior to the state-of-the-art methods. They also state that the proposed DCT regularizer can be independently embedded into most image processing tasks e.g., image in painting and image denoising. Implement this DCT regularizer and confirm the results shown in this paper.

DCT-P36. See [DCT39] The authors have developed recursive approximation of DCT that satisfies the three features (see pages 449–450). Using Figure 1 write down the SMF for C_8hat. Compute the approximated DCT for N = 8. Using Equation (10), compute D_8 the diagonal matrix. The recursive relationship for the approximated DCT is shown in Equation (9).

DCT-P37. See DCT-P36 Using Figure 2 write down the SMF for C_{16} hat. Compute the approximated DCT for N=16. Using Equation(10), compute D_{16} the diagonal matrix.

DCT-P38. See [DCT39] The proposed reconfigurable architecture for approximate DCT of lengths N = 8 and 16 is shown in Figure 3. Verify this. (Hint: use Figure 2).

DCT-P39. See DCT-P37 Sketch the block diagram of the proposed SCT for N=32, similar to that shown in Figure 2 for N = 16. Compute D_{32} the diagonal matrix.

DCT-P40. See DCT-P39. Verify the proposed reconfigurable architecture for approximate DCT of lengths N = 8, 16 and 32 shown in Figure 4.

DCT-P41. See [DCT39]. The performance of the approximated DCT in HEVC (main profile all intra) is compared with other DCT approximations in HEVC (See Figures 5–7). Verify this comparison. This requires extensive simulations.

DCT-P42. See [DCT39]. As future work the authors state "the coding loss could be reduced by an appropriate modification of context models used in CABAC coder." Investigate this.

DCT-P43. See [DCT20] In the conclusion the authors state "As future work we plan to apply our DCT-BASED PVC where perceptual redundancy is effectively removed thus improving the compression efficiency of HEVC based PVC encoders." Explore this.

DCT-P44. In HEVC integer DCTs of sizes 4×4, 8×8, 16×16 and 32×32 are specified. The smaller size INTDCTs are embedded in larger size DCTs [E202]. Develop 64×64 INTDCT maintaining the embedded property.

DCT-P45. See DCT-P44. For the 64×64 INTDCT, obtain the sparse matrix factors so that it can be implemented with reduced complexity.

DCT-P46. See [DCT-P43] An efficient architecture for 2-D inverse INTDCT for sizes (16×16) and (32×32) have been developed. Extend this to size (64×64). See also DCT-P44 and DCT-P45.

DCT-P47. See [DCT9] In the conclusion section the authors state "For future work, we may use the proposed sharpness/blurriness metrics for other imaging applications, such as image edge enhancement." Explore this.

DCT-P48. See [DCT49] Go through this paper in detail and confirm the results shown in all the tables. The implementation is extensive and may require group effort.

DCT-P49. See [DCT51] and the abstract. Carry out the improved FPGA implementation of the FVC 2D transform and confirm the results described in Figure 4 and Tables III and IV.

DCT-P50. See [DCT52] and "Conclusion." Design and implement the proposed two different high performance FVC 2D transform hardware using Verilog HDL. Confirm the results shown in Figure 10 and in Tables V and VI.

DCT-P51. See Figure 4 in [DCT49]. For the adaptive multiple transforms etc., used at the CTU level in inter predicted random access (RA) configuration the encoding time increases substantially with bit rate savings from 0.2% to nearly 1.5%. Can this be improved by applying fast algorism for the DTT. Explore this.

DCT-P52. See [DCT49]. Can the adaptive multiple transforms proposed in JVET joint exploration test model (JEM) be incorporated in AV1 finalized by AOM? If so explore the bit rate savings, computational complexity and impact on hardware realization.

DCT-P53. See [DCT55]. Explore the future work suggested in the conclusion section.

DCT-P54. See [DCT56] High performance FPGA based architecture for adaptive multiple transforms for different sizes (4×4 to 32×32) has been developed. Extend this to size 64×64.

DCT-P55. See [DCT65] and the conclusion section. In Figure 1 structure of modes in the proposed scheme at the TU stage for 4×4, 8×8, 16×16 and 32×32 is shown. Extend this to 64×64 (proposed in JEM. See reference 4 cited in this paper.)

DCT-P56. See [DCT65] and DCT-P55. In Section V Experimental results for SDT modes for various test sequences are based on luma only. Extend these results for SDT modes for all the test sequences based on all colour components.

DCT-P57. See [DCT65] A paragraph from this section Is repeated here.

"Performance in Joint Exploration Test Model HEVC has high coding efficiency. However, there is still industry needs to further reduce the video bandwidth. ITU-T VCEG (Q6/16) and ISO/IEC MPEG (JTC 1/SC 29/WG 11) are studying the potential need for standardization of future video coding technology. More than 20 compression technology designs, such as larger CTU up to 256×256, 64×64, explicit multiple core transform, mode dependent non-separable secondary transforms, were adopted\and integrated to the Joint Exploration Test Model [36],which has significant performance improvement over HEVC.

Our proposed SDTs have beeen adopted to JEM in October 2015 thanks to the good performance. We integrate the proposed intra and inter SDTs to JEM-1.0 in a similar way as that in HM-12.0. Recently, some new efficient techniques have been adopted to JEM-4.0 [37], such as quadtree plus binary tree (QTBT) block structure, mode dependent non-separable secondary transform on 8×8 blocks. In JEM-4.0, we enable 4×4 to 64×64 intra and inter SDTs on the square TUs. Table X shows the BD-rate savings (%), with SDTs disabled versus SDTs enabled. We observe that the performance improvement is smaller than that on HM-12.0, since many other effective tools have reduced the remaining redundancy. Note that we use the default setting in JEM-4.0 with other adopted tools enabled on JEM-4.0. Even so, we can still achieve up to 12.3% BD-rate savings. Table XI shows the ratios of the geometry mean time complexity of JEM-4.0 with SDTs enabled to that with SDTs disabled (from Classes C, D, E, F). Our fast algorithm makes it feasible to utilize large block size SDTs. But the further exploration of fast algorithms for fast searching and derivation of bases are still expected and will be left as future work." Explore this future work.

DCT-P58. See [DCT65] and DCT-P57. Implement JEM-4.0.
DCT-P59. See [DCT67]. Using Eqs. 1 through 13 the butterfly structure of A8 is shown in Figure 6. Note that A8 has only integer values (Eq. 12). Extend this technique and develop A16 (only integer values). Develop the butterfly structure of A16 similar to that shown in Figure 6 for A8. List the number of additions and shifts needed for A8.
DCT-P60. See [DCT67] and DCT-P59. Table II approximation error (eapp) comparisons are listed for N = 8. Extend these comparisons for N = 16.
DCT-P61. See DCT-P59. Extend this to A32. Note that in JEM large size integer DCTs are proposed.
DCT-P62. See [DCT67] Table IV lists the comparisons of RD performances in terms of BDBR (%) in terms of BDBR for low delay P and random access. Extend this to low delay B. This is a major project.

MPA.P8: Transform Coding (IEEE ICIP 2016)

Session Type: Poster
Time: Monday, September 26, 14:00–15:20
Location: Room 301 CD: Poster Area 8
Session Chair: Onur Guleryuz, LG Electronics Mobile Research Lab

MPA.P8.1: A Staircase Transform Coding Scheme for Screen Content Video Coding

Cheng Chen; *University of Iowa*
Jingning Han; *Google Inc.*
Yaowu Xu; *Google Inc.*
James Bankoski; *Google Inc.*

MPA.P8.2: Fast MCT Optimization for the Compression of Whole-Slide Images

Miguel Hernandez-Cabronero; *University of Warwick*
Victor Sanchez; *University of Warwick*
Francesc Auli-LLinas; *Universitat Autónoma de Barcelona*
Joan Serra-Sagristá; *Universitat Autónoma de Barcelona*

MPA.P8.3: GBST: Separable Transforms Based Online Graphs for Predictive Video Coding

Hilmi E. Egilmez; *University of Southern California*
Yung-Hsuan Chao; *University of Southern California*
Antonio Ortega; *University of Southern California*
Bumshik Lee; *LG Electronics Inc.*
Sehoon Yea; *LG Electronics Inc.*

MPA.P8.4: H.264 Intra Coding with Transforms Based on Prediction Inaccuracy Modeling

Xun Cai; *Massachusetts Institute of Technology*
Jae Lim; *Massachusetts Institute of Technology*

MPA.P8.5: Row-Column Transforms: Low-Complexity Approximation of Optimal Non-Separable Transforms

Hilmi E. Egilmez; *University of Southern California*
Onur G. Guleryuz; *LG Electronics Inc.*
Jana Ehmann; *LG Electronics Inc.*
Sehoon Yea; *LG Electronics Inc.*

MPA.P8.6: Extended Block-Lifting-Based Lapped Transforms

Taizo Suzuki; *University of Tsukuba*
Hiroyuki Kudo; *University of Tsukuba*

MPA.P8.7: Transform-Coded Pel-Recursive Video Compression

Jana Ehmann; *LG Electronics Inc.*
Onur G. Guleryuz; *LG Electronics Inc.*
Sehoon Yea; *LG Electronics Inc.*

MPA.P8.8: Rate-Constrained Successive Elimination Of Hadamard-Based Satds

Ismael Seidel; *Federal University of Santa Catarina*
Luiz Cancellier; *Federal University of Santa Catarina*
José Luís Güntzel; *Federal University of Santa Catarina*
Luciano Agostini; *Federal University of Pelotas*

HEVC (High Efficiency Video Coding) References

E1. K.H. Lee et al., "Technical considerations for Ad Hoc Group on new challenges in video coding standardization," ISO/IEC MPEG 85th meeting, M15580, Hanover, Germany, July 2008.

E2. E. Alshina et al., "Technical considerations on new challenges in video coding standardization," ISO/IEC MPEG 86th meeting, M15899, Busan, South Korea, Oct. 2008.

E3. Y. Ye and M. Karczewicz, "Improved H.264 intra coding based on bi-directional intra prediction, directional transform, and adaptive coefficient scanning," IEEE Int'l Conf. Image Process.'08 (ICIP08), pp. 2116–2119, San Diego, CA, Oct. 2008.

E4. G.J. Sullivan, "The high efficiency video coding (HEVC) standardization initiative," Power Point slides, 7th June 2010.

E5. G.J. Sullivan and J.-R. Ohm, "Recent developments in standardization of high efficiency video coding (HEVC)," Proc. SPIE, Applications of Digital Image Processing XXXIII, vol. 7798, pp. 77980V-1 through V-7, San Diego, CA, 1–3 Aug. 2010.

E6. R. Joshi, Y.A. Reznik and M. Karczewicz, "Efficient large size transforms for high-performance video coding," Proc. SPIE, vol. 7798, San Diego, CA, Aug. 2010.

E7. IEEE Trans. on CSVT, vol. 20, Special section on HEVC (several papers), Dec. 2010.

E8. M. Karczewicz et al., "A hybrid video coder based on extended macroblock sizes, improved interpolation and flexible motion representation," IEEE Trans. CSVT, vol. 20, pp. 1698–1708, Dec. 2010.

E9. T. Wiegand, B. Bross, W.-J. Han, J.-R. Ohm, and G. J. Sullivan, WD3: Working Draft 3 of High-Efficiency Video Coding, Joint Collaborative Team emerging HEVC standard on Video Coding (JCT-VC) of ITU-T VQEG and ISO/IEC MPEG, Doc. JCTVC-E603, Geneva, CH, March 2011.

E10. S. Jeong et al., "High efficiency video coding for entertainment quality," ETRI Journal, vol. 33, pp. 145–154, April 2011.

E11. K. Asai et al., "New video coding scheme optimized for high-resolution video sources," IEEE Journal of Selected Topics in Signal Processing, vol. 5, no. 7, pp. 1290–1297, Nov. 2011.

E12. IEEE Journal of Selected Topics in Signal Processing, vol. 5, no. 7, Nov. 2011 (several papers on HEVC), Introduction to the issue on emerging technologies for video compression.

E13. Z. Ma, H. Hu, and Y. Wang, "On complexity modeling of H.264/AVC video decoding and its application for energy efficient decoding," IEEE Trans. Multimedia, vol. 13, no. 6, pp. 1240–1255, Nov. 2011.

E14. J. Dong and K.N. Ngan, "Adaptive pre-interpolation filter for high efficiency video coding," J. VCIR, vol. 22, pp. 697–703, Nov. 2011.

E15. W. Ding et al., "Fast mode dependent directional transform via butterfly-style transform and integer lifting steps," J. VCIR, vol. 22, pp. 721–726, Nov. 2011.

E16. F. De Simone et al., "Towards high efficiency video coding; subjective evaluation of potential coding technologies," J. VCIR, vol. 22, pp. 734–748, Nov. 2011.

E17. B. Bross et al., "High efficiency video coding (HEVC) text specification draft 6," Joint Collaborative Team on Video Coding (JCT-VC) of ITU-T SG16 WP3 and ISO/IEC JTC1/SC29/WG11, 7th Meeting: Geneva, CH, 21–30 Nov. 2011.

E18. Z. Ma and A. Segall, "Low resolution decoding for high-efficiency video coding," IASTED SIP-2011, Dallas, TX, Dec. 2011.

E19. Y-J. Yoon et al., "Adaptive prediction block filter for video coding," ETRI Journal, vol. 34, pp. 106–109, Feb. 2012. (See the references) http://etrij.etri.re.kr.

E20. B. Li, G.J. Sullivan and J. Xu, "Compression performance of high efficiency video coding (HEVC) working draft 4," IEEE ISCAS, pp. 886–889, session B1L-H, Seoul, Korea, May 2012.

E21. N. Ling, "High efficiency video coding and its 3D extension: A research perspective," Keynote Speech, ICIEA, Singapore, July 2012.

E22. C. Fogg, "Suggested figures for the HEVC specification," ITU-T & ISO/IEC JCTVC-J0292r1, July 2012.

E23. J.–R. Ohm, G. J. Sullivan, and T. Wiegand, "Developments and trends in 2D and 3D video coding standardization: HEVC and more," IEEE ICIP 2012, Tutorial, Orlando, FL, 30 Sept. 2012.

E24. A. Saxena and F.C. Fernandes, "On secondary transforms for prediction residual," IEEE ICIP 2012, Orlando, FL, pp. 2489–2492, Sept. 2012.

E25. "High efficiency video coding" MA.P2, IEEE ICIP 2012, Poster session, (several papers), Orlando, FL, Sept.-Oct., 2012. These are listed below (E26–E38).

E26. S. Ma, J. Si and S. Wang, "A study on the rate distortion modeling for high efficiency video coding," IEEE ICIP 2012, pp. 181–184, Orlando, FL, Oct. 2012.

E27. A. Gabriellini et al., "Spatial transform skip in the emerging high efficiency video coding standard," IEEE ICIP 2012, pp. 185–188, Orlando, FL, Oct. 2012.

E28. H. Vianna et al., "High performance hardware architectures for the inverse rotational transform of the emerging HEVC standard," IEEE ICIP 2012, pp. 189–192, Orlando, FL, Oct. 2012.

E29. H. Aoki, "Efficient quantization parameter coding based on intra/inter prediction for visual quality conscious video coders," IEEE ICIP 2012, pp. 193–196, Orlando, FL, Oct. 2012.

E30. X. Zhang et al., "New chroma intra prediction modes based on linear model for HEVC," IEEE ICIP 2012, pp. 197–200, Orlando, FL, Oct. 2012.

E31. D. Palomino, "A memory aware and multiplierless VLSI architecture for the complete intra prediction of the HEVC emerging standard," IEEE ICIP 2012, pp. 201–204, Orlando, FL, Oct. 2012.

E32. H. Schwarz, "Extension of high efficiency video coding (HEVC) for multiview video and depth data," IEEE ICIP 2012, pp. 205–208, Orlando, FL, Oct. 2012.

E33. M. Budagavi and V. Sze, "Unified forward + inverse transform architecture for HEVC," IEEE ICIP 2012, pp. 209–212, Orlando, FL, Oct. 2012.

E34. C.C. Chi, "Improving the parallelization efficiency of HEVC decoding," IEEE ICIP 2012, pp. 213–216, Orlando, FL, Oct. 2012.

E35. G. Correa et al., "Motion compensated tree depth limitation for complexity control of HEVC encoding," IEEE ICIP 2012, pp. 217–220, Orlando, FL, Oct. 2012.

E36. M. Zhang, C. Zhao and J. Xu, "An adaptive fast intra mode decision in HEVC," IEEE ICIP 2012, pp. 221–224, Orlando, FL, Oct. 2012.

E37. J. Stankowski et al., "Extensions of the HEVC technology for efficient multiview video coding," IEEE ICIP 2012, pp. 225–228, Orlando, FL, Oct. 2012.

E38. L. Kerofsky, K. Misra and A. Segall, "On transform dynamic range in high efficiency video coding," IEEE ICIP 2012, pp. 229–232, Orlando, FL, Oct. 2012.

E39. M.E. Sinangil et al., "Memory cost vs. coding efficiency trade-offs for HEVC motion estimation engine," IEEE ICIP 2012, pp. 1533–1536, Orlando, FL, Sept.–Oct., 2012.

E40. M.E. Sinangil et al., "Hardware – aware motion estimation search algorithm development for HEVC standard," IEEE ICIP 2012, pp. 1529–1532, Orlando, FL, Sept.–Oct., 2012.

E41. M. Zhang, C. Zhao and J. Xu, "An adaptive fast intra mode decision in HEVC," IEEE ICIP 2012, pp. 221–224, Orlando, FL, Sept.–Oct., 2012.

E42. H. Zhang and Z. Ma, "Fast intra prediction for high efficiency video coding," Pacific Rim Conf. on Multimedia, PCM2012, Singapore, Dec. 2012. See [E147].

E43. Special issue on emerging research and standards in next generation video coding, IEEE Trans. CSVT, vol. 22, pp. 1646–1909, Dec. 2012.

E44. B. Shrestha, "Investigation of image quality of Dirac, H.264 and H.265," Report for EE 5359: Multimedia Processing, EE Dept., UTA, Arlington, TX, 2012.
http://www.uta.edu/faculty/krrao/dip/

E45. B. Li, G.J. Sullivan, and J. Xu, "Compression performance of high efficiency video coding working draft 4," IEEE ISCAS 2012, Session on HEVC, pp. 886–889, Seoul, Korea, May 2012.

E46. http://www.h265.net has info on developments in HEVC NGVC – Next generation video coding.

E47. JVT KTA reference software http://iphome.hhi.de/suehring/tml/download/KTA/

E48. JM11KTA2.3 (http://www.h265.net/2009/04/kta-software-jm11kta23.html).

E49. Permanent Link: Mode-Dependent Directional Transform (MDDT) in JM/KTA TMUC HEVC Software – http://hevc.kw.bbc.co.uk/svn/jctvc-tmuc/.

E50. F. Bossen, D. Flynn and K. Sühring (July 2011), "HEVC reference software manual" Online available: http://phenix.int-evry.fr/jct/doc_end_user/documents/6_Torino/wg11/JCTVC-F634-v2. zip

E51. http://phenix.it-sudparis.eu/jct/doc_end_user/documents/7_Geneva/wg11/JCTVC-G399-v3.zip

E52. http://www.itu.int/ITU-T/studygroups/com16/jct-vc/index.html.

E53. JCT-VC documents are publicly available at http://ftp3.itu.ch/av-arch/jctvc-site and http://phenix.it-sudparis.eu/jct/.

E54. HEVC Model 9.0, available for download at https://hevc.hhi.fraunhofer.de/svn/svn_HEVCSoftware/tags/HM-9.0/ The latest version available at the time this chapter was prepared is HM-9.0. Version HM 8.0 is available at: http://hevc.kw.bbc.co.uk/trac/browser/tags/HM-8.0

E55. J. Nightingale, Q. Wang and C. Grecos, "HEVStream; A framework for streaming and evaluation of high efficiency video coding (HEVC) content in loss-prone networks," IEEE Trans. Consumer Electronics, vol. 59, pp. 404–412, May 2012.

E56. D. Marpe et al., "Improved video compression technology and the emerging high efficiency video coding standard," IEEE International Conf. on Consumer Electronics, pp. 52–56, Berlin, Germany, Sept. 2011.

E57. HM9 High efficiency video coding (HEVC) Test Model 9 (HM 9) encoder description JCTVC-K1002v2, Shanghai meeting, Oct. 2012.

E58. B. Bross et al., "High efficiency video coding (HEVC) text specification draft 8," JCTVC-J1003, July 2012. http://phenix.int-evry.fr/jct/doc_end_user/current_document.Php?id=5889

E59. G.J. Sullivan et al., "Overview of the high efficiency video coding (HEVC) standard," IEEE Trans. CSVT, vol. 22, pp. 1649–1668, Dec. 2012.

E60. J.-R. Ohm et al., "Comparison of the coding efficiency of video coding standards – including high efficiency video coding (HEVC)," IEEE Trans. CSVT, vol. 22, pp. 1669–1684, Dec. 2012. Software and data for reproducing selected results can be found at ftp://ftp.hhi.de/ieee-tcsvt/2012

E61. F. Bossen et al., "HEVC complexity and implementation analysis," IEEE Trans. CSVT, vol. 22, pp. 1685–1696, Dec. 2012.

E62. B. Bross et al., "Block merging for quadtree based partitioning in HEVC," SPIE Applications of digital image processing XXXV, vol. 8499, paper 8499-26, Aug. 2012.

E63. P. Hanhart et al., "Subjective quality evaluation of the upcoming HEVC video compression standard," SPIE Applications of digital image processing XXXV, vol. 8499, paper 8499-30, Aug. 2012.

E64. M. Horowitz et al., "Informal subjective quality comparison of video compression performance of the HEVC and H.264/MPEG-4 AVC standards for low delay applications," SPIE Applications of digital image processing XXXV, vol. 8499, paper 8499-31, (W84990-1), Aug. 2012.

E65. V. Sze and M. Budagavi, "High throughput CABAC entropy coding in HEVC," IEEE Trans. CSVT, vol. 22, pp. 1778–1791, Dec. 2012.

E66. V. Sze and A.P. Chandrakasan, "Joint algorithm-architecture optimization of CABAC to increase speed and reduce area cost," IEEE ISCAS, Proc. pp. 1577–1580, 2011.

E67. H. Schwartz, D. Marpe and T. Wiegand, "Overview of the scalable video coding extension of the H.264/AVC standard," IEEE Trans. CSVT, vol. 17, pp. 1103–1120, Sept. 2007. This is a special issue on scalable video coding in H.264/AVC.0

E68. JSVM – joint scalable video model reference software for scalable video coding. (online) http://ip.hhi.de/imagecom_GI/savce/downloads/SVC-Reference-software.htm

E69. H. Lakshman, H. Schwarz and T. Wiegand, "Generalized interpolation based fractional sample motion compensation," IEEE Trans. CSVT, vol. 23, pp. 455–466, March 2013. A sample C++ implementation (that can be tested using the first test model of HEVC, HM1.0) can be found in http://phenix.int-evry.fr/jct/doc_end_user/documents/4_Daegu/wg11/JCTVC-D056-v2.zip

E70. K. Ugur et al., "High performance low complexity video coding and the emerging high efficiency video coding standard," IEEE Trans. CSVT, vol. 20, pp. 1688–1697, Dec. 2010.

E71. A. Norkin et al., "HEVC deblocking filter," IEEE Trans. CSVT, vol. 22, pp. 1746–1754, Dec. 2012. Corrections to "HEVC deblocking filter," IEEE Trans. CSVT, vol. 23, pp. 21–41, Dec. 2013.

E72. J. Sole et al., "Transform coefficient coding in HEVC," IEEE Trans. CSVT, vol. 22, pp. 1765–1777, Dec. 2012.

E73. JSVM9 Joint scalable video model 9 http://ip.hhi.de/imagecom_GI/savce/downloads

E74. Z. Shi, X. Sun and F. Wu, "Spatially scalable video coding for HEVC," IEEE Trans CSVT, vol. 22, pp. 1813–1826, Dec. 2012.

E75. A. Krutz et al., "Adaptive global motion temporal filtering for high efficiency video coding," IEEE Trans. CSVT, vol. 22, pp. 1802–1812, Dec. 2012.

E76. J. Lainema et al., "Intra coding of the HEVC standard," IEEE Trans. CSVT, vol. 22, pp. 1792–1801, Dec. 2012.

E77. H. Li, B. Li and J. Xu, "Rate distortion optimized reference picture management for high efficiency video coding," IEEE Trans. CSVT, vol. 22, pp. 1844–1857, Dec. 2012.

E78. Special issue on video coding: HEVC and beyond, IEEE Journal of selected topics in signal processing, vol. 7, Dec. 2013.

E79. G. Bjontegaard, "Calculation of average PSNR differences between RD-Curves," ITU-T SG16, Doc. VCEG-M33, 13th VCEG meeting, Austin, TX, April 2001. http://wfpt3.itu.int/av-arch/video-site/0104_Aus/VCEG-M33.doc

A graphical explanation of BD-PSNR and BS-bitrate is shown in [E202]

E80. G. Bjontegaard, "Improvements of the BD-PSNR model," ITU-T SG16 Q.6, Doc. VCEG-AI11, Berlin, Germany, July 2008.

E81. T. Schierl et al., "System layer integration of HEVC," IEEE Trans. CSVT, vol. 22, pp. 1871–1884, Dec. 2012.

E82. G. Correa et al., "Performance and computational complexity assessment of high efficiency video encoders," IEEE Trans. CSVT, vol. 22, pp. 1899–1909, Dec. 2012.

E83. M. Zhou et al., "HEVC lossless coding and improvements," IEEE Trans. CSVT, vol. 22, pp. 1839–1843, Dec. 2012.

E84. C. Yeo, Y. H. Tan and Z. Li, "Dynamic range analysis in high efficiency video coding residual coding and reconstruction," IEEE Trans. CSVT, vol. 23, pp. 1131–1136 , July 2013.

E85. C.-M. Fu et al., "Sample adaptive offset in the HEVC standard," IEEE Trans. CSVT, vol. 22, pp. 1755–1764, Dec. 2012.

E86. C.C. Chi et al., "Parallel scalability and efficiency of HEVC parallelization approaches," IEEE Trans. CSVT, vol. 22, pp. 1827–1838, Dec. 2012.

E87. J. Vanne et al., "Comparative rate-distortion complexity analysis of HEVC and AVC video codecs," IEEE Trans. CSVT, vol. 22, pp. 1885–1898, Dec. 2012.

E88. Y. Yuan et al., "Quadtree based non-square block structure for interframe coding in high efficiency video coding," IEEE Trans. CSVT, vol. 22, pp. 1707–1719, Dec. 2012.

E89. R. Sjoberg et al., "Overview of high-level syntax and reference picture management," IEEE Trans. CSVT, vol. 22, pp. 1858–1870, Dec. 2012.

E90. P. Helle et al., "Block merging for quadtree-based partitioning in HEVC," IEEE Trans. CSVT, vol. 22, pp. 1720–1731, Dec. 2012.

E91. I.-K. Kim et al., "Block partitioning structure in HEVC standard," IEEE Trans. CSVT, vol. 22, pp. 1697–1706, Dec. 2012.

E92. T. Lin et al., "Mixed chroma sampling-rate high efficiency video coding for full-chroma screen content," IEEE Trans. CSVT, vol. 23, pp. 173–185, Jan. 2013.

E93. E. Peixoto and E. Izquierdo, "A new complexity scalable transcoder from H.264/AVC to the new HEVC codec," IEEE ICIP 2012, session MP.P1.13 (Poster), Orlando, FL, Sept. Oct. 2012.

E94. K. Anderson, R. Sjobetg and A. Norkin, "BD measurements based on MOS," (online), ITU-T Q6/SG16, document VCEG-AL23, Geneva, Switzerland, July 2009.Available: http://wfpt3.itu.int/av-arch/video-site/0906_LG/VCEG-AL23.zip

E95. HEVC open source software (encoder/decoder)

https://hevc.hhi.fraunhofer.de/svn/svn_HEVCSoftware/tags/HM-10.1 JCT-VC (2014) Subversion repository for the HEVC test model reference software
https://hevc.hhi.fraunhofer.de/svn/svn_HEVCSoftware/
http://hevc.kw.bbc.co.uk/svn/jctvc-al24 (mirror site)

E96. T. Shanableh, E. Peixoto and E. Izquierdo, "MPEG-2 to HEVC transcoding with content based modeling," IEEE Trans. CSVT, vol. 23, pp. 1191–1196, July 2013.

References [E97] through [E101] are from VCIP 2012. Please access http://www.vcip2012.org. There are also some papers on HEVC in the Demo session.

E97. G.J. Sullivan, "HEVC: The next generation in video compression." Keynote speech, Visual Communications and Image Processing, VCIP 2012, San Diego, CA, 27–30 Nov. 2012.

E98. Two oral sessions on HEVC (W03 and F01). Also panel discussion on "All about HEVC" Moderator: Shipeng Li, Microsoft Research Asia.

E99. X. Zhang, S. Liu and S. Lei, "Intra mode coding in HEVC standard," Visual Communications and Image Processing, VCIP 2012, pp. 1–6, San Diego, CA, 27–30, Nov. 2012.

E100. L. Yan et al., "Implementation of HEVC decoder on X86 processors with SIMD optimization," Visual Communications and Image Processing, VCIP 2012, San Diego, CA, 27–30, pp. 1–6, Nov. 2012.

E101. W. Dai and H. Xiong, "Optimal intra coding of HEVC by structured set prediction mode with discriminative learning," Visual Communications and Image Processing, VCIP 2012, pp. 1–6, San Diego, CA, 27–30, Nov. 2012.

E102. G. Li et al., "Uniform probability model for deriving intra prediction angular table in high efficiency video coding," Visual Communications and Image Processing, VCIP 2012, pp. 1–5, San Diego, CA, 27–30, Nov. 2012.

E103. Y. Duan, "An optimized real time multi-thread HEVC decoder," Visual Communications and Image Processing, VCIP 2012, p. 1, San Diego, CA, 27–30, Nov. 2012.

References [E99] through [E105] are from VCIP 2012. Please access http://www.vcip2012.org. There are also some papers on HEVC in the Demo session.

E104. S. Gangavathi, "Complexity reduction of H.264 using parallel programming," M.S. Thesis, EE Dept., UTA, Arlington, TX. Dec. 2012.

E105. M.T. Pourazad et al., "HEVC: The new gold standard for video compression," IEEE CE magazine, vol. 1, issue 3, pp. 36–46, July 2012.

E106. Y. Zhang, Z. Li and B. Li, "Gradient-based fast decision for intra prediction in HEVC," Visual Communications and Image Processing, VCIP 2012, San Diego, CA, 27–30, Nov. 2012.

E107. H. Lv et al., "A comparison of fractional-pel interpolation filters in HEVC and H.264/AVC," Visual Communications and Image Processing, VCIP 2012, San Diego, CA, 27–30, Nov. 2012. Several references leading to the selection of DCTIF are listed at the end.

E108. J. Wang et al., "Multiple sign bits hiding for high efficiency video coding," Visual Communications and Image Processing, VCIP 2012, San Diego, CA, 27–30, Nov. 2012.

E109. C.-M. Fu et al., "Sample adaptive offset for HEVC," in Proc. Intl. workshop on multimedia signal processing, (MMSP), Hangzhou, China, Oct. 2011.

E110. S. Subbarayappa, "Implementation and Analysis of Directional Discrete Cosine Transform in Baseline Profile in H.264," M.S. Thesis, EE Dept., UTA, Arlington, Texas, 2012.
http://www.uta.edu/faculty/krrao/dip/

E111. P. Anjanappa, "Performance analysis and implementation of mode dependent DCT/DST in H.264/AVC," M.S. Thesis, EE Dept., UTA, Arlington, Texas, 2012.
http://www.uta.edu/faculty/krrao/dip/

E112. Q. Cai et al., "Lossy and lossless intra coding performance evaluation: HEVC, H.264/AVS, JPEG 2000 and JPEG LS," APSIPA, Los Angeles, CA, Dec. 2012.

E113. K. Chen et al., "Efficient SIMD optimization of HEVC encoder over X86 processors," APSIPA, Los Angeles, CA, Dec. 2012.

E114. H. Lv et al., "An efficient NEON-based quarter-pel interpolation method for HEVC," APSIPA, Los Angeles, CA, Dec. 2012.

E115. Q. Yu et al., "Early termination of coding unit splitting for HEVC," APSIPA, Los Angeles, CA, Dec. 2012.

E116. J.M. Nightingale, Q. Wang and C. Grecos, "Priority-based methods for reducing the input of packet loss on HEVC encoded video streams," SPIE/EI, 8656–15, Feb. 2013.

E117. Joint preliminary call for proposals on scalable video coding extensions of high efficiency video coding (HEVC), Geneva, Switzerland, May 2012.

E118. S. Kamp and M. Wien, "Decoder-side motion vector derivation for block-based video coding," IEEE Trans. CSVT, vol. 22, pp. 1732–1745, Dec. 2012.

E119. A. Saxena, F. Fernandes and Y. Reznik, "Fast transforms for Intra-prediction-based image and video coding," in Proc. IEEE Data Compression Conference (DCC'13), Snowbird, UT, March 20–23, 2013.

E120. R. Chivukula and Y. Reznik, "Fast computing of discrete cosine and sine transforms of types VI and VII," in Applications of Digital Image Processing XXXIV, Proc. SPIE, A. G. Tescher, Ed., vol. 8135, pp. 813505–1–14, 2011.

E121. M. Zhou et al., "HEVC lossless coding and improvements," SPIE/EI, vol. 8666-10, Burlingame, CA, Feb. 2013.

E122. R. Garcia and H. Kalva, "Subjective evaluation of HEVC in mobile devices," SPIE/EI, vol. 8667-19, Burlingame, CA, Feb. 2013. See also "Human mobile-device interaction on HEVC and H.264 subjective evaluation for video use in mobile environment," IEEE ICCE, pp. 639–640, Jan. 2013. See [E210].

E123. B. Bross et al., "High efficiency video coding (HEVC) text specification draft 10 (for FDIS & Consent)," JCTVC-L1003v13, Jan. 2013.

E124. X. Xiu et al., "Improved motion prediction for scalable extensions of HEVC," SPIE/EI, vol. 8666-5, Burlingame, CA, Feb. 2013.

E125. T. Hinz et al., "An HEVC extension for spatial and quality scalable video coding," SPIE/EI, vol. 8666-6, Burlingame, CA, Feb. 2013.

E126. K. Misra et al., "Scalable extensions of HEVC for next generation services," SPIE/EI, vol. 8666-7, Burlingame, CA, Feb. 2013.

E127. ISO/IEC JTC-1/SC29/WG11 w12957, "Joint call for proposals on scalable video coding extensions of high efficiency video coding," July 2012.

E128. S.G. Deshpande et al., "An improved hypothetical reference decoder for HEVC," SPIE/EI, vol. 8666-9, Burlingame, CA, Feb. 2013.

E129. J. Han et al., "Towards jointly optimal spatial prediction and adaptive transform in video/image coding," IEEE ICASSP, pp. 726–729, Mar. 2010.

E130. A. Saxena and F. Fernandes, "CE7: Mode-dependent DCT/DST for intra prediction in video coding," ITU-T/ISO-IEC Document: JCTVC-D033, Jan. 2011.

E131. A. Saxena and F. Fernandes, "Mode dependent DCT/DST for intra prediction in block-based image/video coding," IEEE ICIP, pp. 1685–1688, Sept. 2011.

E132. Y. Kim et al., "A fast intra-prediction method in HEVC using rate-distortion estimation based on Hadamard transform," ETRI Journal, vol. 35, # 2, pp. 270–280, April 2013.

E133. D. Flynn, G. Martin-Cocher and D. He, "Decoding a 10-bit sequence using an 8-bit decoder," JCTVC-M0255, April 2013.

E134. J. -R. Ohm, T. Wiegand and G. J. Sullivan, "Video coding progress; The high efficiency video coding (HEVC) standard and its future extensions," IEEE ICASSP, Tutorial, Vancouver, Canada, June 2013.

E135. D. Springer et al., "Robust rotational motion estimation for efficient HEVC compression of 2D and 3D navigation video sequences," IEEE ICASSP 2013, pp. 1379–1383, Vancouver, Canada, 2013.

E136. H. Li, Y. Zhang and H. Chao, "An optimally scalable and cost-effective fractional motion estimation algorithm for HEVC," IEEE ICASSP 2013, pp. 1399–1403, Vancouver, Canada, June 2013.

E137. Several papers on HEVC in the poster session IVMSP-P3: Video coding II, IEEE ICASSP 2013, Vancouver, Canada, June 2013.

E138. Z. Pan et al., "Early termination for TZSearch in HEVC motion estimation," IEEE ICASSP 2013, pp. 1389–1392, June 2013.

E139. D.-K. Kwon, M. Budagavi and M. Zhou, "Multi-loop scalable video codec based on high efficiency video coding (HEVC)," IEEE ICASSP 2013, pp. 1749–1753, June 2013.

E140. H.R. Tohidpour, M.T. Pourazad and P. Nasiopoulos, "Content adaptive complexity reduction scheme for quality/fidelity scalable HEVC," IEEE ICASSP 2013, pp. 1744–1748, June 2013.

E141. Y.H. Tan, C. Yeo and Z. Li, "Residual DPCM for lossless coding in HEVC," IEEE ICASSP 2013, pp. 2021–2025, June 2013.

E142. M. Wien, "HEVC – coding tools and specifications," Tutorial, IEEE ICME, San Jose, CA, July 2013.

E143. T.L. da Silva, L.V. Agostini and L.A. da Silva Cruz, "HEVC intra coding acceleration based on tree intra-level mode correlation," IEEE Conference 2013 Signal Processing: Algorithms, Architectures, Arrangements, and Applications (SPA 2013), Poznan, Poland, Sept. 2013.

E144. E. Peixoto et al., "An H.264/AVC to HEVC video transcoder based on mode mapping," IEEE ICIP 2013, WP.P4.3 (poster), Melbourne, Australia, 15–18 Sept. 2013,

E145. E. Wige et al., "Pixel-based averaging predictor for HEVC lossless coding," IEEE ICIP 2013, WP.P311 (poster), Melbourne, Australia, 15–18 Sept. 2013.

E146. E. Peixoto, T. Shanableh and E. Izquierdo, "H.264/AVC to HEVC video transcoder based on dynamic thresholding and content modeling," IEEE Trans. on CSVT, vol. 24, pp. 99–112, Jan. 2014.

E147. H. Zhang and Z. Ma, "Fast intra mode decision for high-efficiency video coding," IEEE Trans. CSVT, Vol. 24, pp. 660–668, April 2014. See [E42].

E148. Y. Tew and K.S. Wong, "An overview of information hiding in H.264/AVC compressed video," IEEE Trans. CSVT, vol. 24, pp. 305–319, Feb. 2014. (repeated see E173).

E149. M. Zhou, V. Sze and M. Budagavi, "Parallel tools in HEVC for high-throughput processing," Applications of digital image processing, SPIE, vol. 8499, pp. 849910-1 through 849910-12, 2012.

E150. W.-H. Peng and C.-C. Chen, "An interframe prediction technique combining template matching and block motion compensation for high efficiency video coding," IEEE Trans. CSVT, vol. 23, pp. 1432–1446, Aug. 2013.

E151. S. Blasi, E. Peixoto and E. Izqueirdo, "Enhanced Inter – Prediction using merge prediction transformation in the HEVC codec," IEEE ICASSP 2013, pp. 1709–1713, Van Couver, CA, June 2013.

E152. L.-L. Wang and W.-C. Siu, "Novel Adaptive Algorithm for Intra Prediction With Compromised Modes Skipping and Signaling Processes in HEVC," IEEE Trans. CSVT, Vol. 23, pp. 1686–1694, Oct. 2013.

E153. D. Flynn, J. Sole and T. Suzuki, "High efficiency video coding (HEVC) range extensions text specification," Draft 4, JCT-VC. Retrieved 2013-08-07.

E154. M. Budagavi and D.-Y. Kwon, "Intra motion compensation and entropy coding improvements for HEVC screen content coding," IEEE PCS, pp. 365–368, San Jose, CA, Dec. 2013.

E155. M. Naccari et al., "Improving inter prediction in HEVC with residual DPCM for lossless screen content coding," IEEE PCS, pp. 361–364, San Jose, CA, Dec. 2013.

E156. E. Alshina et al., "Inter-layer filtering for scalable extension of HEVC," IEEE PCS, pp. 382–385, San Jose, CA, Dec. 2013.

E157. J. Zhao, K. Misra and A. Segall, "Content-adaptive upsampling for scalable video coding," IEEE PCS, pp. 378–381, San Jose, CA, Dec. 2013.

E158. G.J. Sullivan et al., "Standardized extensions of high efficiency video coding (HEVC)," IEEE Journal of selected topics in signal processing, vol. 7, pp. 1001–1016, Dec. 2013.

E159. F. Pescador et al., "On an Implementation of HEVC Video Decoders with DSP Technology," IEEE ICCE, pp. 121–122, Las Vegas, NV, Jan. 2013.

E160. F. Pescador et al., "A DSP HEVC decoder implementation based on open HEVC," IEEE ICCE, pp. 65–66, Las Vegas, NV, Jan. 2014.

E161. F. Pescador et al., "Complexity analysis of an HEVC decoder based on a digital signal processor," IEEE Trans. on Consumer Electronics, vol. 59, pp. 391–399, May 2013.

E162. K. Iguchi et al., "HEVC encoder for super hi-vision," IEEE ICCE, pp. 61–62, Las Vegas, NV, Jan. 2014.

E163. K. Miyazawa et al., "Real-time hardware implementation of HEVC encoder for 1080p HD video," IEEE PCS 2013, pp. 225–228, San Jose, CA, Dec. 2013.

E164. Z. Lv and R. Wang, "An all zero blocks early detection method for high efficiency video coding" SPIE, vol. 9029, pp. 902902-1 through 902902-7, IS&T/SPIE Electronic Imaging, San Francisco, CA, Feb. 2014. See [E169, E184].

E165. S. Vasudevan and K.R. Rao, "Combination method of fast HEVC encoding," IEEE ECTICON 2014, Korat, Thailand, May 2014.

http://www.uta.edu/faculty/krrao/dip/ click on courses and then click on EE5359 Scroll down and go to Thesis/Project Title and click on S. Vasudevan.

E166. S. Kim et al., "A novel fast and low-complexity motion estimation for UHD HEVC," IEEE PCS 2013, pp. 105–108, San Jose, CA, Dec. 2013.

E167. H. Zhang, Q. Liu and Z. Ma, "Priority classification based intra mode decision for high efficiency video coding," IEEE PCS 2013, pp. 285–288, San Jose, CA, Dec. 2013.

E168. X. Wang et al., "Paralleling variable block size motion estimation of HEVC on CPU plus GPU platform," IEEE ICME workshop, 2013.

E169. P-T. Chiang and T.S. Chang, "Fast zero block detection and early CU termination for HEVC video coding," IEEE ISCAS 2013, pp. 1640–1643, Beijing, China, May 2013. See [E164, E184]

E170. V. Sze and M. Budagavi, "Design and implementation of next generation video coding systems," Sunday 1 June 2014 (half day tutorial), IEEE ISCAS 2014, Melbourne, Australia, 1–5 June 2014.

E171. T. Nguyen et al., 'Transform coding techniques in HEVC," IEEE Journal of selected topics in signal processing, vol. 7, pp. 978–989, Dec. 2013.

E172. E. Peixoto, T. Shanableh and E. Izquierdo, "H.264/AVC to HEVC transcoder based on dynamic thresholding and content modeling', IEEE Trans. CSVT, vol. 24, pp. 99–112, Jan. 2014.

E173. Y. Tew and K.S. Wong, "An overview of information hiding in H.264/AVC compressed video," IEEE Trans. CSVT, vol. 24, pp. 305–319, Feb. 2014.

E174. T. Na and M. Kim, "A novel no-reference PSNR estimation method with regard to deblocking filtering effect in H.264/AVC bitstreams," IEEE Trans. CSVT, vol. 24, pp. 320–330, Feb. 2014.

E175. D. Grois et al., "Performance comparison of H.265/MPEG-HEVC, VP9 and H.264/MPEG-AVC encoders," IEEE PCS 2013, pp. 394–397, San Jose, CA, Dec. 2013.

E176. F. Bossen, "Common HM test conditions and software reference configurations," document JCTVC-L1100 of JCT-VC, Geneva, CH, Jan. 2013.

E177. A. Abdelazim, W. Masri and B. Noaman, "Motion estimation optimization tools for the emerging high efficiency video coding (HEVC)," SPIE, vol. 9029, paper #1, IS&T/SPIE Electronic Imaging, pp. 902905-1 through 902905-8, San Francisco, CA, Feb. 2014.

E178. H.R. Wu and K.R. Rao, "Digital video image quality and perceptual coding," Boca Raton, FL: CRC Press, 2006.

E179. T. Shen et al., "Ultra fast H.264/AVC to HEVC transcoder," IEEE DCC, pp. 241–250, 2013.

E180. M. Tikekar et al., "A 249-Mpixel/s HEVC video-decoder chip for 4k Ultra-HD applications," IEEE J. of Solid-State Circuits, vol. 49, pp. 61–72, Jan. 2014.

E181. K.R. Rao and J.J. Hwang, "Techniques and standards for image/video/audio coding," Prentice Hall, 1996. Translated into Japanese and Korean.

E182. B. Bross et al., "HEVC performance and complexity for 4K video," 3rd IEEE ICCE-Berlin, pp. 44–47, Sept. 2013.

E183. F. Bossen et al., "HEVC complexity and implementation analysis," IEEE Trans. CSVT, vol. 22, pp. 1685–1696, Dec. 2012.

E184. K. Lee et al., "A novel algorithm for zero block detection in high efficiency video coding," IEEE Journal of selected topics in signal processing, vol. 7, #6, pp. 1124–1134, Dec. 2013. See [E164, E169].

E185. K. Shah, "Reducing the complexity of inter prediction mode decision for high efficiency video coding," M.S. Thesis, EE Dept., University of Texas at Arlington, Arlington, Texas, May 2014. http://www.uta.edu/faculty/krrao/dip click on courses and then click on EE5359 Scroll down and go to Thesis/Project Title and click on K. Shah

E186. HEVC Fraunhofer site containing all the information on HEVC- http://hevc.info/ (software/overview papers/documents etc.)

E187. HEVC decoder for handheld devices implemented by Ace Thought- http://www.acethought.com/index.php/products/hevc-decoder/

E188. M. Revabak and T. Ebrahimi, "Comparison of compression efficiency between HEVC/H.265 and VP9 based on subjective assessments," SPIE Optical Engineering, Applications of digital image processing, Vol. 9217, San Diego, California, USA, August 18–21, 2014. https://infoscience.epfl.ch/record/200925/files/article-vp9-submited-v2.pdf

E189. D. Grois et al., "Comparative assessment of H.265/MPEG-HEVC, VP9, and H.264/MPEG-AVC encoders for low-delay video applications," SPIE Optical Engineering, Applications of digital image processing, Vol. 9217, San Diego, California, USA, August 18–21, 2014. (This has several useful web sites.)

E190. B. Bross et al., "HEVC performance and complexity for 4K video," 3rd IEEE ICCE-Berlin, Berlin, Germany. Sept. 2013.

E191. A. Heindel, E. Wige and A. Kaup, "Analysis of prediction algorithms for residual compression in a lossy to lossless scalable video coding system based on HEVC," SPIE Optical Engineering, Applications of digital image processing, Vol. 9217, San Diego, California, USA, August 18–21, 2014.

E192. A. Heindel, E. Wige and A. Kaup, "Sample-based weighted prediction for lossless enhancement layer coding in HEVC," IEEE PCS 2013, Grand Compression Challenge, 2013.

E193. J. Vanne, M. Viitanen and T.D. Hamalainen, "Efficient mode decision schemes for HEVC inter prediction," IEEE Trans. CSVT, vol. 24, pp. 1579–1593, Sept. 2014.

E194. L. Wang et al., "Lossy to lossless image compression based on reversible integer DCT," IEEE ICIP, pp. 1037–1040, San Diego, CA, Oct. 2008.

E195. K. Gruneberg, T. Schierl and S. Narasimhan, "ISO/IEC 13818-1:2013/FDAM 3, Transport of high efficiency video coding (HEVC) video over MPEG-2 systems, 2013." (FDAM: final draft amendment)

E196. F. Bossen, "Excel template for BD-rate calculation based on piecewise cubic interpolation," JCT-VC Reflector.

E197. L. Shen, Z. Zhang and Z. Liu, "Adaptive inter-mode decision for HEVC jointly utilizing inter-level and spatiotemporal correlations," IEEE Trans. CSVT, vol. 24, pp. 1709–1722, Oct. 2014.

E198. Y.-T. Peng and P.C. Cosman, "Weighted boundary matching error concealment for HEVC using block partition decisions," IEEE 48th Asilomar Conference on Signals, Systems and Computers, Pacific Grove, CA, Nov. 2014.

E199. F. Zou et al., "View synthesis prediction in the 3-D video coding extensions of AVC and HEVC," IEEE Trans. CSVT, vol. 24, pp. 1696–1708, Oct. 2014.

E200. V. Sze, M. Budagavi and G.J. Sullivan (Editors), "High efficiency video coding: Algorithms and architectures," Springer 2014.

E201. M. Warrier, "Multiplexing HEVC video with AAC audio bit streams, de multiplexing and achieving lip synchronization," M.S. Thesis, EE Dept., University of Texas at Arlington, Arlington, Texas, Dec. 2014. http://www.uta.edu/faculty/krrao/dip click on courses and then click on EE5359 Scroll down and go to Thesis/Project Title and click on M. Warrier

E202. P. Mehta, "Complexity reduction for intra mode selection in HEVC using open MP," M.S. Thesis, EE Dept., University of Texas at Arlington, Arlington, Texas, Dec. 2014. http://www.uta.edu/faculty/krrao/dip click on courses and then click on EE5359 Scroll down and go to Thesis/Project Title and click on P. Mehta

E203. J. Dubhashi, "Complexity reduction of H.265 motion estimation using compute unified device Architecture," M.S. Thesis, EE Dept., University of Texas at Arlington, Arlington, Texas, Dec. 2014. http://www.uta.edu/faculty/krrao/dip click on courses and then click on EE5359 Scroll down and go to Thesis/Project Title and click on J. Dubhashi

E204. Y. Umezaki and S. Goto, "Image segmentation approach for realizing zoomable streaming HEVC video," IEEE ICICS 2013.

E205. K. Suresh, "Application of open MP on complexity reduction in inter frame coding in HEVC" M.S. Thesis, EE Dept., University of Texas at Arlington, Arlington, Texas, Dec. 2014. http://www.uta.edu/faculty/krrao/dip click on courses and then click on EE5359 Scroll down and go to Thesis/Project Title and click on K. Suresh

E206. Y. Ye and P. Andrivon, "The scalable extensions of HEVC for ultra-high definition video delivery," IEEE Multimedia magazine, vol. 21, issue, 3, pp. 58–64, July-Sept. 2014. See the references at the end. These are highly useful for further research in SHVC – scalable HEVC.

E207. Shiba Kuanar, Christopher Conly, Kamisetty Rao, "Deep Learning based HEVC in-loop Filtering for Decoder Quality Enhancement," IEEE Picture Coding Symposium (PCS), June 2018, San Francisco, California.

E208. Shiba Kuanar, Kamisetty Rao, Christopher Conly, "Fast Mode Decision in HEVC Intra Prediction using Region wise CNN Feature Classification," IEEE International Conference on Multimedia and Expo (ICME Expo), July 2018, San Diago USA.

Software Repository; Scalable Extensions of HEVC

The source code for the software is available in the following SVN repository.
  https://hevc.hhi.fraunhofer.de/svn/svn_SHVCSoftware/
For tool integration branch for a company can be obtained by contacting:
  Seregin, Vadim (vseregin@qti.qualcomm.com)

Build System

The software can be built under Linux using make. For Windows, solutions for different versions of Microsoft Visual Studio are provided.

Software Structure

The SHVC Test Model Software inherits the same software structure from the HEVC test model HM software, which includes the following applications and libraries for encoding, decoding and down sampling process:

- Applications:
    - TAppEncoder, executable for bit stream generation.
    - TAppDecoder, executable for reconstruction.
    - TAppCommon, common functions for configuration file parsing.
    - TAppDownConvert, down sampling functionalities.
- Libraries:
    - TLibEncoder, encoding functionalities
    - TLibDecoder, decoding functionalities
    - TLibCommon, common functionalities
    - TLibVideoIO, video input/output functionalities

ATSC Advanced Television Systems Committee
TV Tomorrow: ATSC 3.0 Advances

ATSC 3.0 will be a radical departure from the current standard ATSC 1.0 was developed around 20 years ago, when cellphones were analog and streaming was unheard of. It relies 8-VSB modulation capable of delivering 19.39 Mbps in a 6 MHz TV channel - enough to carry a high – definition program compressed by a factor of 50 using MPEG-2 to a fixed receiver.

With ATSC 3.0, the committee seeks to increase that data rate by 30 percent, or roughly 25.2 Mbps. The overall intent of 3.0 is to enable seamless transmission of HD, 4K, 22.2 audio and other data streams to fixed, mobile and handheld devices in all types of terrain.

The most differentiating characteristic of ATSC 3.0 is that it will not be backward – compatible with 1.0 or even 2.0, which is now in development. In other words, televisions now capable of processing over-the-air TV signals will not be able to decode ATSC 3.0 signals. It is also being developed with a global perspective in mind, meaning that modulations schemes other than 8-VSB – particularly co-orthogonal frequency division multiplexing, or COFDM – will likely be on the table.

In [E206], it is stated that "SHVC is a candidate technology in ATSC and it is expected to be proposed soon to DVB (digital video broadcasting) for inclusion in their codec toolbox." www.atsc.org

[E207] J. Chen et al., "Scalable HEVC (SHVC) Test Model 6 (SHM 6), JCTVC-Q1007_v1, 17th JCT-VC meeting," Valencia, ES, 27 March–4 April 2014.

[E208] L. Shen, Z. Zhang and P. An, "Fast CU size decision and mode decision algorithm for HEVC intra coding," IEEE Trans. on consumer electronics, vol. 59, pp. 207–213, Feb. 2013.

[E209] E. Kalali, Y. Adibelli and I. Hamzaoglu, "A high performance deblocking filter hardware for high efficiency video coding," IEEE Trans. on consumer electronics, vol. 59, pp. 714–720, Aug. 2013.

[E210] R. Garcia and H. Kalva, "Subjective evaluation of HEVC and AVC/H.264 in mobile environments," IEEE Trans. on Consumer Electronics, vol. 60, pp. 116–123, Feb. 2014. See also [E122].

[E211] (ATSC) Washington, D.C. 3D-TV terrestrial broadcasting, part2-SCHC using real time delivery; Doc a/104 2012, Dec 26, 2012.

[E212] S. Wang, D. Zhou and S. Goto "Motion compensation architecture for 8K HEVC decoder," IEEE ICME, Chengdu, China, 14–18 July 2014.

[E213] A. Aminlou et al., "Differential coding using enhanced inter-layer reference picture for the scalable extension of H.265/HEVC video codec," IEEE Trans. CSVT, vol. 24, pp. 1945–1856, Nov. 2014. There are several papers on SHVC in the references.

[E214] W. Hamidouche, M. Raulet and O. Deforges, "Parallel SHVC decoder: Implementation and analysis," IEEE ICME, Chengdu, China, 14–18 July 2014.

[E215] H. Wang, H. Du and J. Wu, "Predicting zero coefficients for high efficiency video coding," IEEE ICME, Chengdu, China, 14–18 July 2014.

[E216] G. Braeckman et al., "Visually lossless screen content coding using HEVC base-layer," IEEE VCIP 2013, pp. 1–6, Kuching, China, 17–20, Nov. 2013.

[E217] W. Zhu et al., "Screen content coding based on HEVC framework," IEEE Trans. Multimedia, vol. 16, pp. 1316–1326 Aug. 2014 (several papers related to MRC) MRC: mixed raster coding.

[E218] ITU-T Q6/16 Visual Coding and ISO/IEC JTC1/SC29/WG11 Coding of Moving Pictures and Audio
Title: Joint Call for Proposals for Coding of Screen Content
Status: Approved by ISO/IEC JTC1/SC29/WG11 and ITU-T SG16 Q6/16 (San Jose, 17 January 2014)
Final draft standard expected: late 2015.

[E219] S. Cho et al., "HEVC hardware decoder implementation for UHD video applications," IEEE ICASSP, Florence, Italy, 4–9 May 2014.

[E220] H. Kim et al., "Multi-core based HEVC hardware decoding system," IEEE ICME workshop 2014, Chengdu, China, 14–18, July 2014.

[E221] C. Yan et al., "Efficient parallel framework for HEVC motion estimation on many core processors," IEEE Trans. CSVT, vol. 24, pp. 2077–2089, Dec. 2014.

[E222] S.-F. Tsai et al., "A 1062 Mpixels/s 8192 × 4320p high efficiency video coding (H.265) encoder chip," 2013 Symposium on VLSIC, pp. 188–189, 2013.

[E223] Y. Fan, Q. Shang and X. Zeng, "In-block prediction-based mixed lossy and lossless reference frame recompression for next-generation video encoding," IEEE Trans. CSVT, vol. 25, pp. 112–124, Jan. 2015.

The following papers are presented in IEEE ICCE, Las Vegas, NV, Jan. 2015. In the ICCE proceedings, each paper is limited to 2 pages. The authors can be contacted by email for detailed papers, if any.

[E224] Real-time Video Coding System for Up to 4K 120P Videos with Spatio-temporal Format Conversion

Toshie Misu (NHK, Japan); Yasutaka Matsuo (Japan Broadcasting Corporation (NHK), Japan); Shunsuke Iwamura and Shinichi Sakaida (NHK, Japan)

A real-time UHDTV video coding system with spatio-temporal format conversion is proposed. The system pre-reduces the UHDTV source resolution to HDTV before HEVC/H.265 encoding. The decoded video is reconstructed to UHDTV in the receiver. We have developed a hardware format converter with super-resolution techniques, and have tested the coding performance. This involves down sampling (decimation), HEVC codec and upsampling (interpolation).

[E225] Rate-Distortion Optimized Composition of HEVC-Encoded Video

Weiwei Chen (Vrije Unniversiteit Brussel, Belgium); Jan Lievens and Adrian Munteanu (Vrije Universiteit Brussel, Belgium); Jrgen Slowack and Steven Delputte (Barco N. V., Belgium)

A rate control scheme is proposed to maximize the encoding quality of a video composition based on HEVC. By solving an R-D optimization problem, this scheme can optimally allocate rate for each video source and maximize the video quality of the composition. The experiments show 14.3% bitrate reduction on average.

[E226] Improvement of 8K UHDTV Picture Quality for H.265/HEVC by Global Zoom Estimation

Ryoki Takada and Shota Orihashi (Waseda University, Japan); Yasutaka Matsuo (Waseda University & Japan Broadcasting Corporation (NHK), Japan); Jiro Katto (Waseda University, Japan)

In this paper, to handle zooming in 8K video sequences, we propose a method for improving the picture quality by global zoom estimation based on motion vector analysis extracted by block matching.

[E227] Improving the Coding Performance of 3D Video Using Guided Depth Filtering

Sebastiaan Van Leuven, Glenn Van Wallendael, Robin Bailleul, Jan De Cock and Rik Van de Walle (Ghent University - iMinds, Belgium)

Autostereoscopic displays visualize 3D scenes based on encoded texture and depth information, which often lack the quality of good depth maps. This paper investigates the coding performance when applying guided depth filtering as a pre-processor. This content adaptive filtering shows gains of 5.7% and 9.3% for 3D-HEVC and MV-HEVC, respectively.

[E228] Watermarking for HEVC/H.265 Stream

Kazuto Ogawa and Go Ohtake (Japan Broadcasting Corporation, Japan)

We propose a watermarking method for HEVC/H.265 video streams that embeds information while encoding the video.

[E229] Video Copy Detection Based on HEVC Intra Coding Features

Kai-Wen Liang, Yi-Ching Chen, Zong-Yi Chen and Pao-Chi Chang (National Central University, Taiwan)

This work utilizes the coding information in HEVC for video copy detection. Both directional modes and residual coefficients of the I-frames are employed as the texture features for matching. These features are robust against different quantization parameters and different frame sizes. The accuracy is comparable with traditional pixel domain approaches.

[E230] Fast Intra Mode Decision Algorithm Based on Local Binary Patterns in High Efficiency Video Coding (HEVC)

Jong-Hyeok Lee (SunMoon University, Korea); Kyung-Soon Jang (Sunnoon University, Korea); Byung-Gyu Kim (SunMoon University, Korea); Seyoon Jeong and Jin Soo Choi (ETRI, Korea)

This paper presents a fast intra mode decision algorithm using the local binary pattern of the texture of an encoded block in HEVC RExt. The Experimental result shows 14.59% of time-saving factor on average with 0.69% of of bit-rate and 0.19 dB of PSNR loss, respectively.

[E231] Fast Thumbnail Extraction Algorithm for HEVC

Wonjin Lee (University of Hanyang, Korea); Gwanggil Jeon (University of Ottawa, Canada); Jechang Jeong (Hanyang University, Korea)

The proposed method only reconstructs 4×4 boundary pixels which are needed for thumbnail image. After partial decoding process, it could be concluded that the proposed method significantly reduces decoding time. In addition, the visual quality of obtained thumbnail images was almost identical to thumbnail images extracted after full decoding process.

[E232] A Probabilistic-Based CU Size Pre-Determination Method for Parallel Processing of HEVC Encoders

Taeyoung Na and Sangkwon Na (Samsung Electronics, Korea); Kiwon Yoo (Samsung Electronics, Korea)

In this paper, CU sizes to be considered for one are pre-determined according to a probabilistic decision model instead of checking all CU size candidates before inter prediction. Thus, motion estimation (ME) with selected CU sizes can be performed in parallel and this is advantageous for H/W encoder design.

[E233] A Multicore DSP HEVC Decoder Using an Actor-Based Dataflow Model

Miguel Chavarrias, Fernando Pescador, Matias J Garrido, Eduardo Juarez and Cesar Sanz (Universidad Politécnica de Madrid, Spain)

The OpenRVC-CAL compiler framework has been used along with the OpenMP-API to implement an HEVC video decoder based on multicore DSP technology. Currently, two DSP-cores have been used, though the technique may be applied to the development of N-core based implementations. The two-core decoder outperforms a single-core implementation by 70%

[E234] A New Mode for Coding Residual in Scalable HEVC (SHVC)

Hamid Reza Tohidypour (University of British Columbia, Canada); Mahsa T Pourazad (TELUS Communications Company, Canada); Panos Nasiopoulos

(University of British Columbia, Canada); Jim Slevinsky (TELUS Communications, Canada)

Recently, an effort on standardizing the scalable extension of High Efficiency Video Coding (HEVC), known as SHVC, has been initiated. To improve the compression performance of SHVC, we propose a new mode for coding residual information, which reduces the bit-rate of the enhancement layer by up to 3.13%.

[E235] Coding Efficiency of the Context-based Arithmetic Coding Engine of AVS 2.0 in the HEVC Encoder

Hyunmin Jung, Soonwoo Choi and Soo-ik Chae (Seoul National University, Korea)

This paper evaluates coding efficiency of the HEVC encoder before and after the CABAC is replaced with the CBAC while the rate errors are controlled by varying the bit depth of the fractional part for estimated bits in rate estimation tables. Experimental result shows the CBAC provides 0.2% BD-rate reduction.

[E236] Machine Learning for Arbitrary Downsizing of Pre-Encoded Video in HEVC

Luong Pham Van, Johan De Praeter, Glenn Van Wallendael, Jan De Cock and Rik Van de Walle (Ghent University - iMinds, Belgium)

This paper proposes a machine learning-based transcoding scheme for arbitrarily downsizing a pre-encoded High Efficiency Video Coding video. Machine learning exploits the correlation between input and output coding information to predict the split-flag of coding units. The experimental results show that the proposed techniques significantly reduce the transcoding complexity.

[E237] Accelerating H.264/HEVC Video Slice Processing Using Application Specific Instruction Set Processor

Dipan Kumar Mandal and Mihir N. Mody (Texas Instruments, India); Mahesh Mehendale (Texas Instruments Inc., India); Naresh Yadav and Chaitanya Ghone (Texas Instruments Ltd., India); Piyali Goswami (Texas Instruments India Pvt. Ltd., India); Hetul Sanghvi (Texas Instruments Inc, India); Niraj Nandan (Texas Instruments, USA)

The paper presents a programmable approach to accelerate slice header decoding in H.264, H.265 using an Application Specific Instruction Set Processor (ASIP). Purpose built instructions accelerate slice processing by 30%–70% for typical to worst case complexity slices. This enables real time universal video decode for all slice-complexity-scenarios without sacrificing programmability.

[E238] Fast Coding Unit Size Selection for HEVC Inter Prediction

Thanuja Mallikarachchi (University of Surrey & University of Surrey, United Kingdom); Anil Fernando (Center for Communications Research. University of Surrey, United Kingdom); Hemantha Kodikara Arachchi (University of Surrey, United Kingdom)

Determining the best partitioning structure for a given CTU is a time consuming operation within the HEVC encoder. This paper presents a fast CU size selection algorithm for HEVC using a CU classification technique. The proposed algorithm achieves an average of 67.83% encoding time efficiency improvement with negligible rate-distortion impact.

[E239] S.-J. You, S.-J. Chang and T.-S. Chang, "Fast motion estimation algorithm and design for real time QFHD high efficiency video coding," IEEE Trans. CSVT, vol. 25, pp. 1533–1544, Sept. 2015. Also many good references are listed at the end.

[E240] W. Zhao, T. Onoye and T. Song, "Hierarchical structure based fast mode decision for H.265/HEVC," IEEE Trans. CSVT, vol. 25, pp. 1651–1664, Oct. 2015. Several valuable references related to fast mode decision approaches resulting in reduced HEVC encoder complexity are cited at the end.

[E241] Y. Umezaki and S. Goto, "Image segmentation approach for realizing zoomable streaming HEVC video," ICICS 2013.

[E242] G. Correa et al., "Fast HEVC encoding decisions using data mining," IEEE Trans. CSVT, vol. 25, pp. 660–673, April 2015.

[E243] Access to HM 16.9 Software Manual: https://hevc.hhi.fraunhofer.de/svn/svn_HEVCSoftware/tags/HM-16.9+SCM-4.0rc1/doc/software-manual.pdf

[E244] HM Encoder Description: http://mpeg.chiariglione.org/standard/mpeg-h/high-efficiency-video-coding/n14703-high-efficiency-video-coding-hevc-encoder

[E245] T. Nguyen and D. Marpe, "Objective Performance Evaluation of the HEVC Main Still Picture Profile," IEEE Trans. CSVT, vol. 25, pp. 790–797, May 2015.

[E246] C.C. Chi et al., "SIMD acceleration for HEVC decoding," IEEE Trans. CSVT, vol. 25, pp. 841–855, May 2015.

[E247] B. Min and C.C. Cheung, "A fast CU size decision algorithm for the HEVC intra encoder," IEEE Trans. CSVT, vol. 25, pp. 892–896, May 2015.

[E248] H.-S. Kim and R.-H. Park, "Fast CU partitioning algorithm for HEVC using online learning based Bayesian decision rule," IEEE Trans. CSVT, (Early access).

[E249] Y.- H. Chen and V. Sze, "A deeply pipelined CABAC decoder for HEVC supporting level 6.2 high-tier applications," IEEE Trans. CSVT, vol. 25, pp. 856–868, May 2015.

[E250] D. Zhou et al., "Ultra-high-throughput VLSI architecture of H.265/HEVC CABAC encoder for UHDTV applications," IEEE Trans. CSVT, vol. 25, pp. 497–507, March 2015.

[E251] D. Grois, B. Bross and D. Marpe, "HEVC/H.265 Video Coding Standard (Version 2) including the Range Extensions, Scalable Extensions, and Multiview Extensions," (Tutorial) Monday 29 June 2015 11.30am–3:00pm , IEEE ICME 2015, Torino, Italy, 29 June–3 July, 2015.

[E252] L. Hu et al., "Hardware – oriented rate distortion optimization algorithm for HEVC Intra – frame encoder," IEEE International Workshop on multimedia computing, in conjunction with IEEE ICME 2015, Torino, Italy, 29 June–3 July, 2015. (Same as [[E258]])

[E253] I. Hautala et al., "Programmable low-power multicore coprocessor architecture for HEVC/H.265 in-loop filtering," IEEE Trans. CSVT, vol. 25, pp. 1217–1230, July 2015.

[E254] J. KIM, S.-Ho and M. Kim, "An HEVC complaint perceptual video coding scheme based on JND models for variable block-sized transform kernels," IEEE Trans. CSVT, (early access).

[E255] F. Zou et al., "Hash Based Intra String Copy for HEVC Based Screen Content Coding, "IEEE ICME (workshop), Torino, Italy, June 2015. (In references, there are several papers on Screen Content Coding).

[E256] HEVC SCC extension reference software: https://hevc.hhi.fraunhofer.de/svn/svn_HEVCSoftware/tags/HM-14.0+RExt-7.0+SCM-1.0/

[E257] C. Chen et al., "A New Block- Based Coding Method for HEVC Intra Coding," IEEE ICME (Workshop), Torino, Italy, June 2015. Also published in IEEE CSVT Trans. vol. 27, pp. 1727–1736, Aug. 2017.

[E258] L. Hu et al., "Hardware-Oriented Rate-Distortion Optimization Algorithm for HEVC Intra-Frame Encoder," IEEE ICME, Torino, Italy, June 2015. (Same as [E252])

[E259] Y. Zhang et al., "Machine Learning-Based Coding Unit Depth Decisions for Flexible Complexity Allocation in High Efficiency Video Coding," IEEE Trans. on Image Processing, Vol. 24, No. 7, pp. 2225–2238, July 2015.

[E260] D. Grois, B. Bross and D. Marpe, "HEVC/H.265 Video Coding Standard including the Range Extensions, Scalable Extensions, and Multiview Extensions," (Tutorial), IEEE ICCE , Berlin, Germany, 6–9 Sept. 2015.

[E261] S. J. Kaware and S. K. Jagtap, "A Survey: Heterogeneous Video Transcoder for H.264/AVC to HEVC," IEEE International Conference on Pervasive Computing (ICPC), pp. 1–3, Pune, India, 8–10 Jan. 2015.

[E262] Several papers related to HDR, HEVC, Beyond HEVC and SCC in "Applications of Digital Image Processing XXXVIII," HDR I session, HDR II, Applications and Extensions for High Efficiency Video Coding (HEVC), vol. 9599, SPIE. Optics + photonics, San Diego, California, USA, 9–13, Aug. 2015. Website : www.spie.org/op

[E263] E. Peixoto et al., "Fast H.264/AVC to HEVC Transcoding based on Machine Learning," IEEE International Telecommunications Symposium (ITS), pp. 1–4, 2014.

[E264] J. M. Boyce et al., "Overview of SHVC: Scalable Extensions of the High Efficiency Video Coding (HEVC) Standard," IEEE Trans. on CSVT. vol. 26, pp. 20–34, Jan. 2016.

[E265] X265: Documentations on wavefront parallel Processing. [Online] Available: http://x265.readthedocs.org/en/latest/threading.html#wavefront-parallel-processing, accessed Aug. 1, 2015

[E266] J. Kim, S.-H. Bae and M. Kim, "An HEVC – Compliant Perceptual Video Coding Scheme based on JND Models for Variable Block-sized Transform Kernels," IEEE Trans. on CSVT. EA.

[E267] Y. Chen et al., "Efficient Software H.264/AVC to HEVC Transcoding on Distributed Multicore Processors," IEEE Trans. on CSVT, vol. 25, pp. 1423–1434, Aug. 2015.

[E268] H. Lee et al., "Early Skip Mode Decision for HEVC Encoder with Emphasis on Coding Quality," IEEE Trans. on Broadcasting. EA.

[E269] K. Lim et al., "Fast PU skip and split termination algorithm for HEVC Intra Prediction," IEEE Trans. on CSVT, vol. 25, No. 8, pp. 1335–1346, Aug. 2015.

[E270] K. Won and B. Jeon, "Complexity – Efficient Rate Estimation for Mode Decision of the HEVC Encoder," IEEE Trans. on Broadcasting. EA.

[E271] L. F. R. Lucas et al., "Intra predictive Depth Map Coding using Flexible Block partitioning," IEEE Trans. on Image Processing. EA

[E272] K. Chen et al., "A Novel Wavefront – Based High parallel Solution for HEVC Encoding," IEEE Trans. on CSVT. EA

[E273] J. Zhang. B. Li and H. Li, "An Efficient Fast Mode Decision Method for Inter Prediction in HEVC," IEEE Trans. on CSVT, vol. 26, pp. 1502–1515, Aug. 2016.

[E274] N. Hu and E.-H. Yang, "Fast Mode Selection for HEVC intra frame coding with entropy coding refinement based on transparent composite model," IEEE Trans. on CSVT, vol. 25, No.9, pp. 1521–1532, Sept. 2015. ERRATUM: (early access)

[E275] S.-Y. Jou, S.-J. Chang and T.-S. Chang, "Fast Motion Estimation Algorithm and Design For Real Time QFHD High Efficiency Video Coding," IEEE Trans. on CSVT, vol. 25, pp. 1533–1544, Sept. 2015. The papers [E276] through [E284] were presented in IEEE ICCE Berlin, 6–9 Sept. 2015 and can be accessed from IEEEXPLORE.

[E276] N. Dhollande et al., "Fast block partitioning method in HEVC Intra coding for UHD video," IEEE ICCE , Berlin, Germany, 6–9 Sept. 2015.

[E277] W. Shi et al., "Segmental Downsampling Intra Coding Based on Spatial Locality for HEVC," IEEE ICCE, Berlin, Germany, 6–9 Sept. 2015.

[E278] S.–H. Park et al., "Temporal Correlation-based Fast Encoding Algorithm in HEVC Intra Frame Coding," IEEE ICCE, Berlin, Germany, 6–9 Sept. 2015.

[E279] R. Skupin, Y. Sanchez and T. Schirel, "Compressed Domain Processing for Stereoscopic Tile Based Panorama Streaming using MV-HEVC," IEEE ICCE, Berlin, Germany, 6–9 Sept. 2015.

[E280] A. S. Nagaraghatta et al., "Fast H.264/AVC to HEVC transcoding using Mode Merging and Mode Mapping," IEEE ICCE, Berlin, Germany, 6–9 Sept. 2015.

[E281] A. Mikityuk et al., "Compositing without Transcoding for H.265/HEVC in Cloud IPTV and VoD services," IEEE ICCE, Berlin, Germany, 6–9 Sept. 2015.

[E282] F. Chi et al., "Objective identification of most distorted coding units over UHD HEVC sequences," IEEE ICCE, Berlin, Germany, 6–9 Sept. 2015.

[E283] M. Tomida and T. Song, "Small area VLSI architecture for deblocking filter of HEVC," IEEE ICCE, Berlin, Germany, 6–9 Sept. 2015.

[E284] T. Biatek et al., "Toward Optimal Bitrate Allocation in the Scalable HEVC Extension: Application to UHDTV," IEEE ICCE, Berlin, Germany, 6–9 Sept. 2015. References [E285] through [E319], [E323–E324] are presented in IEEE ICIP 2015, Quebec City, Canada

[E285] J. Mir et al., "Rate Distortion Analysis of High Dynamic Range Video Coding Techniques," IEEE ICIP, Quebec City, Canada, 27–30 Sept. 2015.

[E286] S. G. Blasi et al., "Context adaptive mode sorting for fast HEVC mode decision," IEEE ICIP, Quebec City, Canada, 27–30 Sept. 2015.

[E287] N. Hu and E.-H. Yang, "Fast inter mode decision for HEVC based on transparent composite model," IEEE ICIP, Quebec City, Canada, 27–30 Sept. 2015.

[E288] T. Laude and J. Ostermann, "Copy mode for static screen content coding with HEVC," IEEE ICIP, Quebec City, Canada, 27–30 Sept. 2015.

[E289] A. Eldeken et al., "High throughput parallel scheme for HEVC deblocking filter," IEEE ICIP, Quebec City, Canada, 27–30 Sept. 2015.

[E290] X. Shang et al.," Fast cu size decision and pu mode decision algorithm in HEVC intra coding," IEEE ICIP, Quebec City, Canada, 27–30 Sept. 2015.

[E291] Y. Chen and H. Liu, "Texture based sub-pu motion inheritance for depth coding," IEEE ICIP, Quebec City, Canada, 27–30 Sept. 2015.

[E292] Y.-C. Sun et al., "Palette mode - a new coding tool in screen content coding extensions of HEVC," IEEE ICIP, Quebec city, Canada, 27–30 Sept. 2015.

[E293] D. Grois, B. Bross and D. Marpe, "HEVC/H.265 Video Coding Standard (Version 2) including the Range Extensions, Scalable Extensions, and Multiview Extensions," (Tutorial) Sunday 27 Sept 2015, 9:00 am to 12:30 pm, IEEE ICIP, Quebec city, Canada, 27–30 Sept. 2015.

[E294] D. Schroeder et al., "Block structure reuse for multi-rate high efficiency video coding," IEEE ICIP, Quebec city, Canada, 27–30 Sept. 2015.

[E295] Y. Yan, M. M. Hannuksela and H. Li, "Seamless switching of H.265/HEVC-coded dash representations with open GOP prediction structure," IEEE ICIP, Quebec city, Canada, 27–30 Sept. 2015.

[E296] A. Sohaib and A. R. Kelly, "Radiometric calibration for HDR imaging," IEEE ICIP, Quebec city, Canada, 27–30 Sept. 2015.

[E297] J. Liu et al., "Chromatic calibration of an HDR display using 3d octree forests," IEEE ICIP, Quebec city, Canada, 27–30 Sept. 2015.

[E298] S. Patel, D. Androutsos and M. Kyan, "Adaptive exposure fusion for high dynamic range imaging," IEEE ICIP, Quebec city, Canada, 27–30 Sept. 2015.

[E299] Y. Zang, D. Agrafiotis and D. Bull, "High dynamic range content calibration for accurate acquisition and display," IEEE ICIP, Quebec city, Canada, 27–30 Sept. 2015.

[E300] J.-F. Franche and S. Coulombe, "Fast H.264 to HEVC transcoder based on post-order traversal of quadtree structure," IEEE ICIP, Quebec city, Canada, 27–30 Sept. 2015.

[E301] A. Arrufat, P. Philippe and O. Deforges, "Mode-dependent transform competition for HEVC," IEEE ICIP, Quebec city, Canada, 27–30 Sept. 2015.

[E302] W.-K. Cham and Q. Han, "High performance loop filter for HEVC," IEEE ICIP, Quebec city, Canada, 27–30 Sept. 2015.

[E303] A. Arrufat, A.-F. Perrin and P. Philippe, "Image coding with incomplete transform competition for HEVC," IEEE ICIP, Quebec city, Canada, 27–30 Sept. 2015.

[E304] Y. Liu and J. Ostermann, "Fast motion blur compensation in HEVC using fixed-length filter," IEEE ICIP, Quebec city, Canada, 27–30 Sept. 2015.

[E305] L. Yu and J. Xiao, "Statistical approach for motion estimation skipping (SAMEK)," IEEE ICIP, Quebec city, Canada, 27–30 Sept. 2015.

[E306] H.-B. Zhang et al., "Efficient depth intra mode decision by reference pixels classification in 3D-HEVC," IEEE ICIP, Quebec city, Canada, 27–30 Sept. 2015.

[E307] M. Joachimiak et al., "Upsampled-view distortion optimization for mixed resolution 3D video coding," IEEE ICIP, Quebec city, Canada, 27–30 Sept. 2015.

[E308] X. Yu et al., "VLSI friendly fast cu/pu mode decision for HEVC intra encoding: leveraging convolution neural network," IEEE ICIP, Quebec city, Canada, 27–30 Sept. 2015.

[E309] M. Hannuksela et al., "Overview of the multiview high efficiency video coding (MV-HEVC) standard," IEEE ICIP, Quebec city, Canada, 27–30 Sept. 2015.

[E310] W.-N. Lie and Y.-H. Lu, "Fast encoding of 3D color-plus-depth video based on 3D-HEVC," IEEE ICIP, Quebec city, Canada, 27–30 Sept. 2015.

[E311] M. Meddeb, M. Cagnazzo and B.-P. Popescu "ROI-based rate control using tiles for an HEVC encoded video stream over a lossy network," IEEE ICIP, Quebec city, Canada, 27–30 Sept. 2015.

[E312] W. Gueder, P. Amon and E. Steinbach, "Low-complexity block size decision for HEVC intra coding using binary image feature descriptors," IEEE ICIP, Quebec city, Canada, 27–30 Sept. 2015.

[E313] L. Du and Z. Liu, "3.975MW 18.396GBPS 2R2W SRAM for SBAC context model of HEVC," IEEE ICIP, Quebec city, Canada, 27–30 Sept. 2015.

[E314] M. Pettersson, R. Sjoberg and J. Samuelsson, "Dependent random access point pictures in HEVC," IEEE ICIP, Quebec city, Canada, 27–30 Sept. 2015.

[E315] V. Sanchez, "Lossless screen content coding in HEVC based on sample-wise median and edge prediction," IEEE ICIP, Quebec city, Canada, 27–30 Sept. 2015.

[E316] H. Maich et al., "A multi-standard interpolation hardware solution for H.264 and HEVC," IEEE ICIP, Quebec city, Canada, 27–30 Sept. 2015.

[E317] Y. Wang, F. Duanmu and Z. Ma, "Fast CU partition decision using machine learning for screen content compression," IEEE ICIP, Quebec City, Canada, 27–30 Sept. 2015.

[E318] M. Wang and K. N. Ngan, "Optimal bit allocation in HEVC for real-time video communications," IEEE ICIP, Quebec City, Canada, 27–30 Sept. 2015.

[E319] X. lian et al., "Pixel-grain prediction and k-order ueg-rice entropy coding oriented lossless frame memory compression for motion estimation in HEVC," IEEE ICIP, Quebec City, Canada, 27–30 Sept. 2015.

[E320] Y. Shishikui, "The coming of age of 8K super hi-vision (ultra high definition television) – The meaning of the ultimate," IEEE PCS 2015, Cairns, Australia, June 2015.

[E321] D. Flynn et al., "Overview of the Range Extensions for the HEVC Standard: Tools, Profiles and Performance," IEEE Trans. CSVT, vol. 26, pp. 4–19, Jan. 2016.

[E322] J. Xu, R. Joshi and R. A. Cohen, "Overview of the Emerging HEVC Screen Content Coding Extension," IEEE Trans. CSVT, vol. 26, pp. 50–62, Jan. 2016.

[E323] Y. Yan, M. M. Hannuksela and H. Li, "Seamless switching of H.265/HEVC-coded dash representations with open GOP prediction structure," IEEE ICIP, Quebec City, Canada, 27–30 Sept. 2015.

[E324] D. Grois, B. Bross and D. Marpe, "HEVC/H.265 Video Coding Standard (Version 2) including the Range Extensions, Scalable Extensions, and Multiview Extensions," (Tutorial) Sunday 27 Sept. 2015, 9:00 am to 12:30 pm, IEEE ICIP, Quebec city, Canada, 27–30 Sept. 2015.

[E325] J.M. Boyce et al., "Overview of SHVC: Scalable extensions of the high efficiency video coding (HEVC) standard," IEEE Trans. CSVT. vol. 26, pp. 20–34, Jan. 2016.

[E326] G. Tech et al., "Overview of the Multiview and 3D extensions of high efficiency video coding," IEEE Trans. CSVT. vol. 26, pp. 35–49, Jan. 2016.

[E327] M. M. Hannuksela, J. Lainema and V. K. M. Vadakital, "The high efficiency image file format standards," IEEE Signal Processing Magazine, vol. 32, pp. 150–156, IEEE Trans. CSVT. Vol. (Early Access)

[E328] C. Li, H. Xiong and D. Wu, "Delay–Rate–Distortion Optimized Rate Control for End-to-End Video Communication Over Wireless Channels," IEEE Trans. CSVT, vol. 25, no. 10, pp. 1665–1681, Oct 2015.

[E329] W. Hamidouche, M. Raulet and O. Deforges, "4K Real-Time and Parallel Software Video Decoder for Multi-layer HEVC Extensions," IEEE Trans. CSVT. vol. 26, pp. 169–180, Jan. 2016.

[E330] H.R. Tohidpour, M.T. Pourazad and P. Nasiopoulos, "Online Learning-based Complexity Reduction Scheme for 3D-HEVC," IEEE Trans. CSVT. EA.

[E331] X. Deng et al.," Subjective-driven Complexity Control Approach for HEVC," IEEE Trans. CSVT. (Early access).

[E332] S.-H. Jung and H. W. Park, "A fast mode decision method in HEVC using adaptive ordering of modes," IEEE Trans. CSVT. Vol. 26, pp. 1846–1858, Oct. 2016.

[E333] A. J. D. Honrubia et al., "Adaptive Fast Quadtree Level Decision Algorithm for H.264 to HEVC Video Transcoding," IEEE Trans. CSVT. vol. 26, pp. 154–168, Jan. 2016.

[E334] K. L. Chung et al., "Novel Bitrate-Saving and Fast Coding for Depth Videos in 3D-HEVC," IEEE Trans. CSVT. EA.

[E335] C.-H. Lin, K.-L. Chung and C.-W. Yu, "Novel Chroma Subsampling Strategy based on Mathematical Optimization for Compressing Mosaic Videos with Arbitrary RGB Color Filter Arrays in H.264/AVC and HEVC," IEEE Trans. CSVT. (Early access).

[E336] G. Correa et al., "Pareto-Based Method for High Efficiency Video Coding with Limited Encoding Time," IEEE Trans. CSVT. EA.

[E337] T. Vermeir et al., "Guided Chroma Reconstruction for Screen Content Coding," IEEE Trans. CSVT. EA.

[E338] M. Abeydeera et al., "4K Real Time HEVC Decoder on FPGA," IEEE Trans. CSVT. vol. 26, pp. 236–249, Jan. 2016.

[E339] E. Francois et al., "High Dynamic Range and Wide Color Gamut Video Coding in HEVC: Status and Potential Future Enhancements," IEEE Trans. CSVT. vol. 26, pp. 63–75, Jan. 2016.

[E340] H. Lee et al., "Fast Quantization Method with Simplified Rate-Distortion Optimized Quantization for HEVC Encoder," IEEE Trans. CSVT. vol. 26, pp. 107–116, Jan. 2016.

[E341] T.-Y. Huang and H. H. Chen, "Efficient Quantization Based on Rate-Distortion Optimization for Video Coding," IEEE Trans. CSVT. EA

[E342] G. Pastuszak and A. Abramowski, "Algorithm and Architecture Design of the H.265/HEVC Intra Encoder," IEEE Trans. CSVT. vol. 26, pp. 210–222, Jan. 2016.

[E343] M. Kim et al., "Exploiting Thread-Level Parallelism on HEVC by Employing Reference Dependency Graph," IEEE Trans. CSVT. (Early access).

[E344] A. Aminlou et al., "A New R-D Optimization Criterion for Fast Mode Decision Algorithms in Video Coding and Transrating," IEEE Trans. CSVT. vol. 26, pp. 696–710, April 2016.

[E345] P.-T. Chiang et al., "A QFHD 30 fps HEVC Decoder Design," IEEE Trans. CSVT, vol. 26, pp. 724–735, April 2016.

[E346] W. Xiao et al., "HEVC Encoding Optimization Using Multicore CPUs and GPUs," IEEE Trans. CSVT. EA.

[E347] HEVC Demystified: A primer on the H.265 Video Codec – https://www.elementaltechnologies.com/resources/white-papers/hevc-h265-demystified-primer

[E348] N.S. Shah, "Reducing Encoder Complexity of Intra – Mode Decision Using CU Early Termination Algorithm," M.S. Thesis, EE Dept., University of Texas at Arlington, Arlington, Texas, Dec. 2015. http://www.uta.edu/faculty/krrao/dip click on courses and then click on EE5359 Scroll down and go to Thesis/Project Title and click on N.S. Shah.

[E349] D, Hingole, "H.265 (HEVC) bitstream to H.264 (MPEG 4 AVC) bitstream transcoder" M.S. Thesis, EE Dept., University of Texas at Arlington, Arlington, Texas, Dec. 2015. http://www.uta.edu/faculty/krrao/dip click on courses and then click on EE5359 Scroll down and go to Thesis/Project Title and click on D. Hingole.

[E350] U.S.M. Dayananda, "Investigation of Scalable HEVC & its bitrate allocation for UHD deployment in the context of HTTP streaming," M.S. Thesis, EE Dept., University of Texas at Arlington, Arlington, Texas, Dec. 2015. http://www.uta.edu/faculty/krrao/dip click on courses and then click on EE5359 Scroll down and go to Thesis/Project Title and click on U.S.M. Dayananda.

[E351] V. Vijayaraghavan, "Reducing the Encoding Time of Motion Estimation in HEVC using parallel programming," M.S. Thesis, EE Dept., University of Texas at Arlington, Arlington, Texas, Dec. 2015. http://www.uta.edu/faculty/krrao/dip click on courses and then click on EE5359 Scroll down and go to Thesis/Project Title and click on V. Vijayaraghavan

[E352] S. Kodpadi, "Evaluation of coding tools for screen content in High Efficiency Video Coding," M.S. Thesis, EE Dept., University of Texas at Arlington, Arlington, Texas, Dec. 2015.

http://www.uta.edu/faculty/krrao/dip click on courses and then click on EE5359 Scroll down and go to Thesis/Project Title and click on S. Kodpadi.

[E353] N.N. Mundgemane, *"Multi-stage prediction scheme for Screen Content based on HEVC,"* M.S. Thesis, EE Dept., University of Texas at Arlington, Arlington, Texas, Dec. 2015.

http://www.uta.edu/faculty/krrao/dip click on courses and then click on EE5359 Scroll down and go to Thesis/Project Title and click on N. Mundgemane.

[E354] E. de la Torre, R.R.Sanchez and J.L. Martinez, "Fast video transcoding from HEVC to VP9," IEEE Trans. Consumer Electronics, vol. 61, pp. 336–343, Aug. 2015.

[E355] W.-S. Kim et al., "Cross-component prediction in HEVC," IEEE Trans. CSVT, vol. 27, EA.

(Related papers: T. Nguyen et al., **"Extended Cross-Component Prediction in HEVC"**, *Proc. Picture Coding Symposium (PCS 2015)*, Cairns, Australia, May 31–Jun 03, 2015.

A. Khairatet et al., '**Adaptive Cross-Component Prediction for 4:4:4 High Efficiency Video Coding**', *Proc. IEEE International Conference on Image Processing (ICIP 2014)*, Paris, Oct. 27–30, 2015. Conclusion section in [E355] is reproduced below for ready reference.

Conclusion

"Coding in the RGB colour space provides better colour fidelity compared to the YCbCr coding, while YCbCr coding can achieve better coding efficiency. In this paper, CCP is proposed where the correlation among colour components are efficiently removed by applying prediction between luma and chroma residual signals using the linear regression. Experimental results show that the proposed method achieves great coding efficiency improvements for RGB coding and maintains colour fidelity. CCP also provides good coding gain when applied to YCbCr coding, as the proposed scheme can adapt to the given colour space through the signalling of the scale factor. The computational and implementation complexities are also analyzed to show that the proposed CCP can be implemented and used in real-time processing systems. In the future, more study can be made on how to facilitate software and hardware implementation. We also leave it as a further study how to efficiently apply CCP to various applications including high dynamic range video coding."

[E356] C.-K. Fong, Q. Han and W.-K. Cham, "Recursive integer cosine transform for HEVC and future video coding standards," IEEE Trans. CSVT, vol. 27, pp. 326–336, Feb. 2017.

[E357] P. Hsu and C. Shen, "The VLSI architecture of a highly efficient deblocking filter for HEVC systems," IEEE Trans. CSVT, vol. 26, Jan 2016. [early access] (Several references related to deblocking and SAO filters in HEVC and deblocking filters in H.264/AVC are listed at the end.)

242 Appendix

[E358] H. Lee et al., "Fast quantization method with simplified rate-distortion optimized quantization for an HEVC encoder" IEEE Trans. CSVT, vol. 26, pp. 107–116, Jan. 2016.

[E359] S. Wang et al., "VLSI implementation of HEVC motion compensation with distance biased direct cache mapping for 8K UHDTV applications," IEEE Trans. CSVT, (early access)

[E360] G. Georgios, G. Lentaris and D. Reisis, "Reduced complexity super resolution for low-bitrate compression," TEEE Trans. CSVT, vol. 26, pp. 342–345, Feb. 2016. This paper can lead to several valuable and challenging projects.

[E361] C.-H. Kuo, Y.-L. Shih and S.-C. Yang, "Rate control via adjustment of Lagrangian multiplier for video coding," IEEE Trans. CSVT, vol. 26, pp. 2069–2078, Nov. 2016.

[E362] J. Zhang, B. Li and H. Li, "An efficient fast mode decision method for inter prediction in HEVC," IEEE Trans. CSVT, vol. 26, pp. 1502–1515, Aug.2016.

[E363] C.-P. Fan, C.-W. Chang and S.-J. Hsu, "Cost effective hardware sharing design of fast algorithm based multiple forward and inverse transforms for H.264/AVC, MPEG-1/2/4, AVS and VC-1 video encoding and decoding applications," IEEE Trans. CSVT, vol. 26, (early access)

[E364] J.-S. Park et al., "2-D large inverse transform (16×16, 32×32) for HEVC (high efficiency video coding)," J. of Semiconductor Technology and Science, vol. 12, pp. 203–211, June 2012.

[E365] A. Ilic, et al., "Adaptive scheduling framework for real-time video encoding on heterogeneous systems." IEEE Trans. CSVT, vol. 26, (early access).

[E366] IEEE DCC , Snow Bird, Utah, 29 March–1 April 2016 Panel Presentation

Video Coding: Recent Developments for HEVC and Future Trends

Abstract

This special event at DCC 2016 consists of a keynote talk by Gary Sullivan (co-chair of the MPEG & VCEG Joint Collaborative Team on Video Coding) followed by a panel discussion with key members of the video coding and standardization community. Highlights of the presentation include HEVC

Screen Content Coding (SCC), High Dynamic Range (HDR) video coding, the Joint Exploration Model (JEM) for advances in video compression beyond HEVC, and recent initiatives in royalty-free video coding.

Panel Members:

Anne Aaron
Manager, Video Algorithms - Netflix

Arild Fuldseth
Principal Engineer, Video Coding - Cisco Systems

Marta Karczewicz
VP, Technology, Video R&D and Standards - Qualcomm

Jörn Ostermann
Professor - Leibniz Universität Hannover Institute for Information Processing

Jacob Ström
Principal Researcher - Ericsson

Gary Sullivan
Video Architect - Microsoft

Yan Ye
Senior Manager, Video Standards – InterDigital

Slides can be accessed from:

http://www.cs.brandeis.edu/~dcc/Programs/Program2016Keynoteslides-Sullivan.pdf

[E367] M. Jridi and P.K. Meher, "A scalable approximate DCT architectures for efficient HEVC compliant video coding" IEEE Trans. CSVT, vol. 27, pp. 1815–1825, Aug. 2017. See also P.K. Meher et al., "Generalized architecture for integer DCT of lengths N = 8, 16 and 32," IEEE Trans. CSVT, vol. 24, pp. 168–178, Jan. 2014.

[E368] T. Zhang et al., "Fast intra mode and CU size decision for HEVC," IEEE Trans. CSVT, vol. 27, pp. 1714–1726, Aug. 2017.

[E369] X. Liu et al., "An adaptive mode decision algorithm based on video texture characteristics for HEVC intra prediction," IEEE Trans. CSVT, vol. 27, pp. 1737–1748, Aug. 2017.

Abstract is reproduced here.

Abstract

The latest High Efficiency Video Coding (HEVC) standard could achieve the highest coding efficiency compared with the existing video coding standards. To improve the coding efficiency of Intra frame, a quadtree-based variable block size coding structure which is flexible to adapt to various texture characteristics of images and up to 35 Intra prediction modes for each prediction unit (PU) is adopted in HEVC. However, the computational complexity is increased dramatically because all the possible combinations of the mode candidates are calculated in order to find the optimal rate distortion (RD) cost by using Lagrange multiplier. To alleviate the encoder computational load, this paper proposes an adaptive mode decision algorithm based on texture complexity and direction for HEVC Intra prediction. Firstly, an adaptive Coding Unit (CU) selection algorithm according to each depth levels' texture complexity is presented to filter out unnecessary coding block. And then, the original redundant mode candidates for each PU are reduced according to its texture direction. The simulation results show that the proposed algorithm could reduce around 56% encoding time in average while maintaining the encoding performance efficiently with only 1.0% increase in BD-rate compared to the test mode HM16 of HEVC.

There are also several references related to intra coding both in HEVC and H.264/AVC.

[E370] Z. Pan et al., "Early termination for TZSearch in HEVC motion estimation," IEEE ICASSP 2013, pp. 1389–1392, June 2013.

[E371] R. Shivananda, "Early termination for TZSearch in HEVC motion estimation," project report, Spring 2016. http://www.uta.edu/faculty/krrao/dip click on courses and then click on EE5359 Scroll down and go to projects Spring 2016 and click on R. Shivananda final report. See[E370]. Following this technique, Shivananda is able to reduce HEVC encoding time by 38% with negligible loss in RD performance for LD, RA and RA early profiles. See Tables 3 through 14 for the results.

[E372] S.K. Rao, "Performance comparison of HEVC intra, JPEG, JPEG 2000, JPEG XR, JPEG LS and VP9 intra," project report, Spring 2016. http://www.uta.edu/faculty/krrao/dip click on courses and then click on EE5359 Scroll down and go to projects Spring 2016 and click on Swaroop Krishna Rao final report.

[E373] H.N. Jagadeesh, "Sample Adaptive Offset in the HEVC standard," project report, Spring 2016. http://www.uta.edu/faculty/krrao/dip

click on courses and then click on EE5359 Scroll down and go to projects Spring 2016 and click on Jagadeesh final report. He has implemented HM software using HEVC test sequences with and without the SAO filter. In general SAO filter improves the subjective and objective qualities of the video output at the cost of slight increase in implementation complexity.

[E374] N. Thakur, "Fast Intra Coding Based on Reference Samples Similarity in HEVC," project report Spring 2016. http://www.uta.edu/faculty/krrao/dip click on courses and then click on EE5359 Scroll down and go to projects Spring 2016 and click on Nikita Thakur final report.

[E375] M.N. Sheelvant, "Performance and Computational Complexity Assessment of High-Efficiency Video Encoders," project report Spring 2016. http://www.uta.edu/faculty/krrao/dip click on courses and then click on EE5359 Scroll down and go to projects Spring 2016 and click on M.N. Sheelvant final report.

[E376] Y. Tew, K.S. Wong and R.C.-W. Phan, "Region of interest encryption in high efficiency video coding compressed video," IEEE- ICCE Taiwan, May 2016.

[E377] S.-H. Bae, J. Kim and M. Kim, "HEVC-Based perceptually adaptive video coding using a DCT-Based local distortion detection probability," IEEE Trans. Image Processing, vol. 25, pp. 3343–3357, July 2016.

[E378] C. Diniz et al., "A deblocking filter hardware architecture for the high efficiency video coding standard," Design, Automation and Test in Europe conference and exhibition (DATE), 2015.

[E379] D.S. Pagala, "Multiplex/demultiplex of HEVC video with HE-AAC v2 audio and achieve lip synch.," M.S. Thesis, EE Dept., University of Texas at Arlington, Arlington, Texas, Aug. 2016. http://www.uta.edu/faculty/krrao/dip click on courses and then click on EE5359 Scroll down and go to Thesis/Project Title and click on D.S. Pagala. See (arib std-b32 version 3.9-E1, "Video coding, Audio coding and Multiplexing Specifications for Digital Broadcasting," 2016.)

[E380] Sessions related to HEVC in IEEE ICIP, Phoenix, Arizona, Sept. 2016:

MA-L3: SS: Image compression grand challenge,
Several sessions on Image quality assessment
MPA.P3: Image and Video Coding.

MPA.P8: Transform Coding.
TA-L1: HEVC Intra Coding.
TP-L1: HEVC optimization
WPB-P2 HEVC processing and coding
Panel Session: Is compression dead or are we wrong again?

[E381] M. Masera, M. Martina and G. Masera, "Adaptive approximated DCT architectures for HEVC," IEEE Trans. CSVT, (Early access) This has several references related to integer DCT architectures.

[E382] B. Min, Z. Xu and C.C. Cheung, "A fully pipelined hardware architecture for intra prediction of HEVC," IEEE Trans. CSVT (early access). (Same as [E397]) The fully pipelined design for intra prediction in HEVC can produce 4 pels per clock cycle. Hence the throughput of the proposed architecture is capable of supporting 3840x2160 videos at 30 fps.

[E383] G. Correa et al., "Pareto-based method for high efficiency video coding with limited encoding time," IEEE Trans. CSVT, vol. 26, pp. 1734–1745, Sept. 2016.

[E384] C.-H. Liu, K.-L. Chung and C.-YW. Yu, "Novel Chroma Sub-sampling Strategy Based on Mathematical Optimization for Compressing Mosaic Videos with Arbitrary RGB Color Filter Arrays in H.264/AVC and HEVC," IEEE Trans. CSVT, vol. 26, pp. 1722–1733, Sept. 2016.

[E385] C. Herglotz et al., "Modeling the energy consumption of HEVC decoding process," IEEE Trans. CSVT, vol. 28, pp. 217–229, Jan. 2018.

[E386] IEEE DATA COMPRESSION CONFERENCE (DCC) Snowbird, Utah, April 4–7, 2017. http://www.cs.brandeis.edu/~dcc

Keynote address

"Video Quality Metrics," S. Daly, Senior Member Technical Staff, Dolby Laboratories

Special sessions:

"Video coding," G. Sullivan and Y. Ye, Session Chairs

"Genome Compression," J. Wen and K. Sayood, Session Chairs

(note there is Session 10 on Genome Compression)

"Quality Metrics and Perceptual Compression" Y. Reznik and T. Richter, Session Co-Chairs

Web site below.

http://www.cs.brandeis.edu/~dcc/Programs/Program2016KeynoteSlides-Sullivan.pdf

There are also several papers related to HEVC.

[E387] H. Siddiqui, K. Atanassov and S. Goma, "Hardware-friendly universal demosaick using non-iterative MAP reconstruction," IEEE ICIP 2016 (MA.P.3.4), Phoenix, AZ, Sept. 2016.

[E388] J. Seok et al., "Fast Prediction Mode Decision in HEVC Using a Pseudo Rate-Distortion Based on Separated Encoding Structure", ETRI J. vol. 38, #5, pp. 807–817, Oct. 2016. Abstract is reproduced below.

A novel fast algorithm is suggested for a coding unit (CU) mode decision using pseudo rate-distortion based on a separated encoding structure in High Efficiency Video Coding (HEVC). A conventional HEVC encoder requires a large computational time for a CU mode prediction because prediction and transformation procedures are applied to obtain a rate-distortion cost. Hence, for the practical application of HEVC encoding, it is necessary to significantly reduce the computational time of CU mode prediction. As described in this paper, under the proposed separated encoder structure, it is possible to decide the CU prediction mode without a full processing of the prediction and transformation to obtain a rate-distortion cost based on a suitable condition. Furthermore, to construct a suitable condition to improve the encoding speed, we employ a pseudo rate-distortion estimation based on a Hadamard transformation and a simple quantization. The experimental results show that the proposed method achieves a 38.68% reduction in the total encoding time with a similar coding performance to that of HEVC reference model.

This is an excellent paper which describes the complexity involved in implementing the HM software and develops techniques that reduce the encoder complexity significantly with minimal loss in PSNR and BD rate supported by simulation results (4K and 2K test sequences).

[E389] T. Mallikarachchi et al., "Content-Adaptive Feature-Based CU Size Prediction for Fast Low-Delay Video Encoding in HEVC," IEEE Trans. CSVT, vol. 28, pp. 693–705, March 2018.

[E390] J. Zhou et al., "A Variable-Clock-Cycle-Path VLSI Design of Binary Arithmetic Decoder for H.265/HEVC" IEEE Trans. CSVT, EA.

[E391] A. Zheng et al., "Adaptive Block Coding Order for Intra Prediction in HEVC," IEEE Trans. CSVT, vol. 26, pp. 2152–2158, Nov. 2016.

[E392] G. Fracastoro, S.M. Fosson and E. Magli, "Steerable discrete cosine transform," IEEE Trans. IP, vol. 26, pp. 303–314, Jan. 2017. This has several references related to DDT, MDDT and graph based transforms. See also G. Fracastoro and E. Magli, "Steerable discrete cosine transform," in Proc. IEEE International Workshop on Multimedia Signal Processing, 2015 (MMSP), 2015.

[393] Q. Huang et al., "Understanding and Removal of False Contour in HEVC Compressed Images," IEEE Trans. CSVT, (Early access).

Abstract is repeated here.

Abstract

A contour-like artefact called false contour is often observed in large smooth areas of decoded images and video. Without loss of generality, we focus on detection and removal of false contours resulting from the state-of-the-art HEVC codec. First, we identify the cause of false contours by explaining the human perceptual experiences on them with specific experiments. Next, we propose a precise pixel-based false contour detection method based on the evolution of a false contour candidate (FCC) map. The number of points in the FCC map becomes fewer by imposing more constraints step by step. Special attention is paid to separating false contours from real contours such as edges and textures in the video source. Then, a decontour method is designed to remove false contours in the exact contour position while preserving edge/texture details. Extensive experimental results are provided to demonstrate the superior performance of the proposed false contour detection and removal (FCDR) method in both compressed images and videos.

[E394] F. Chen et al., "Block-composed background reference for high efficiency video coding," IEEE Trans. CSVT, vol. 27, pp. 2639–2651, Dec. 2017.

[E395] D. Schroeder et al., "Efficient multi-rate video encoding for HEVC-based adaptive HTTP streaming," in IEEE Trans. CSVT, vol. 28, pp. 143–157. Jan. 2018.

Part of abstract is repeated here:

"We propose methods to use the encoding information to constrain the rate-distortion-optimization of the dependent encodings, so that the encoding complexity is reduced, while the RD performance is kept high. We additionally show that the proposed methods can be combined, leading to an efficient multi-rate encoder that exhibits high RD performance and substantial complexity reduction. Results show that the encoding time for 12 representations at different spatial resolutions and signal qualities can be reduced on average by 38%, while the average bitrate increases by less than 1%"

[E396] L. Li et al., "Lambda domain optimal bit allocation algorithm for HEVC," IEEE Trans. CSVT, vol. 28, pp. 130–142, Jan. 2018.

[E397] B. Min, Z. Xu and R.C.C Cheung, "A fully pipelined architecture for intra prediction of HEVC," IEEE Trans. CSVT, EA. (Same as [E382])

[E398] P.-K. Hsu and C.-A. Shen, "The VLSI architecture of a highly efficient deblocking filter for HEVC systems," IEEE Trans. CSVT, vol. 27, pp. 1091–1103, May 2017. There are many other references related to deblocking filter.

[E399] R. Wang et al., "MPEG internet video coding standard and its performance evaluation," IEEE Trans. CSVT, vol. 28, pp. 719–733, March 2018.

MPEG Internet Video Coding

This standard approved as ISO/IEC (MPEG-4 Internet Video Coding) in June 2015 is highly beneficial to video services over the internet. This paper describes the main coding tools used in IVC and evaluates the objective and subjective performances compared to web video coding (WVC), video coding for browsers (VCB) and AVC high profile.

[E400] M. Xu et al., "Learning to detect video saliency with HEVC features," IEEE Trans. IP, vol. 26, pp. 369–385, Jan. 2017.

Introduction is repeated here.

According to the study on the human visual system (HVS) [1], when a person looks at a scene, he/she may pay much visual attention on a small region (the fovea) around a point of eye fixation at high resolutions. The other regions, namely the peripheral regions, are captured with little attention at low resolutions. As such, humans are able to avoid is therefore a key to perceive the world around humans, and it has been extensively studied in psychophysics, neurophysiology, and even computer vision societies [2]. Saliency detection is an effective way to predict the amount of human visual attention attracted by different regions in images/videos, with computation on their features. Most recently, saliency detection has been widely applied in object detection [3, 4], object recognition [5], image retargeting [6], image quality assessment [7], and image/video compression [8, 9].

[E401] T. Mallikarachchi et al., "A feature based complexity model for decoder complexity optimized HEVC video encoding," IEEE ICCE, Las Vegas, Jan. 2017. See also: T. Mallikarachchi et al., "Decoding-complexity-aware HEVC encoding using a complexity-rate-distortion model," IEEE Trans. CE, Early access.

[E402] F. Pescador et al., "Real-time HEVC decoding with openHEVC and openMP," IEEE ICCE, Las Vegas, Jan. 2017.

[E403] G. Xiang et al., "An improved adaptive quantization method based on perceptual CU early splitting for HEVC," IEEE ICCE, Las Vegas, Jan. 2017.

[E404] H.M. Maung, S. Aramvith and Y. Miyanaga, "Error resilience aware rate control and mode selection for HEVC video transmission," IEEE ICCE, Las Vegas, Jan. 2017

[E405] B. Boyadjis et al., "Extended selective encryption of H.264/AVC (CABAC) and HEVC encoded video streams," IEEE Trans. CSVT, vol. 27, pp. 892–906, April 2017.

Conclusion is repeated here.

Conclusion

This paper proposes a novel method for SE of H.264/AVC (CABAC) and HEVC compressed streams. Our approach tackles the main security challenge of SE: the limitation of the information leakage through protected video streams.

The encryption of luma prediction modes for Intra blocks/units is proposed in addition to the traditional ES in order to significantly improve the *structural* deterioration of video contents. Especially on I frames, it brings a solid complement to previous SE schemes relying on residual encryption (mainly affecting the *texture* of objects) and motion vector encryption (disrupting the *movement* of objects). The proposed scheme shows both numerical and visual improvements of the scrambling performance regarding state-of-the-art SE schemes. Based on the encryption of *regular*-mode treated syntax element, the approach exhibits a loss in compression.

See also IEEE DCC 2018, March 2018, Snow Bird, UTAH

Special sessions:

"Toward a Next-Generation International Video Coding Standard"

Gary Sullivan, Microsoft Research

Yan Ye, Interdigital, Inc.

Session Coordinators

[E407] J. An et al., "Block partitioning structure for next generation video coding," ITU-T SG16, Doc. COM16-C966, Sept. 2015.

[E408] H. Huang et al., "EE2.1: Quadtree plus binary structure integration in JEM tools," JVET-C0024, Geneva, May 2016.

[E409] S.-H. Park and E.S. Jang, "An efficient motion estimation method for QTBT structure in JVET," IEEE DCC, April 2017.

[E410] P.-L. Cabarat, W. Hamidouche and O. Deforges, "Real-time and parallel SHVC hybrid codec AVC to HEVC decoder," IEEE ICASSP 2017, pp. 3046–3050, New Orleans, March 2017.

[E411] K. Senzaki et al., "BD-PSNR/Rate Computation tool for five data points," document JCTVC-B055, Geneva, Switzerland, July 2010.

[E412] Y. Fan et al., "A hardware oriented IME algorithm for HEVC and its hardware implementation," IEEE Trans. CSVT, vol. 28, pp. 2048–2057, Aug. 2018. Same as [E449].

[E413] T. Zhang et al., "Signal dependent transform based on SVD for HEVC intra coding," IEEE Trans. Multimedia, EA.

Section VI conclusions is repeated here.

This paper investigated the residuals characteristics of intra prediction and the deficiency of DCT/DST for residual blocks. Based on the residual analysis, a signal dependent transform based on SVD (SDT-SVD) which is derived from the synthetic block is proposed to compact the residuals from angular intra prediction. The proposed SDT-SVD is also extended to template matching prediction. Experimental results suggest that the proposed coding method with SDT-SVD for AIP outperforms the latest HEVC reference software with a bit rate reduction of 1.0% on average and it can be up to 2.1%. When SDT-SVD is extended to TMP based intra coding, the overall bit rate reduction is 2.7% on average and can be up to 5.8%. In the future, fast algorithms, such as early skip for SDT-SVD transform and fast template searching method, can be exploited to reduce the complexity of the proposed method. To further improve the efficiency of the SDT-SVD transform, more work can be done to design an efficient block with more similar structures as the residual block for deriving SVD transforms.

[E414] X. Lu et al., "Fast intra coding implementation for high efficiency video coding," IEEE DCC 2017.

[E415] H. Ryu et al., "Load balancing based on transform unit partition information for high efficiency video coding deblocking filter," ETRIJ, vol. 39, #3, pp. 301–309, June 2017.

[E416] MD. Asikuzzaman and M. R. Pickering, "An overview of digital video watermarking," IEEE Trans. CSVT, EA. Same as [E452]. (several papers related to watermarking in HEVC)

[E417] Y. Gong et al., "Rate-Distortion-Optimization-Based Quantization Parameter Cascading Technique for Random-Access Configuration in H.264/HEVC," IEEE Trans. CSVT. vol. 27, pp. 1304–1312, June 2017.

[E418] S.-H. Yang and P.B. Vo, "Adaptive bit allocation for consistent video quality in scalable high efficiency video coding," IEEE Trans. CSVT, vol. 27, pp. 1555–1567, July 2017.

[E419] X. Hoang Van, J. Ascenso and F. Pereira, "Adaptive scalable video coding: a high efficiency video coding based framework combining the predictive and distributed paradigms," IEEE Trans. CSVT, Early access.

[E420] R. Parois et al., "Efficient parallel architecture for a real-time UHD scalable HEVC encoder," EUSIPCO 2017, KOS Island, Greece, Aug. 2017.

[E421] M. Kozir et al., "Heuristics for tile parallelism in HEVC," EUSIPCO 2017, KOS Island, Greece, Aug. 2017.

[E422] D.S. Silveira, M. Porto and S. Bampi, "Performance and energy consumption analysis of X265 video encoder," EUSIPCO 2017, KOS Island, Greece, Aug. 2017.

[E423] N. Sidaty, "Reducing computational complexity in HEVC decoder for mobile energy saving," EUSIPCO 2017, KOS Island, Greece, Aug. 2017.

[E424] I. Dvir et al., "Trapezoidal block split using orthogonal C2 transforms for HEVC video coding," EUSIPCO 2017, KOS Island, Greece, Aug. 2017.

[E425] H. Ryu et al., "Load balancing based on transform unit partition information for HEVC deblocking filter," ETRI J, vol. 39, #3 pp. 301–309, June 2017

[E426] IEEE DATA COMPRESSION CONFERENCE (DCC)

Snowbird, Utah

March 27–March 30, 2018

http://www.cs.brandeis.edu/~dcc

Keynote address: "Inpainting-Based Data Compression"

   Prof. Joachim Weickert

   Saarland University, Germany

Special sessions:

   "Toward a Next-Generation International Video Coding Standard"

   Gary Sullivan, Microsoft Research

   Yan Ye, Interdigital, Inc.

Session Coordinators:

   "The Emerging AV1 Video Codec" AV1 by alliance for open media.

   Jingning Han, Google

   Iraj Sodagar, Microsoft

   Session Coordinators

   "Computation over Compressed Data"

   Travis Gagie, University of Helsinki

   Gonzalo Navarro, University of Chile

   Session Coordinators

[E427] M. Bilal et al., "A low complexity pedestrian detection framework for smart video surveillance systems," IEEE Trans. CSVT, vol. 27, pp. 2260–2273, Oct. 2017. See also X. Jim, L. An and B. Bhanu, "Group structure preserving pedestrian tracking in a multicamera video network," IEEE Trans. vol. 27, pp. 2165–2176, Oct.2017.

[E428] X. Zhang et al., "Low-rank-based nonlocal adaptive loop filter for high efficiency video compression," IEEE Trans. CSVT, vol. 27, pp. 2177–2188, Oct. 2017.

[E429] E.M.-Enriquez et al., "Directional transforms for video coding based on lifting on graphs," IEEE Trans. CSVT, vol. 28, pp. 933–846, April 2018.

[E430] T.S. Kim et al., "Fast integer motion estimation with bottom-up motion vector prediction for an HEVC encoder," IEEE Trans. CSVT, EA.

Last section Conclusion is repeated here:

This paper presents a fast IME algorithm designed for HEVC that supports a highly flexible block partitioning structure. This new algorithm reduces its computational complexity significantly by searching narrow regions around multiple accurate MVPs. The accurate MVPs are derived from the MVs of the PUs in the CUs in the lower layer in the coding block hierarchy. Thanks to the accuracy of the bottom-up MV prediction, the reduced search range does not significantly affect the degradation of the coding efficiency. Experimental results show that the proposed IME reduces computational complexity up to 81.95% with 0.39% BD-BR increase. Furthermore, the proposed IME is designed to maintain compatibility with other fast algorithms that are based on top-down encoding order [37–50], which is a crucial issue when designing a new scheme because many fast algorithms can conflict with each other, thereby not being able to achieve performance improvement when they are used together. In order to keep compatibility with HEVC standard, the bottom-up MVPs are utilized for only IME and bitstream is generated by AMVP.

A possible future work of this paper can be extension of the bottom-up motion vector prediction to FME search and merge mode decision, which simply exploits correlation among motion vectors of blocks in different hierarchical depths of the block. However, the proposed bottom-up MV prediction may not be effective if it is applied to FME as it is. Because the MV correlation between different depths is not large enough to be able to predict the fractional MV. Therefore, additional study on bottom-up encoding order is required for FME.

[E431] J. Wang et al., "On BD rate calculation." Doc JCTVC-F270 6th meeting Joint Collaborative Team on Video Coding (JCT-VC) of ITU-T SG 16 WP 3 and ISO/IEC JTC 1/SC 20/WG 11, Torino, Italy, 2011.

[E432] S.K. Rao, "Bit-rate reduction using superpixel for 360-degree videos – VR application," M.S. Thesis, EE Dept., Univ. of Texas at Arlington, Arlington, TX, Dec. 2017.

[E433] H.N. Jagadish "implementation and comparison of serial and parallel algorithms of simple adaptive offset in HEVC," M.S. Thesis, EE Dept., Univ. of Texas at Arlington, Arlington, TX, Dec. 2017.

[E434] N.C. Vayalil, M. Paul and Y. Kong, "A residue number system hardware design of full search variable motion estimation accelerator," IEEE Trans. CSVT, EA.

This paper proposes a modified algorithm based on a test zone (TS) search algorithm for hardware implementation for encoding ultra high density (UHD) videos in real time.

[E435] F. Kamisli, "Lossless image and intra-frame compression with integer-to-integer DST," IEEE Trans. CSVT, EA. See related papers by this author in the references.

[E436] X. Meng et al., "A new HEVC in-loop filter based on multi-channel long short-term dependency residual networks," IEEE DCC 2018, Snow Bird, UTAH, March 2018.

[E437] K. Thiyagarajan et al., "Energy-Aware Encryption for Securing Video Transmission in Internet of Multimedia Things." IEEE Trans. CSVT, EA.

[E438] C. Ma et al., "Reference clip for inter prediction in video coding" IEEE Trans. CSVT, Early access.

The last section Conclusion is repeated here.

Conclusion

In this paper, we propose to optimize the reference content based on clips. Compared with the traditional multiple reference pictures, the proposed reference clips can collect the reference content more flexibly. We design a complete video coding framework, which features reference clip generation and management, using reference clip for inter prediction, and bit-rate allocation among reference clips. The experimental results have shown that our scheme achieves on average 5.1% and 5.0% BD-rate reduction than the HEVC anchor, for LDB and LDP configurations respectively, without significant increase of computational complexity. Therefore, we believe reference clip opens up a new dimension for optimizing inter prediction in video coding. The techniques of reference frames have been developed for decades and proved to improve the coding efficiency substantially. Analogously, our proposed reference clip is worthy of further study because of its great potential shown in this paper. The future work on reference clip may proceed in several directions. First, integrating reference clip into random-access setting is necessary. Second, how to allocate memory between reference frames and reference clips for different videos is important to improve coding efficiency. Third, using external data like a prebuilt dictionary of clips in video coding is very interesting to study.

[E439] Y. Zhang et al., "Effective data driven coding unit size decision approaches for HEVC intra coding" IEEE Trans. CSVT, EA.
[E440] X. Liu et al., "An adaptive coding unit size decision algorithm for HEVC intra prediction based on complexity classification using machine learning," IEEE Trans. CSVT, EA.

Conclusions section is repeated here:

Conclusions

In this paper, an adaptive fast CU size decision algorithm, named as CCML, is proposed for HEVC Intra frame coding. Firstly, we analyse the quadtree-based CU size decision process and the relationship between CU partitions and image features. And then, a three-output classification model is constructed based on the CU complexity by using SVM technique, which guarantees a relatively higher classification accuracy than a simple binary classifier. Finally, the classification model is loaded into the HEVC reference encoder (HM15.0) to assist in deciding the optimal CU size in advance without performing the full RDO process. The experimental results show that the proposed CCML algorithm succeeds to reduce around 59.66% encoding time on average compared with the original test model HM15.0, while efficiently maintaining the similar coding performance. It could also achieve better coding performance compared with the other existing fast CU size decision algorithms referred in this paper. We believe that the proposed CCML algorithm has the great potential to be widely applied for real time multimedia communication systems."

[E441] Y. Gong et al., "Temporal-layer-motivated lambda domain picture level rate control for random access configuration in H. 265/HEVC," IEEE Trans. CSVT, Early access
[E442] K. Rouis, C. Larabi and J.B. Tahar, "Towards perceptually guided rate distortion optimization for JHEVC," IEEE ICASSP2018, Calgary, Canada, April 2018.
[E443] R. Peesapati et al., "Design of streaming deblocking filter for HEVC decoder," IEEE Trans. CE, vol. 62, #3, pp. 225–233, Aug. 2017.
This has several references related to hardware architecture for H.264/AVC and HEVC deblocking filters.

[E444] S. Baldev et al., "Design and implementation of efficient streaming deblocking and SAO filter for HEVC decoder," IEEE Trans. CE, Early access.

This has several references related to hardware implementation of these filters.

[E445] D. Grois, T. Nguyen and D. Marpe, "Performance comparison of AV1, JEM, VP9 and HEVC encoders" Applications of digital image processing XL, (ED: A.G. Tescher), Proc. of SPIE, vol. 10396, San Diego, CA, Aug. 2017.

[E446] Y. Xue and Y. Wang, "A novel video coding framework using self-adaptive dictionary," IEEE Trans. CSVT, Early access.

Part of the conclusion section is repeated here

"Although our proposed coder has only shown limited gain over HEVC P-frame, and falls short to HEVC with B-frame coding, additional gains are likely with further optimization of the various components. First of all, the rate-distortion behaviors of both the first stage for coding the SeAD atoms and the second stage for coding the DCT atoms exhibit a very smooth trend, indicating that some analytical models for the rate-distortion curves of these two stages may be established, which may be utilized to yield a better rate distortion optimized decision on the adaptive switching between the two stages and on the block partition decision. Secondly, how to design a better SeAD dictionary and a better second stage dictionary can be further explored. Specifically, the proposed framework only uses atoms from the previous frame in SeAD, while one could investigate using multiple frames, with both previous and future frames. For the second stage dictionary, although our piloting study on designing a special fixed dictionary through dictionary learning returned similar result as with using the DCT, other new development along this line might be borrowed to design a better dictionary or to adapt the dictionary. Thirdly, we believe significant gains are possible with more careful design of the entropy coder for coding the first stage atom locations. Lastly, although our preliminary attempt at extending the two stage coding frame work to intra-frame coding has not yielded promising results, further optimization of the entropy coder, especially for coding the intra-prediction candidate locations, is likely to make it more competitive than HEVC."

[E447] T.S. Kim et al., "Fast integer motion estimation with bottom-up motion vector prediction for an HEVC encoder" IEEE Trans. CSVT, Early access. Several valuable papers on fast motion estimation are listed at the end.

[E448] H. Jiang et al., "Highly-parallel low-cost embedded HEVC encoder on TI Keystone multicore DSP" IEEE Trans. CSVT, Early access.

[E449] Y. Fan et al., "A hardware oriented integer motion estimation algorithm for HEVC and its hardware implementation," IEEE Trans. CSVT, vol. 28, pp. 2048–2057, Aug. 2018. Same as [E412].

Several papers related to both IME and FME algorithms are listed at the end.

[E450] N. Doan et al., "A hardware oriented concurrent TZ search algorithm for high efficiency video coding," EURASIP J. on advances in signal processing, Nov. 2017.

[E451] J.F.M. Carreira et al., "A two-stage approach for robust HEVC coding and streaming," IEEE Trans. CSVT, vol. 28, pp. 1966–1973, Aug. 2018.

[E452] Md. Asikuzzaman and M.R. Pickering, "An overview of digital video watermarking," IEEE Trans. CSVT, vol. 28, pp. 2131–2153, Sept.2018. Same as [OP14].

A section of this paper related to HEVC is reproduced here.

Over the past two decades, a great deal of research using the compressed domain has been conducted, with most studies focusing on existing compression standards, including MPEG2, MPEG-4 and H.264/AVC. The latest compression standard, H.265/HEVC, is now replacing the others as it contains many advanced features. Although some researchers have embedded the watermark in the H.265/HEVC domain, as very few papers have been published, this field is relatively young. Therefore, it seems that there is a great deal of scope to explore this area for developing a watermarking technique. As the robustness of existing techniques to image or video compression are assessed using any of the existing compression methods, there is potential for further research into watermarking techniques that can deal with H.265/HEVC compression which is the next-generation compression standard.

[E453] C. Lan et al., "Variable block-sized signal dependent transform for video coding," IEEE Trans. CSVT, vol. 28, pp. 1920–1933, Aug. 2018. Same as [DCT65].

[E454] J. He et al., "Adaptive quantization parameter selection for H.265/HEVC by employing inter-frame dependency" IEEE Trans. CSVT, Early access.

[E455] S. Ki et al., "Learning-based just-noticeable-quantization-distortion for perceptual video coding" IEEE Trans. IP, vol. 27, pp. 3178–3193, July 2018.

[E456] Y. Zhang et al., "Residual highway convolutional neural networks for in-loop filtering in HEVC," IEEE Trans. IP, vol. 27, pp. 3827–3841, Aug. 2018

[E457] D.-H. Kim, Y.-J. Kwon and K.-H. Choi, "Motion-vector refinement for video error concealment using downhill simplex approach," ETRI J., vol. 40, issue 2, pp. 266–274, April 2018.

[E458] M. Jiang et al., "On Derivation of Most Probable Modes for Intra Prediction in Video Coding", IEEE ISCAS 2018, Florence, Italy, May 2018.

Abstract is repeated here.

High Efficiency Video Coding (HEVC) standard introduced 35 intra prediction modes in order to improve intra prediction performance. To signal mode of the current block to the decoder a list of three most probable modes (MPM) is derived based on the modes of the two neighbouring blocks. Similarly, six MPMs are used in the recent Joint Exploration Model (JEM) to accommodate the increased number to 67 intra prediction modes. One method to achieve this is to use an extension of HEVC approach. The other one employs the modes of the five neighbouring blocks, together with their derived modes and default modes, in an ordered MPM candidate list. Our contribution is to use case-based multiple candidate lists. For each defined case, the corresponding list is sorted based on the probability of each candidate. The index of a case is computed based on the defined status values. Experimental results show 0.22%, 0.19%, 0.18% BD rate savings for Y, U, and V components for all-intra configuration, respectively.

[E459] R. Fan, Y. Zhang and B. LI, "Motion Classification-Based Fast Motion Estimation for High-Efficiency Video Coding," IEEE ISCAS 2018, Florence, Italy, May 2018.

Abstract is repeated here.

HEVC achieves excellent coding performance, at the cost of substantially high complexity, preventing it from real-time applications. This paper address the problem of fast motion estimation, the most time-consuming module therein, and proposes a motion classification-based fast ME algorithm. The prediction units are first categorized into three classes and different search strategies are carefully designed accordingly. Furthermore, fast search priority-based partial internal termination schemes are presented to skip impossible positions for speedup. Experimental results demonstrate the proposed algorithm can achieve significant complexity reductions with negligible efficiency degradation, and outperforms state-of-the-art algorithms in terms of both coding performance and complexity reduction.

[E460] M. Grellert et al., "Fast coding unit partition decision for HEVC using support vector machines," IEEE Trans. CSVT EA.

Conclusion section is repeated here.

This paper presented a fast CU partitioning decision solution using Support Vector Machines. An extensive analysis of coding features was initially conducted, showing that the features introduced in this work, combined with ones borrowed from references in the literature, are very effective to predict the correct CU partitioning. The methodology designed to train our SVM classifiers proved to be very useful, as it reduced the number of features to be used and produced simpler classifiers, using only a subset of the training samples. In addition, an adaptation to increase the rate-distortion performance was implemented by adjusting the decision threshold of SVMs. In this analysis we concluded that it is possible to reduce the rate-distortion penalties significantly with simple changes in the decision functions. The compression results showed that significant time savings can be achieved with tolerable rate-distortion performance penalties. For instance, for the smallest tested decision threshold, we achieved a 34.9% time savings with a 0.13% increase in BD-BR on average. Additional savings can be obtained by increasing the decision threshold used. We also identified the possibility to achieve different complexity reduction points by adjusting this threshold parameter, which hints at the possibility of building an adaptive system that provides the smallest rate-distortion performance loss for the required throughput, or conversely the highest time savings. The major conclusion from all the results is that the designed solution is able to effectively reduce the HEVC encoding complexity with competitive results in terms of rate-distortion. Furthermore, the methodology we applied can also be used for other HEVC encoding decisions, such as the PU and the TU partitioning.

[E461] J. Begaint et al., "Region-based models for motion compensation in video coding," IEEE PCS 2018, San Francisco, CA, June 2018.

The section conclusions is repeated here.

"In this paper we present a novel video prediction mode to describe complex motion in video sequences. The efficiency of the proposed solution was demonstrated against state-of-the-art video coding tools on multiple sequences, with an average gain of −3.37% over HEVC. The complexity of the prediction mode is limited, especially on the decoder side, with respect to the gains that can be obtained. Although the region- based prediction is currently limited to one reference frame, it could be extended to use more frames from the reference pictures buffer. Improving the speed and the robustness of the prediction also constitutes an important future work."

[E462] Y. Sugito et al., "A study on the required video bit-rate for 8K 120-Hz HEVC/H.265 temporal scalable coding," IEEE PCS2018, San Francisco, CA, June 2018. At the end the authors propose how to extend this study and also how to improve the image quality at the encoder.

[E463] J.-R. Ohm and G.J. Sullivan, "Versatile video coding-towards the next generation video coding," INVITED TALK, IEEE PCS2018, San Francisco, CA, June 2018.

[E463] H Tang et al., "Hadamard transform based optimized HEVC video coding," IEEE CSVT, Early access.

[E464] Y. Dar, M. Elad and A.M. Druckstein, "Optimized pre-compensating compression," IEEE Trans. IP, vol. 27, pp. 4798–4809, Oct. 2018.

The last section Conclusion is repeated here.v
In this paper, we studied an image/video compression problem where a linear degradation affects the decompressed signal. We handled the difficult rate-distortion optimization using the ADMM approach, resulting in an iterative procedure relying on a standard compression technique. Our method was shown to be effective in experiments of adjusting HEVC's image and video compression to post-decompression blur degradations. We consider the proposed method as a promising paradigm for addressing complicated rate-distortion optimizations arising in intricate systems and applications. Future work may extend our approach for optimizing systems where several display types are possible. Another interesting process to examine includes a degradation that precedes the compression (e.g., an acquisition stage) in addition to the post-decompression degradation (e.g., rendering). Other intriguing research directions may explore the utilization of our idea for optimizing compression with respect to various perceptual distortion metrics or to nonlinear systems.

[E465] M. Xu et al., "Reducing complexity of HEVC: A deep learning approach," IEEE Trans. IP, EA.

Appendix

The last section Conclusion is repeated here.

In this paper, we have proposed a deep learning approach to reduce the encoding complexity of intra- and inter-mode HEVC, which learns to predict the optimal CU partition instead of using conventional brute-force RDO search. To deepen the networks of our deep learning approach, the CPH-Intra and CPH-Inter database was established, consisting of large-scale data of HEVC CU partition at intra- and intermodes, respectively. Then, two deep learning architectures, i.e., ETH-CNN and ETH-LSTM, were proposed to predict the CU partition for reducing the HEVC complexity at the intra- and inter-modes. The output of these architectures is HCPM, which hierarchically represents the CU partition in a CTU. Upon the representation of HCPM, an early termination mechanism was introduced in ETH-CNN and ETH-LSTM to save the computational time. The experimental results show that our deep learning approach performs much better than other state-of-the-art approaches in terms of both complexity reduction and RD performance.

There are three promising directions for future works. Our work, at the current stage, mainly focuses on predicting the CU partition to reduce the HEVC encoding complexity. Other components, such as PU and TU prediction, can also be replaced by deep learning models to further reduce the encoding complexity of HEVC. This is an interesting future work. In the deep learning area, various techniques have been proposed to accelerate the running speed of deep neural networks. Our deep learning approach may be sped up by applying these acceleration techniques, which is seen as another promising future work. Beyond the CPU implementation of the current work, our approach may be further implemented in the FPGA device in future, for practical applications.

[E466] J. Cui et al., "Hybrid all zero soft quantized block detection for HEVC," IEEE Trans. IP, EA.

[E467] D. Wang et al., "Efficient multi-strategy intra prediction for quality scalable high efficiency video coding," IEEE Trans. IP, EA.

Conclusion section is repeated here.

Conclusion

In this paper, a novel and effective Intra prediction algorithm for QS is proposed. The encoding procedure includes four fast strategies: "CBDP" allows skipping depths with low probabilities; "DB-IMD" permits skipping unnecessary Intra prediction; "HCB-IMP" allows checking only a subset of IMs instead of all 35 IMs; and "SDB-DET" allows early termination of depth selection. Therefore, the proposed algorithm can significantly improve the coding speed with negligible losses in coding efficiency. It is noted that some of the proposed strategies may be extended to inter QS and spatial SHVC (SS), such as CBDP and SDB-DET, with some possible adaptations. In addition, DB-IMD and HCB-IMP can also be extended to Intra SS by taking into account the changed inter-layer correlation due to the differences between QS (same resolution but different QP among layers) and SS (same or similar QP but different resolutions among layers). In other words, for inter QS and SS, the proposed schemes need to be adapted according to the statistics change of inter-layer correlation, as well as the partitions in CBDP inter modes. This is our ongoing work.

[E468] M. Jamali and S. Coulombe, "Fast HEVC intra mode decision based on RDO cost prediction," IEEE Trans. Broadcasting, EA

[E469] H. Guo et al., "Optimal bit allocation at frame level for rate control in HEVC," IEEE Trans. Broadcasting, EA.

Abstract is repeated here.

Abstract

Rate control plays an important role in video coding systems, which makes the output bitrate of a video encoder equal to the target bitrate while minimizing the distortion of the compressed video. However, most of the existing rate control schemes achieve accurate bitrate control at the loss of rate-distortion (R-D) performance. This paper proposes an effective frame level bit allocation method to improve the R-D performance whilst maintaining the high accuracy of the bitrate control in HEVC. First, an improved R-D model is presented at the frame level, which, by making use of the information of encoded frames more completely, achieves higher accuracy with lower bitrate mismatch rate. Second, a group of picture (GOP) level bit allocation approach is introduced with the consideration of the temporal R-D dependency among different GOPs, which can further enhance the R-D performance. Finally, to achieve optimal R-D performance globally, a formulation for optimal frame level bit allocation is developed with a GOP level Lagrange multiplier introduced, which takes into account the coding effects of a current frame on the other frames within the same GOP. A scheme of recursive Taylor expansion is employed to find the GOP level Lagrange multiplier. Experimental results demonstrate that the proposed method can achieve considerable

R-D performance improvement and accurate bitrate control. Specifically, our method shows 3.9% and 3.8% BD-rate saving in average compared against the HEVC reference software HM16.7 with rate control in the low-delay B and low-delay P coding structures, respectively.

[E470] L. Trudeau, S. Coulombe and C. Desrosiers, "Cost-based search ordering for rate-constrained motion estimation applied to HEVC," IEEE Trans. Broadcasting, EA.

Conclusion section is repeated here.

Conclusion

In this paper, we exposed two problems related to the increasing rate rule: asymmetric distribution of MV costs and off-centered search areas. These problems cannot be solved efficiently by static search orderings based on geometric patterns, such as spiral searches. We proposed a new model to build dynamic search orderings for BMA. From this model, we developed a fast algorithm capable of producing cost-based search orderings that are unaffected by asymmetric distributions of MV costs and off-centered search areas. These orderings not only improve elimination, but they allow for early termination, a novelty for optimal algorithms Our experiments show that the proposed solution is more than five times faster than the HEVC HM encoder in full search mode, with the same BD-PSNR. When compared to an HEVC HM encoder that would implement RCSEA, the proposed solution remains superior, with a $14\times$ speedup. Furthermore, the proposed early termination only requires performing 36% and 46% of block-matching loop iterations for Random Access and Low Delay, respectively, and the number of SAD operations also decreased by approximately 3%. Finally, we measured an average of $134\times$ speedup when we replaced the semi-exhaustive fallback in the TZ-Search by a suboptimal cost-based refinement. These results demonstrate the potential of our findings to be tailored to suboptimal algorithms. In an upcoming paper, we will show that when extending the proposed techniques to the context of TZ-Search even more significant speedups can be reached.

[E471] H. Azgin et al., "Reconfigurable intra prediction hardware for future video coding," IEEE Trans. CE, vol. 63, issue 4, pp. 419–425, Nov. 2017.

[E472] M. Cheon and J.-S. Lee, "Subjective and objective quality assessment of 4K UHD videos for immersive experience," IEEE Trans. vol. 28, pp. 1467–1480, July 2018.

The last section Conclusion is repeated here.

In this paper, we have presented subjective and objective quality assessment of 4K UHD videos targeting an immersive viewing environment using a large sized screen. The perceptual quality of 4K UHD and FHD videos upscaled to 4K UHD was investigated for three codecs, i.e., AVC, HEVC, and VP9. The added value of 4K UHD over FHD was clearly observed for high bitrate conditions, which was more prominent for contents having high spatial complexity. The benchmarking study of objective quality metrics was also conducted, where the superiority of ADM, VQM, FSIM, and OSS-PSNR was shown. The necessity of developing objective quality metrics optimized for 4K immersive environments and the latest compression methods was raised. There are several related research issues to be resolved in the future. Further investigation of perceptual experience in immersive visual environments will be necessary, e.g., visual attention. In addition, as aforementioned, developing new objective quality metrics for such environments will be desirable. We hope that our 4K database is useful for these future studies.

[E473] S.-H. Park et al., *"Efficient Reference Frame Selection Method for Lightweight HEVC Encoder"*, IEEE ICCE West (Berlin), 3–5 Sept. 2018. Abstract is reproduced here. As a demand of video resolution increases highly, video encoder can face difficulties on the computational complexity to process huge pixel data with real-time environment. Many researches have focused on the complexity of high efficiency video coding (HEVC) test model (i.e., HM) which is far from real-time encoding for even high-definition video. However, it has been rarely studied in literature to establish a lightweight HEVC test model (LHM) and to decrease the complexity of LHM further. In this paper, LHM that equips fast encoding options is newly established, and a fast motion estimation (ME) method is proposed to challenge decreasing the encoding complexity of LHM. The proposed method exploits the intriguing correlation of the motion information of merge/skip mode with the information of other motion partitions to minimize the required number of reference frames in ME process. Experimental results showed that the proposed method could substantially decrease the complexity of LHM in view of encoding time as well as memory access.

[E474] R. Kumar et al., *"Retargeting UHD 4k Video for Smartphones,"* IEEE ICCE West (Berlin), 3–5 Sept. 2018.

Abstract is reproduced here.

This paper evaluates a new video-retargeting algorithm jointly with two others previously published in the literature. Four Ultra High Definition (UHD) 4K resolution video sequences were used for subjective assessment. Subjective testing results show that the proposed algorithm yields the best performance when UHD 4K video is retargeted to 1920 × 1080 resolution. The main application of these retargeting algorithms is to increase the quality of experience (QoE) provided to consumers when playing UHD videos on small screen devices like smartphones.

[E475] F. Golaghazadeh et al., "Checksum-filtered list decoding applied to H.264 and H.265 to video error correction," IEEE Trans. CSVT, vol. 28, pp. 1993–2006, Aug. 2018.

This is a valuable paper related to error concealment, error detection/ correction, error control strategy as applied to coded bit streams (H.264 and H.365).

Conclusion section is repeated here.

Conclusion

In this paper, we proposed a new method that exploits the receiver side UDP checksum information to dramatically reduce the number of candidates that need to be considered by list decoding. For one bit in error, the method allows the removal of 97% of the candidates. For two bits in error, this reduction reaches 99.6%. Such a filtering of the candidates as proposed, supplemented by checksum information dramatically reduces the complexity of the list decoding approach. Simulations results showed that the H.264 baseline, using CAVLC, is more robust to desynchronization due to errors, as compared to HEVC, using CABAC. This led to better error correction performance for H.265 as corrupted packets are more likely to cause desynchronization errors which invalidate the erroneous packets. Our simulation results revealed that 79% of the H.264 could be corrected perfectly, compared to 91% for HEVC. The proposed approach provides, on average, a 2.79 dB gain over FC-EC using JM, and a 3.57 dB gain over our implementation of FC-EC in HM. Although, current applications do not typically have access to soft information, the proposed CFLD approach can also be applied to that context, allowing it to perform even better by enabling it to exploit the soft information to rank the candidate bitstreams in each BEE. We also expect a further increase in performance by exploiting pixel domain information to select the best decodable candidate rather than selecting the first decodable candidate. This will be the subject of future research.

[E476] A.S. Dias et al., "Time-constrained video delivery using adaptive coding parameters," IEEE Trans. EA.

Conclusion section is repeated here.

"This paper proposed a joint encoding and uploading time control scheme based on an adaptive QP selection algorithm that relies on accurate encoding time and bit-rate estimation techniques performed during the encoding process. The overall scheme targets video compression applications where video content needs to be encoded and transmitted within given overall time constraints. The reported experimental results show that the proposed method is able to accurately meet the overall time constraints for different overall target times and different bandwidths considered for transmission. Future research work can be performed to refine the proposed technique, for example by defining a method to select an appropriate initial base QP depending on the overall time constraints and characteristics of the video content.

[E477] N. Yan et al., "Convolutional neural network-basefd fractional-pIxel motion compensation" IEEE Trans. EA.

Conclusion section is repeated here.

Conclusion

We have proposed FRCNN, a CNN-based method for fractional-pixel MC to improve the efficiency of inter prediction in video coding. We have formulated the fractional-pixel MC as an inter-picture regression problem instead of interpolation, and proposed to utilize CNN to solve the regression problem. Specifically, we have adopted CNN models to build the relationship between fractional references and integer-pixel values, each model accounts for one proper fractional MV. We also proposed to train a second set of CNN models for bi-directional MC. We have addressed the issue of generating training data for FRCNN, in which we proposed to train QP-dependent models. We have designed algorithms to integrate FRCNN into HEVC, including the ME and MC algorithms, the FRFT selection, and the FRFT merge mode. We have performed extensive experiments to verify the effectiveness of FRCNN, using different network structures and different training data. Experimental results show that our scheme achieves signification bits saving compared to HEVC. The work presented in this paper is still preliminary. There are a lot of open issues for further study, some of which have been mentioned before. Most importantly, how to design a good network structure to achieve better compression efficiency and lower complexity is indispensable. It is also interesting to study the joint optimization of CNN for compression artefact reduction [11] and FRCNN, since the latter also performs reduction of compression noise. In addition, we may extend the basic idea of formulating fractional-pixel MC as an inter-picture regression problem to the case of integer-pixel MC or even general MC without requiring MV."

[E478] S. Ryu and J.-W. Kang, "Machine learning-based fast angular prediction mode decision technique in video coding," IEEE Trans. IP, vol. 27, pp. 5525–5538, Nov. 2018.

[E479] F. Zhang et al., "BVI-HD: A video quality database for HEVC compressed and texture synthesized content," IEEE Trans. MM, vol. 20, pp. 2620–2630, Oct. 2018.

Part of the conclusion section is repeated here.

"The BVI-HD video quality database reported here has been made available online1 for public testing. Its content diversity and representativeness make it a reliable test database for objective video metric evaluation, and the reference sequences can also be used for video compression performance testing. It should be noted that the number of reference and test sequences in the BVI-HD database is still limited and only full HD resolution has been tested. Future work should focus on using more immersive video formats as test content, including higher dynamic ranges, higher frame rates and higher spatial resolutions."

[E480] D. Zang et al., "Efficient multi-strategy intra prediction for quality scalable high efficiency video coding," IEEE Trans. IP, EA.

Part of the conclusion is repeated here.

"It is noted that some of the proposed strategies may be extended to inter QS and spatial SHVC (SS), such as CBDP and SDB-DET, with some possible adaptations. In addition, DB-IMD and HCB-IMP can also be extended to Intra SS by taking into account the changed inter-layer correlation due to the differences between QS (same resolution but different QP among layers) and SS (same or similar QP but different resolutions among layers). In other words, for inter QS and SS, the proposed schemes need to be adapted according to the statistics change of inter-layer correlation, as well as the partitions in CBD Pinter modes. This is our ongoing work."

[E481] J.Y. Lee et al., "Efficient inter-view motion vector prediction in multi-view HEVC," IEEE Trans. Broadcasting, vol. 64, pp. 666–680, Sept. 2018.

[E482] L. Jia et al., "A new rate-complexity distortion model for fast motion estimation algorithm in HEVC," IEEE Trans. MM, EA.

Conclusion section is repeated here.

Conclusion

In this paper, we formulated an R-C-D model for ME. This model explicitly explains how the computation complexity affects the R-D performance of the FS ME algorithm. Instead of threshold setting and arbitrary search pattern, our proposed FME is based on an R-C-D optimization framework. We derive a Lagrangian multiplier to set an optimal tradeoff between the complexity and the R-D cost function. The experimental results shows that for the low-delay P configuration in HEVC, our proposed RCDO-FME reduces the complexity of HEVC FME by 87% on average, while it improves the RD performance by −0.2% BDBR reduction. Our proposed fast algorithm is also better than other state-of-the-art FME methods with consistent performance in terms of complexity reduction and R-D performance improvement. Moreover, the proposed method is not limited to software implemented ME. The strategy of joint R-C-D optimization can be extended to an R-C-D optimized mode decision method. The R-C-D optimized frame work is promising in dealing with the ME and mode decision together, which will bring about higher complexity reduction and better coding performance for inter coding. Additionally, it is possible to design a hardware architecture specifically designed to implement the proposed fast algorithm."

[E483] H. Wang et al., "A collaborative scheduling-based parallel solution for HEVC encoding on multicore platforms." IEEE Trans. MM, EA.

Part of the conclusion section is repeated here.

"In the future, it is desirable to investigate the parallelization of HEVC encoding on various parallel processing architectures and to reduce the overhead of parallel task scheduling. In the current work, the proposed CSPS focuses on parallelizing HEVC encoding within frames. However, it is feasible to integrate the proposed solution with GOP-based/frame-based TLP approaches and various ILP methods with minor efforts in order to further speedup HEVC encoding. In addition, the strategies designed for CSPS are not only applicable to HEVC encoding but also advisable for parallelizing other computationally intensive applications especialy for those adopting the wavefront paradigm."

[E484] M. Wang et al, "UHD Video Coding: A Light-weight Learning-based Fast Super-block Approach," IEEE Trans. CSVT, EA.

Conclusions and future directions are repeated here.

Conclusion and Future Directions

This work systematically studied a set of suitable solutions to provide light-weight complexity yet high coding efficiency for UHD video coding in HEVC. A new MCU block structure was proposed to avoid unnecessary recursive brute-force CU partition search in SCU. In addition, the Direct-MCU and SCU-to-MCU modes were proposed to encode a SCU based on the characteristics of UHD video content, where the DirectMCU mode guaranteed the compression efficiency by avoiding extra-size CU partition while the SCU-to-MCU mode saved the time complexity by skipping fine-grained CU partition search. Furthermore, three effective features were extracted to represent SCU of UHD content, and the MLP classifier was trained and used to speed up the SCU encoding progress. As demonstrated in the simulation experiments, the proposed method greatly improved the encoding performance by providing tolerable complexity and excellent compression efficiency. In the future, we plan to migrate the proposed algorithm to the FVC software platform, so more complex prediction coding and larger transform coding may help to improve the coding efficiency of Direct-MCU and SCU-to-MCU methods. In addition, we are trying to use convolutional neural network (CNN) to directly extract the coding features, and integrate the QP feature and train a joint model. Since the complexity of the FVC reference encoder is a still challenge (e.g., 11 times complexity compared to that the HEVC encoder), efficient feature representation can improve the accuracy of the classifier which helps to greatly reduce the encoding time.

[E485] R. Yang et al, "Enhancing quality for HEVC compressed videos," IEEE Trans. CSVAT, EA.

Last paragraph in the conclusion section is repeated here

"In our prototype, at the current stage, four high-end GPUs are adopted to achieve real-time enhancement. This prototype shows an example that our TQEO scheme is able to control the computational time of quality enhancement to a specific target, adjusting to the computational resources. Fortunately, there are plenty of acceleration hardwares to speed up the CNN networks. For example, the Cambricon-X accelerator has more than 7 times speed of GPU with much lower energy [41]. Recently, the Tensor Processing Unit (TPU) [42] developed by Google achieves 15 to 30 times speed compared with GPU, with smaller size and lower power consumption. Applying our QE-CNN method and TQEO scheme on these Application Specific Integrated Circuits (ASIC) can be seen as an interesting further work."

[E486] H. Wang et al., "A Collaborative Scheduling-Based Parallel Solution for HEVC Encoding on Multicore Platforms," (Same as [E483]) IEEE Trans. MM, vol. 20, pp. 2935–2948, Nov. 2018.

The section conclusion and future works is repeated here.

Conclusion and Future Works

In this paper, an efficient solution named CSPS is proposed, which is capable of overcoming the disadvantages of WPP and exploiting the parallelism of HEVC encoding on multicore platforms. Prior to implementing CSPS, a performance model, which provides a theoretical deduction and a simulation algorithm, is developed to formulate WPP for HEVC encoding in terms of parallelization under various conditions. With the aid of the performance model, the shortages and limitations to parallelize HEVC encoding with WPP, which include load imbalance, ramping inefficiency and limited parallelism, can be discerned. The proposed CSPS employs a strategy named adaptive parallel mode decision to defeat load imbalance of WPP and fully explores the parallelism of multicore architectures, where multi-grained parallel tasks are dynamically allocated as required during the RDO process. The multi-grained task scheduling algorithm is designed to efficiently schedule parallel tasks with different granularities. To diminish ramping inefficiency, an asynchronous frame-level interpolation filter is utilized by CSPS during the ramp-up stage of WPP at the cost of minor memory requirements. The extensive experiments have been performed to verify the effectiveness of CSPS in terms of the amount of parallelism, coding efficiency and parallel scalability. In the future, it is desirable to investigate the parallelization of HEVC encoding on various parallel processing architectures and to reduce the overhead of parallel task scheduling. In the current work, the proposed CSPS focuses on parallelizing HEVC encoding within frames. However, it is feasible to integrate the proposed solution with GOP-based/frame-based TLP approaches and various ILP methods with minor efforts in order to further speedup HEVC encoding. In addition, the strategies designed for CSPS are not only applicable to HEVC encoding but also advisable for parallelizing other computationally intensive applications, especially for those adopting the wavefront paradigm."

[E487] T. Zhao et al, "3D reconstruction of human body shape from a single commodity depth camera," IEEE Trans. MM, EA.

[E488] S. Tian et al, "A benchmark of DIBR synthesized view quality assessment metrics on a new database for immersive media applications," IEEE Trans. MM, EA.

Conclusion and perspective is repeated here

Conclusion and Perspective

"DIBR is widely used in FVV, VR, AR, and other popular topics considered as the next generation of 3D broadcasting applications, in order to provide a better QoE to users. In this paper, a new DIBR-synthesized image database which focuses on the distortions induced by different state-of-the-art DIBR view synthesis algorithms is presented. Ten MVD sequences and seven state-of-the-art DIBR view synthesis algorithms are selected to generate the virtual view images. The subjective experiment is conducted following the SAMVIQ protocol in a controlled environment as recommended by ITU-R Rec. BT.1788 [48]. Results show that the inter-view synthesis methods, which have more input information, significantly outperform the single view based synthesis algorithms. Furthermore, several objective measurements were used to assess the quality of synthesized images on this database. Their performance results indicate that further work has to be done to exploit deeply the characteristics of these specific distortions, for new objective metrics with a better correlation with subjective scores. In current database, only the MPEG MVD source images are included; in the future work, more source images, such as the images from Middlebury database [36], will be considered to make the experiment results more benchmark. All the data of this presented database, including images, the ground truth depth maps and their associated DMOS, is publicly accessible (https://vaader-data.insa-rennes.fr/data/stian/ ieeetom/IETR DIBR Database.zip), for the improvement of the QOE of DIBR related applications."

MPEG-DASH

(DASH- Dynamic adaptive streaming technologies over HTTP)

[MD1] C. Cocoloto et al., "Adaptive streaming of HEVC tiled videos using MPEG-DASH," IEEE Trans. CSVT, V0l.28, pp. 1981–1992, Aug. 2018. To explain MPEG-DASH some sections in the introduction are repeated here.

"Adaptive HTTP streaming technologies, such as MPEG-DASH [3], enable the live or on-demand delivery of media streams over HTTP. The quality of the media delivered to the client can be widely adapted: to the user's preferences (e.g., language, point of view), to the client's capabilities (e.g., support for a given codec) and to fluctuating network conditions. This adaptability is client-driven, as the client seamlessly downloads and plays small continuous media files of a given quality and dynamically switches between alternative media files of different qualities offered by HTTP servers. The client only downloads the necessary quality given the situation based on its adaptation algorithm. Streaming of HEVC using MPEG-DASH has been demonstrated [4] and is in deployment within the industrial world.

Streaming of tiled videos has been studied in the academic world [5, 6] for lecture streaming or panoramic video streaming. Combining the concepts of HEVC tiling and adaptive HTTP streaming has not been studied and opens up the way for new streaming possibilities. A client can indeed select and adaptively stream only one or several tiles of interest to reduce the decoding requirements. Even when all tiles are transmitted, new adaptation algorithms can be designed to let the client select different qualities per tile thus reducing the overall bitrate while maintaining the best quality on the tiles of interest. Both approaches introduce some overhead: at the coding level or at the system level. Forcing tiles to be independent over time restricts the coding tools that can be used and therefore reduces the coding efficiency. The ability to efficiently stream tiles independently requires additional signalling at the MPEG-DASH level and additional packaging information, at the ISOBMFF level.

[MD2] D. Yun and K. Chung, "DASH-based multi-view video streaming system," IEEE Trans. CSVT, vol. 28, pp. 1974–1980, Aug. 2018.

MD-P1 See [MD1] Conclusion and future work is repeated here

"This paper has presented a novel tile-based streaming system which relies on a single, compliant HEVC decoder on the client-side, embedded in a standard adaptive streaming system based on ISOBMFF and MPEG-DASH. Using this system, high resolution video content is split into independent tiles using motion-constrained encoding. These tiles are then individually packaged into segmented ISOBMFF files, described with their tiling association within an MPEG-DASH description. Experimentations show that such a description enables an open-source, standard DASH client to download,decode and present some or all the tiles, as desired. A thorough evaluation indicates that this system induces an overhead of at most 7.3% when using 25 tiles, and at most 3.5% for 9 tiles, on UHD content encoded at 10 Mbps or above, compared to when tiles are not used and therefore when no spatial access is provided. Compared to an approach using separate bitstreams, which also provides spatial access, our approach induces only a limited overhead of less than 2% but with the benefit of using standard adaptive streaming and a single decoder instance. With our approach it therefore becomes possible to deliver UHD videos even on networks with limited bandwidth, by delivering only some tiles of interest or by adaptively varying the quality of each tile. This opens the way for more research in the design of tile-based adaptation algorithms." Explore this research.

High dynamc range image/video coding

[HDR1] R. Mukherjee et al., "Uniform colour space based high dynamic range video coding," IEEE Trans. CSVT, EA.

Conclusion and future work are repeated here.

"In this paper, we have proposed a novel HDR video compression algorithm optimized for 10-bit encoding. The algorithm uses the IPT uniform colour opponent space, a novel PATF providing better image reconstruction than existing PTFs and a novel EMF for accurate chroma preservation. Objective evaluation results demonstrate that the proposed algorithm performs at par or better than the four published and/or patented state-of-the-art algorithms. The proposed algorithm can also use existing PTFs albeit at the cost of reconstruction quality. Future work also includes further refinement of the PATF and EMF to further reduce quantization errors. Moreover, the use of other colour spaces such as $L\alpha\beta$ for the colour space transform phase can be also be explored as well as adapting curves per scenario [40]. Furthermore, the algorithm can also be further evaluated across other HDR video sequences especially those processed with the BT.2020 primaries [41] at higher bit-depths such as 12- and 14-bits/pixel/channel when optimized codec support becomes available. Finally, the objective and subjective evaluations can also be extended using two recently developed quality assessment metrics using eye-tracking data such as the no-reference metric proposed in [42] and spatiotemporal correlation data using eye tracking as proposed in [43]. However, their suitability for HDR video content also needs to be verified before they can be used to evaluate the HDR video compression algorithms."

[HDR2] Y. Liu et al., "An adaptive quantizer for high dynamic range content: Application to video coding," IEEE Trans. EA.

[HDR3] P. Hanhart et al., "Benchmarking of Objective Quality Metrics for HDR Image Quality Assessment," EURASIP Journal on Image and Video Processing, vol. 2015, no. 1, pp. 1–18, 2015.

[HDR4] M. Sousa, M. V. Bernardo, and M. G. A. Pinheiro, "HDR Image Compression: A New Challenge for Objective Quality Metrics," International Workshop on Quality of Multimedia Experience - QoMEX, vol. 1, pp. 1–6, Sept. 2014.

[HDR5] C. Mantel, S. C. Ferchiu, and S. Forchhammer, "Comparing Subjective and Objective Quality Assessment of HDR Images Compressed with JPEG-XT," IEEE, 2014.

[HDR6] P. Rerabek et al., "Subjective and Objective Evaluation of HDR Video Compression," 9th International Workshop on Video Processing and Quality Metrics for Consumer Electronics (VPQM), pp. 1–6, Jan.2015.

[HDR7] R. Mantiuk et al., "HDR-VDP2: A Calibrated Visual Metric for Visibility and Quality Predictions in All Luminance Conditions," ACM Trans. Graph., vol. 30, no. 4, pp. 40:1–40:14, July 2011.

[HDR8] M. Narwaria, M. P. Da Silva, and P. Le Callet, "HDR-VQM: An Objective Quality Measure for High Dynamic Range Video," Signal Processing: Image Communication, vol. 35, pp. 46–60, July 2015.

[HDR9] S. Mahmalat et al., "Pipelines for HDR video coding based on luminance independent chromacity preprocessing," IEEE Trans. CSVT, EA.

[HDR10] P. Ledda, "Product review: High dynamic range displays," Presence, vol. 16, no. 1, pp. 119–122, Feb. 2007.

[HDR11] E. Reinhard, "High Dynamic Range Imaging: Acquisition, Display, and Image-Based Lighting.," San Francisco, CA, Morgan Kaufmann Publishers, 2010.

[HDR12] R. Mantiuk et al., "Perception motivated high dynamic range video encoding," ACM SIGGRAPH 2004 Papers, pp. 733–741, ser. SIGGRAPH'04. New York, NY, USA: ACM, 2004.

[HDR13] R. Mantiuk, K. Myszkowski, and H. Seidel, "Lossy compression of high dynamic range images and video," Human Vision and Electronic Imaging XI, Feb. 2006, pp. 60 570V–10., pp. 1–9, Oct. 2012.

[HDR14] D. Touze et al., "HDR video coding based on local LDR quantization," in HDRi2014 - Second International Conference and SME Workshop on HDR imaging, Sarajevo, 2014.

[HDR15] R. Boitard, R. Mantiuk, and T. Pouli, "Evaluation of colour encodings for high dynamic range pixels," Human Vision and Electronic Imaging, vol. 9394, pp. 93 941K–9. 2015.

[HDR16] J. Strom, J. Sole, and Y. He, "Report of HDR CE1," Joint Collaborative Team on Video Coding (JCT-VC), JCT-VC m37605, Feb. 2016.

[HDR17] M. Azimi et al., "Evaluating the performance of existing full-reference quality metrics on High Dynamic Range (HDR) video content," vol. 1, no. 11, p. 788, 2014.

[HDR18] E. Zerman, G. Valenzise, and F. Dufaux, "An extensive performance evaluation of full-reference HDR image quality metrics," Quality and User Experience, vol. 2, no. 1, p. 5, 2017.

[HDR19] HDR-VQM software. [Online]. Available: https://sites.google.com/site/narwariam/hdr-vqm

HDR-P1 See [HDR1]. Explore the future work suggested at the end. This is a major research and may require group effort and extensive lab/testing facilities.

ATSC Advanced Television Systems Committee
TV Tomorrow: ATSC 3.0 Advances

ATSC 3.0 will be a radical departure from the current standard ATSC 1.0 which was developed around 20 years ago, when cellphones were analog and streaming was unheard of. It relies 8-VSB modulation capable of delivering 19.39 Mbps in a 6 MHz TV channel – enough to carry a high – definition program compressed by a factor of 50 using MPEG-2 to a fixed receiver.

With ATSC 3.0, the committee seeks to increase that data rate by 30 percent, or roughly 25.2 Mbps. The overall intent of 3.0 is to enable seamless transmission of HD, 4K, 22.2 audio and other data streams to fixed, mobile and handheld devices in all types of terrain.

The most differentiating characteristic of ATSC 3.0 is that it will not be backward – compatible with 1.0 or even 2.0, which is now in development. In other words, televisions now capable of processing over-the-air TV signals will not be able to decode ATSC 3.0 signals. It is also being developed with a global perspective in mind, meaning that modulation schemes other than 8-VSB – particularly co-orthogonal frequency division multiplexing, or COFDM – will likely be on the table. For details see

L. Fay et al., "An overview of the ATSC 3.0 physical layer specification" IEEE Trans. Broadcasting, vol. 62, no.1, pp. 159–171, March 2016.

J.M. Boyce, "The U.S. digital television broadcast terrestrial standard," IEEE SP Magazine, May 2009.

NABSHOW ATSC 3.0: (National Association of Broadcasters)

https://mentor.ieee.org/802.18/dcn/12/18-12-0011-00-0000-nab-presentation-on-atsc-3.pdf

ATSC Standards:

http://www.atsc.org/cms/standards/a53/A53-Part-1-2013.pdf
http://www.atsc.org/cms/standards/a53/a_53-Part-2-2011.pdf
http://www.atsc.org/cms/standards/a53/A53-Part-3-2013.pdf
http://www.atsc.org/cms/standards/a53/a_53-Part-4-2009.pdf
http://www.atsc.org/cms/standards/a53/a_53-Part-5-2010.pdf
http://www.atsc.org/cms/standards/a53/A53-Part-6-2013.pdf

(parts 1 and 3 are recent updates - 2013).
There are some papers related to ATSC:

http://www.tvnewscheck.com/article/75280/updating-tvs-progress-toward-atsc-30
http://www.telestream.net/pdfs/whitepapers/wp-calm-act-compliance.pdf

(ATSC MOBILE DTV)

ATSC Mobile DTV is being developed to support a variety of services including free (advertiser-supported) television and interactive services delivered in real-time, subscription-based TV, and file-based content download for playback at a later time. The standard may also be used for transmission of new data broadcasting services.

The A/153 document set has been approved by TSG and will go to membership ballot shortly.

278 *Appendix*

The version history of this document set is as follows:
- Initial release: 31 December 2009
- Update #1: 12 February 2009. Editorial and/or substantive changes were made in all Parts.
- Update #2: 13 March 2009. Substantive changes were made in Part 2; editorial changes (correcting typos) were made in Part 3.
- Update #3: 9 April 2009. Substantive changes were made in Part 3, Part 4, and Part 6; editorial changes were made in Part 1 and Part 2.
- Update #4: 29 May 2009. Substantive changes were made in all Parts.
- Update #5: 10 September 2009. Substantive changes were made in Parts 2 and 3.

The ATSC Mobile DTV Candidate Standard (document number A/153) is composed of separate Parts as follows:
- Part 1 – Mobile/Handheld Digital Television System
- Part 2 – RF/Transmission System Characteristics
- Part 3 – Service Multiplex and Transport Subsystem Characteristics
- Part 4 – Announcement
- Part 5 – Presentation Framework
- Part 6 – Service Protection
- Part 7 – Video System Characteristics
- Part 8 – Audio System Characteristics

All eight Parts can also be downloaded as a ZIP file.

Changes relative to the previous revision number are detailed in the redline ZIP file.

Part 1 of A/153 includes an overall system description and serves to tie the other Parts of the document together.

Note that editorial changes are likely to be made in these documents prior to publication as an ATSC Standard.

Audio Measurement Documents
- CS/TSG-757r0, "Candidate Standard: Amendment No. 1 to ATSC Digital Television Standard (A/53) Part 5:2007"
- CS/TSG-758r0, "Candidate Standard: Amendment No. 1 to ATSC Digital Television Standard (A/53) Part 6:2007"

The purpose of these amendments is to change the method of measuring the loudness of the average spoken program dialogue from the original "LAeq" method described in ANSI document ANSI S1.4-1983 (R 2001)

"Specification for Sound Level Meters," with amendment S1.4A-1995, to the method described by ITU-R Recommendation BS.1770, "Algorithms to measure audio program loudness and true-peak

Richard Schwartz, Axcera, LLC, Lawrence, PA, US, "Integrating ATSC M/H into a broadcast transmission system: options and choices – what you need to know.

ATSC3.0 (Next Gen TV) 4K Ultra HD HDR broadcasts ATSC3.0 SYSTEM (A/300) Doc. A/300; 2017.

ATSC1 S. Myung et al., "Offset and normalized min-sum algorithms for ATSC3.0 LDPC decoder," IEEE Trans. Broadcasting, vol. 63, pp. 734–739, Dec. 2017. Several papers related to ATSC3.0 are listed in the references.

ATSC2 S.-Ik Park et al., "Field comparison tests on LDM and TDM in ATSC," IEEE Trans. Broadcasting, EA,

ATSC3 ATSC System Discovery and Signaling document A/321:2016, Adv. Television Syst. Committee, Washington, DC, USA, Mar. 2016.

ATSC4 ATSC Physical Layer Protocol Standard document A/322: 2017, Adv. Television syst. Committee, Washington, DC, USA, Feb. 2017.

ATSC5 L. Fay et al., "An overview of ATSC 3.0 physical layer specification," IEEE Trans. B, vol. 62, pp. 159–171, Mar. 2016.

ATSC6 M. Fuentes et al., "Physical Layer Performance Evaluation of LTE-Advanced Pro Broadcast and ATSC 3.0 Systems," IEEE Trans. B, EA

HEVC Licensing:
HEVC is part of MPEG LA patent pool. However some patent owners are dissatisfied with the low fees being charged and split off to form HEVC Advance – a *pool*.

IEEE ICCE, Las Vegas, NV, 8–11 Jan. 2016 (ICCE international conference on consumer electronics)

Papers on HEVC: (These papers can lead to several projects.)

ICCE1. Error Resilience Aware Motion Estimation and Mode Selection for HEVC Video Transmission

Gosala Kulupana, Dumidu S. Talagala, Hemantha Kodikara Arachchi and **Anil Fernando** (University of Surrey, United Kingdom)

Error concealment techniques such as motion copying require significant changes to HEVC (High Efficiency Video Coding) motion estimation process when incorporated in error resilience frameworks.

This paper demonstrates a novel motion estimation mechanism incorporating the concealment impact from future coding frames to achieve an average 0.73dB gain over the state-of-the-art.

ICCE2. H.721: Standard for Multimedia Terminal Devices Supporting HEVC

Fernando M Matsubara (Mitsubishi Electric Corporation, Japan); **Hideki Yamamoto** (Oki, Japan)

The newest ITU-T H.721 standard "IPTV terminal devices: Basic model" is described. Main goal of this revision is to enhance content quality by adopting latest video coding and adaptive streaming techniques. Selected technologies include HEVC and MPEG-DASH. This paper summarizes the standard and carefully chosen profiles to maximize perceived quality.

ICCE3. A Downhill Simplex Approach for HEVC Error Concealment in Wireless IP Networks

Kyoungho Choi (Mokpo National University, Korea); **Do Hyun Kim** (ETRI, Korea)

In this paper, a novel video error concealment algorithm is presented for HEVC in wireless IP networks. In the proposed approach, a downhill simplex approach is adopted for fine-tuning motion vectors, considering residual errors and block reliability, and minimizing boundary errors along prediction blocks at the same time.

ICCE4. Power Consumption Comparison Between H.265/HEVC and H.264/AVC for Smartphone Based on Wi-Fi, 3G and 4G Networks

Min Xu (California State University, Long Beach, USA); **Xiaojian Cong** (California State University, Long Beach, USA); **Qinhua Zhao** (California State University, Long Beach, USA); **Hen-Geul Yeh** (California State University, Long Beach, USA)

The H.265/HEVC can help reduce much bandwidth for streaming video on mobile networks where wireless spectrum is at a premium. However, it will take more computing power for decoding. This paper is to perform a power consumption evaluation of streaming and decoding H.265/HEVC and H.264/AVC video for smartphones.

ICCE5. A New Scene Change Detection Method of Compressed and Decompressed Domain for UHD Video Systems

Yumi Eom (Seoul National University of Science and Technology, Korea); **Sang-Il Park** (Korea Communications Commission, Korea); **Chung Chang Woo** (Seoul Natl. University of Science and Technology, Korea)

We propose a new method using two layers that can detect the scene change frames in a fast and accurate way. Also, we propose an algorithm that can be applied to the HEVC. Through the experimental results, our proposed method will helps UHD video contents analysis and indexing.

ICCE6. Improvement of H.265/HEVC Encoding for 8K UHDTV by Detecting Motion Complexity

Shota Orihashi (Waseda University & Graduate School of Fundamental Science and Engineering, Japan); **Harada Rintaro** (Waseda University & Fundamental Science and Engineering, Japan); **Yasutaka Matsuo** (Japan Broadcasting Corporation (NHK), Japan); **Jiro Katto** (Waseda University, Japan)

We propose a method to improve H.265/HEVC encoding performance for 8K UHDTV moving pictures by detecting amount or complexity of object motions. The proposed method estimates motion complexity by external process, and selects an optimal prediction mode and search ranges of motion vectors for highly efficient and low computation encoding.

ICCE7. Design of Multicore HEVC Decoders Using Actor-based Dataflow Models and OpenMP

Miguel Chavarrias, Matias J Garrido and **Fernando Pescador** (Universidad Politécnica de Madrid, Spain); **Maxime Pelcat** (INSA Rennes, France); **Eduardo Juarez** (Universidad Politécnica de Madrid, Spain)

This paper explains a backend for the RVC compiler framework. This backend uses OpenMP instead of the pthreads to automatically generate code for multicore architectures. Implementations of an HEVC decoder have been automatically generated for three multicore architectures. The backend doesn't introduce a performance penalty regarding to the C backend.

ICCE8. Perceptual Distortion Measurement in the Coding Unit Mode Selection for 3D-HEVC

Sima Valizadeh (University of British Columbia, Canada); **Panos Nasiopoulos** (The University of British Columbia, Canada); **Rabab Ward** (University of British Columbia, Canada)

In this paper, we propose to integrate a perceptual video quality metric inside the rate distortion optimization process of the 3D-HEVC. Specifically, in the coding unit (CU) mode selection process, PSNR-HVS is used as a measure for distortion. Our proposed approach improves the compression efficiency of the 3D-HEVC.

ICCE9. CTU Level Decoder Energy Consumption Modelling for Decoder Energy-Aware HEVC Encoding

Thanuja Mallikarachchi (University of Surrey & University of Surrey, United Kingdom); **Hemantha Kodikara Arachchi, Dumidu S. Talagala** and **Anil Fernando** (University of Surrey, United Kingdom)

Accurate modelling of the decoding energy of a CTU is essential to determine the appropriate level of quantization required for decoder energy-aware video encoding. The proposed method predicts the number of nonzero DCT coefficients, and their energy requirements with an average accuracy of 4.8% and 11.19%, respectively.

ICCE10. Real-time Encoding/Decoding of HEVC Using CUDA Capable GPUs

Tony James (Smartplay Technologies Pvt Ltd, Georgia)

HEVC is latest video standard from JCT-VC. It offers better compression performance compared to H.264. However, the computational complexity is highly demanding for real-time applications especially at UHD resolutions. In this paper, we propose few optimization methods to achieve real-time encoding/decoding on CUDA capable GPUs. Experimental results validate our proposal.

ICCE11. SATD Based Intra Mode Decision Algorithm in HEVC/H.265

Jongho Kim (ETRI, Korea)

HEVC which is the next generation video coding standard provides up to 35 intra prediction modes to improve the coding efficiency. These various prediction modes bring a high computation burden. In this paper, the fast intra mode decision algorithm is proposed.

ICCE12. Content Dependent Intra Mode Selection for Medical Image Compression Using HEVC

Saurin Parikh (Florida Atlantic University & Nirma University, USA); **Damian Ruiz** (Universitat Politécnica de Valéncia, Spain); **Hari Kalva** (Florida Atlantic University, USA); **Gerardo Fernandez** (University of Castilla La Mancha, Spain)

This paper presents a method for complexity reduction in medical image encoding that exploits the structure of medical images. The HEVC lossless intra coding of medical images of CR modality, shows reduction upto 52.47% in encoding time with a negligible penalty of 0.22%, increase in compressed file size.

ICCE13. Decoding of Main 4:2:2 10-bit Bitstreams in HEVC Main 8-bit Best-Effort Decoders

Woo-Seok Jeong, Hyunmin Jung and **Soo-ik Chae** (Seoul National University, Korea)

This paper describes a new method of best-effort decoding that substantially reduces PSNR loss especially in intra prediction. For intra prediction in the decoder, reference pixels are decoded in the 4:2:2 10-bit format and non-reference pixels are in the 4:2:0 8-bit format.

http://x265.org
   This is the HEVC encoder developed by Multicoreware Inc. The main difference from HM is that x265 is fast compared to HM, they have assembly level optimizations up to avx2 so far.

https://x265.readthedocs.org/en/default/api.html
This is the link to their documentation. Even ffmpeg has x265 integrated in itself, but I don't think we can use x265 in ffmpeg, since it will take a while to find where x265 is implemented in ffmpeg and then modify it and build it.

The following papers are presented in SPIE PHOTONICS WEST, February 2015, San Francisco, CA
Link: http://spie.org/EI/conferencedetails/real-time-image-video-processing

1. **Efficient fast thumbnail extraction algorithm for HEVC**
   Paper 9400-15

Author(s): Wonjin Lee, Hanyang Univ. (Korea, Republic of); Gwanggil Jeon, Univ. of Incheon (Korea, Republic of); Jechang Jeong, Hanyang Univ. (Korea, Republic of)

The proposed algorithm is fast thumbnail extraction algorithm for HEVC. The proposed method only reconstructs 4×4 boundary pixels which are needed for thumbnail image. The proposed method generates the thumbnail image without full decoding. The proposed algorithm can reduce the computational complexity by performing only calculation needed for thumbnail image extraction. The proposed algorithm can significantly reduce the computational complexity for intra prediction and inverse transform. Also, it does not perform de-blocking filter and sample adaptive offset (SAO). After partial decoding process, it could be concluded that the proposed method significantly reduces the decoding time. In addition, the visual quality of obtained thumbnail images was almost identical to thumbnail images extracted after full decoding process.

2. **A simulator tool set for evaluating HEVC/SHVC streaming**
   Paper 9400-22
   Author(s): James M. Nightingale, Tawfik A. Al Hadhrami, Qi Wang, Christos Grecos, Univ. of the West of Scotland (United Kingdom); Nasser Kehtarnavaz, The Univ. of Texas at Dallas (United States)

Since the delivery of version one of H.265, the Joint Collaborative Team on Video Coding have been working towards standardization of a scalable extension (SHVC), a series of range extensions and new profiles. As these enhancements are added to the standard the range of potential applications and research opportunities will expand. For example the use of video is also growing rapidly in other sectors such as safety, security, defense and health with real-time high quality video transmission playing an important role in areas like critical infrastructure monitoring and disaster management, each of which may benefit from the application of enhanced HEVC/H.265 and SHVC capabilities. The majority of existing research into HEVC/H.265 transmission has focused on the consumer domain with the lack of freely available tools widely cited as an obstacle to conducting this type of research. In this paper we present a toolset that facilitates the transmission and evaluation of HEVC/H.265 and SHVC encoded video on an open source emulator. Our toolset provides researchers with a modular, easy to use platform for evaluating video transmission and adaptation proposals on large-scale wired, wireless and hybrid architectures. The proposed toolset would significantly facilitate further research in delivering adaptive video streaming based on the latest video coding standard over wireless mesh and other ad hoc networks.

3. **Subjective evaluation of H.265/HEVC based dynamic adaptive video streaming over HTTP (HEVC-DASH)**
Paper 9400-24
Author(s): Iheanyi C. Irondi, Qi Wang, Christos Grecos, Univ. of the West of Scotland (United Kingdom)

With the surge in Internet video traffic, real-time HTTP streaming of video has become increasingly popular. Especially, applications based on the MPEG Dynamic Adaptive Streaming over HTTP standard (DASH) are emerging for adaptive Internet steaming in response to the unstable network conditions. Integration of DASH streaming technique with the new video coding standard H.265/HEVC is a promising area of research in light of the new codec's promise of substantially reducing the bandwidth requirement. The performance of such HEVC-DASH systems has been recently evaluated using objective metrics such as PSNR by the authors and a few other researchers. Such objective evaluation is mainly focused on the spatial fidelity of the pictures whilst the impact of temporal impairments incurred by the nature of reliable TCP communications is also noted. Meanwhile, subjective evaluation of the video quality in the HEVC-DASH streaming system is able to capture the perceived video quality of end users, and is a new area when compared with the counterpart subjective studies for current streaming systems based on H.264-DASH. Such subjective evaluation results will shed more light on the Quality of Experience (QoE) of users and overall performance of the system. Moreover, any correlation between the QoE results and objective performance metrics will help designers in optimizing system performance. This paper presents a subjective evaluation of the QoE of a HEVC-DASH system implemented in a hardware test bed. Previous studies in this area have focused on using the current H.264/AVC or SVC codecs and have hardly considered the effect of Wide Area Network (WAN) characteristics. Moreover, there is no established standard test procedure for the subjective evaluation of DASH adaptive streaming. In this paper, we define a test plan for HEVC demonstrate the bitrate switching operations in response to various network condition patterns. The test bed consists of real-world servers (web server and HEVC-DASH server), a WAN emulator and a real-world HEVC-DASH client. -DASH with a carefully justified data set taking into account longer video sequences that would be sufficient to We evaluate the QoE by investigating the perceived impact of various network conditions such as different packet loss rates and fluctuating bandwidth, and the perceived quality of using different DASH video stream segment sizes on a video streaming session and using different video content types. Furthermore, we demonstrate the temporal structure and impairments as identified by previous objective quality metrics and capture how they are perceived by the subjects. The Mean Opinion Score (MOS) is employed and a beyond MOS evaluation method is designed based on a questionnaire that gives more insight into the performance of the system and the expectation of the users. Finally, we explore the correlation between the MOS and the objective metrics and hence establish optimal HEVC-DASH operating conditions for different video streaming scenarios under various network conditions.

NAB2014, Las Vegas, NV April 2014 (following 3 papers) (National Association of Broadcasters)

NAB is held every year in Las Vegas, NV generally in April. It is a B2B convention with nearly 10,000 attendees and has innumerable products related to broadcasting.

1. J. Pallett, (Telestream, Inc.) "HEVC; comparing MPEG-2, H.264 and HEVC"
2. S. Kumar, (Interra systems) "Aspects of video quality assessment for HEVC compressed format,"
3. Y. Ye (Interdigital) , "HEVC video coding"

NAB2015, Las Vegas, NV, April 2015 (following are the papers related to HEVC) (National Association of Broadcasters)

1. The New Phase of Terrestrial UHD Services: Live 4K UHD Broadcasting via Terrestrial Channel.
2. The Combination of UHD, MPEG-DASH, HEVC and HTML5 for New User Experiences.
3. Implications of High Dynamic Range on the Broadcast Chain for HD and Ultra-HD Content.
4. Beyond the Full HD (UHD & 8K).

NTT-electronics has developed HEVC/H.265 based ASICs and codec systems in 2015.

Digital video and systems business group.

http://www.ntt-electronics.com/en

Special Issues

[Spl H1] Special issue on emerging research and standards on next generation video coding, IEEE Trans. CSVT, vol. 22, pp. 1646–1909, Dec. 2012.

[Spl H2] Introduction to the issue on video coding: HEVC and beyond, IEEE Journal of selected topics in signal processing, vol. 7, pp. 931–1151, Dec. 2013.

[Spl H3] CALL FOR PAPERS, IEEE TRANSACTIONS ON CIRCUITS AND SYSTEMS FOR VIDEO TECHNOLOGY Special Section on Efficient HEVC Implementation (M. Budagavi, J.R. Ohm, G.J. Sullivan, V. Sze and T. Wiegand Guest editors).

[Spl H4] Special Issue on New Developments in HEVC Standardization and Implementations Approved by the CSVT editorial board, July 2014.

[Spl H5] IEEE Journal on Emerging and Selected Topics in Circuits and Systems (JETCAS) Special Issue on Screen Content Video Coding and Applications. Vol. 6, issue 4, pp. 389–584 Dec. 2016.

[Spl H6] Journal of real-time image processing: Special issue on architectures and applications of high efficiency video coding (HEVC) standard for real time video applications. B.-G. Kim, K. Psannis and D.-S. Jun (Editors).

[Spl H7] Special issue on Broadcasting and Telecommunications Media technology, ETRI journal, Deadline: 12/31/2015 and Publication: Oct. 2016. Website: http://etrij.etri.re.kr

[Spl H8] Special issue on HEVC extensions and efficient HEVC implementations, IEEE Trans. CSVT, vol. 26, pp. 1–249, Jan. 2016.

[Spl H9] Special issue on quality of experience for advanced broadcast services, IEEE Trans. Broadcasting, vol. 64, issue2, part 2, pp. 335–634, June 2018.

Special Sessions

SS1 J. Han, "Recent advances in video compression technology in open codecs," IEEE ICIP 2017, Beijing, China, Sept. 2017.

HEVC Overview Online

S. Riabstev, "Detailed overview of HEVC/H.265," [online]. Available: https://app.box.com/s/ rxxxzr1a1lnh7709yvih (accessed on June 12 2014).

HEVC tutorial by I.E.G. Richardson: http://www.vcodex.com/h265.html

JCT-VC DOCUMENTS can be found in JCT-VC document management system

http://phenix.int-evry.fr/jct (see [E183])

All JCT-VC documents can be accessed. [online]. Available: http://phenix.int-evry.fr/jct/doc_end_user/current_meeting.php?id_meeting=154&type_order =&sql_type=document_number

VCEG & JCT documents available from

http://wftp3.itu.int/av-arch in the video-site and jvt-site folders (see [E183])

HEVC encoded bit streams

ftp://ftp.kw.bbc.co.uk/hevc/hm-11.0-anchors/bitstreams/

Test Sequences/Data Bases

TS.1. http://media.xiph.org/video/derf/ Xiph.org video test media
TS.2. http://trace.eas.asu.edu/yuv/
TS.3. http://media.xiph.org/
TS.4. http://www.cipr.rpi.edu/resource/sequences/
TS.5. HTTP://BASAK0ZTAS.NET
TS.6. www.elementaltechnologies.com – 4K Video Sequences

> 4K (3840 × 2160) UHD video test sequences
> Multimedia communications with SVC, HEVC, and HEVC
> http://r2d2n3po.istory.com/50

TS.7. Elemental 4K Test Clips.: [online] Available: http://www.elementaltechnologies.com/resources/4k-test-sequences, accessed Aug. 1, 2014.
TS.8. Harmonic 4K Test Clips: [online] Available: http://www.harmonicinc.com/resources/videos/4k- video-clip-center, accessed Aug. 1, 2014.
TS.9. Kodak Lossless True Color Image Suite. [online]. Available: http://r0k.
us/graphics/kodak/
TS.10. Test sequences from dropbox
https://www.dropbox.com/sh/87snmt6j70qquyb/AADuHPbf1o9Y-YnB8N4lCt-Na?dl=0
TS.11. F. Bossen, "Common test conditions and software reference configuration," JCT-VC-1100, July 2012 (Access to test sequences and QP values)
TS.12. Test Sequences: ftp://ftp.kw.bbc.co.uk/hevc/hm-11.0-anchors/bitstreams/
TS.13. D. Ghadiyaram, J. Pan, and A.C. Bovik, "A Subjective and Objective Study of Stalling Events in Mobile Streaming Videos," IEEE Trans.

Image Proc., 2017. There are test sequences for Mobile Streaming and many other video quality evolution.
http://live.ece.utexas.edu/research/quality/index.htm

TS.14. Images of camera captured content: http://www.lifeisspiritual.com/wp-content/uploads/2014/06/Peace-Nature-Jogging.jpg

TS.15. Image of a video clip with text overlay http://www.newscaststudio.com/graphics/bloomberg/

TS.16. Video test sequences - http://forum.doom9.org/archive/index.php/t-135034.html

TS.17. European Broadcasting Union "EBU UHD-1 Test Set," 2013 Available online

  1. To access these sequences free there is an order form (to be filled) subject to the rules and regulations. https://tech.ebu.ch/testsequences/uhd-1

Thanks for the hint, it seems that EBU change the location, please check the new location that I found on the EBU website: https://tech.ebu.ch/Jahia/site/tech/cache/offonce/testsequences/uhd-1

BR, Benjamin

On Sep 14, 2014, at 10:54 PM, "Rao, Kamisetty R" <rao@uta.edu> wrote:
In one of your papers you cited : EBU UHD-1 Test Set (online)
http://tech.edu.ch/testsequences/uhd-1

I am unable to access this web site. Please help me on this. I like to access this set for my research group. Thanks.

TS.18. http://mcl.usc.edu/

The MCL-JCI dataset consists of 50 source images with resolution 1920 × 1080 and 100 JPEG-coded images for each source image with the quality factor (QF) ranging from 1 to 100.

Here is the link to some useful videos on HEVC standard

http://www.youtube.com/watch?v=YQs9CZY2MXM
You might also find this useful and may subscribe to the same
http://vcodex.com/h265.html
HEVC/H.265 PRESENTATION (Krit Panusopone)
www.box.com/s/rxxxzr1a1lnh7709yvih

TS.19. "The New Test Images - Image Compression Benchmark," Rawzor-Lossless Compression Software for Camera RAW Images [Online],

Available: http://imagecompression.info/test images/.
Implementation of Intra Prediction in the HM reference encoder: https://hevc.hhi.fraunhofer.de/svn/svn_HEVCSoftware/tags/HM-12.0/

HM 10.1 SOFTWARE
https://hevc.hhi.fraunhofer.de/svn/svn_HEVCSoftware/branches/HM-10.1-dev/
Latest is HM16.9
Software reference manual for HM. https://hevc.hhi.fraunhofer.de/svn/svn_HEVCSoftware/branches/HM-16.9-dev/doc/software-manual.pdf

TS.20. "The new test images – Image compression benchmark," http://imagecompression.info/test_images

TS.21. BOSSbase image data set 512 × 512 test images
IEEE Jo*BossBase-1.01-hugo-alpha=0.4.tar.bz2*. [Online]. Available: http://agents.fel.cvut.cz/stegodata/, accessed Mar. 2014.

TS.22. http://www3.americanradiology.com/pls/web1/wwimggal.vmg

TS.23. K. Seshadrinathan, A. C. Bovik and L. K. Cormack. (2010). *LIVE Video Quality Database*. [Online].
Available: http://live.ece.utexas.edu/ research/quality/live_video.html

TS.24. S. Péchard, R. Pépion, and P. Le Callet. *IRCCyN/IVC Videos 1080i Database*. [Online]. Available: http://www.irccyn.ec-nantes.fr/~lecallet/platforms.htm, accessed Oct. 16, 2015.

TS.25. F. Zhang, S. Li, L. Ma, Y. C. Wong, and K. N. Ngan.
IVP Subjective Quality Video Database. [Online].
Available: http://ivp.ee.cuhk.edu.hk/research/databas/subjective/, accessed Oct. 16, 2015.

TS.26. J.-S. Lee et al., (2010). *MMSP Scalable Video Database*. [Online].
Available:http://mmspg.epfl.ch/svd
http://mmspg.epfl.ch/hdr-eye   HDR data set

TS.27. J. Guo et al., "Response to B1002 Call for Test Materials: Five Test Sequences for Screen Content Video Coding May 2016." Document JVET-C0044, ITU-T SG 16 WP3 and ISO/IEC JTC1/SC29/WG 11, May 2016.

TS.28. [https://testimages.org]
The TESTIMAGES archive is a huge and free collection of sample images designed for analysis and quality assessment of different kinds of displays (i.e. monitors, televisions and digital cinema projectors) and image processing techniques. The archive includes more than

2 million images originally acquired and divided in three different categories: SAMPLING and SAMPLING_PATTERNS (*aimed at testing resampling algorithms*), COLOUR (*aimed at testing colour rendering on different displays*) and PATTERNS (*aimed at testing the rendering of standard geometrical patterns*).

TS.29. Xiph.org Video Test Media (online) Available https://media.xiph.org/video/derf online test video database

TS.30. Screen content images (SCI) database developed by Z. Ni et al., See [SCC49]. This database has 1800 distorted SCIs generated from 40 reference SCIs. Open to public.
http://smartviplab.org/pubilcations/SCID.html

TS.31. HDR images from HD (1920 × 1080) to (6032 × 4018) including computer generated images
http://rit-mcsl.org/fairchild/HDR.html

TS.32. BOSSbase image data set http://www.agents.cz/boss/BOSSFinal IMAGE DATA SET

TS.33. N. Ponomarenko, V. Lukin, A. Zelensky, K. Egiazarian, M. Carli, and F. Battisti, "Tid2008—A database for evaluation of full-reference visual quality assessment metrics," *Adv. Modern Radioelectron.*, vol. 10, no. 4, pp. 30–45, 2009.

TS.34. N. Ponomarenko et al., "Colour image database TID2013: Peculiarities and preliminary results," in Proc. 4th European Workshop Vis. Inf., June 2013, pp. 106–111.

TS.35. R. C. Gonzalez and R. E. Woods. 2008. Standard Test Image Database. [Online]. Available FTP: http://www.imageprocessingplace.com/ downloads_V3/root_downloads/image_databases/standard_test_images.

TS.36. H. Yu, R. Cohen, K. Rapaka, and J. Xu, Common Test Conditions for Screen Content Coding, document JCTVC-T1015, 20th Meeting, Geneva, Switzerland, Feb. 2015.

Test sequences include the text and graphics with motion part specified in the common test condition (CTC). These sequences are FlyingGraphics, Desktop, Console, WebBrowsing, Map, Programming, and SlideShow. To verify our method, we also test some other screen sequences that were used by JCT-VC. They are SlideEditing, WordEditing, CADWaveform, VideoConferencingDocSharing, Web, and PPTDocXLS. All the sequences are in YCbCr 4:4:4 format, except for SlideEditing that is in YCbCr 4:2:0 format.

TS.37. user generated content (UGC): X. Ma et al., "Huawei test sequences of UGC feature for video coding development," in *Proc. 22th JVT-VC Meeting*, Geneva, Switzerland, Oct.

The selected sequences display a wide variety of motion caused by camera zooms, camera rotations, camera shakes, and classical 2D translational motion.

TS.38. N. Asuni and A. Giachetti, "TESTIMAGES: A large-scale archive for testing visual devices and basic image processing algorithms," in Proc. Eurograph. Italian Chapter Conf.pp. 63–70, 2014.

TS.39. G. Schaefer and M. Stich, "UCID: An uncompressed colour image database,"Proc. SPIE, vol. 5307, pp. 472–481, Dec. 2003.

TS.40. D. Martin et al., "A database of human segmented natural images and its application to evaluating segmentation algorithms and measuring ecological statistics," Proc. 8th IEEE Int. Conf. Comput. Vision (ICCV), vol. 2, July 2001.

TS.41. sjtu test images SJTU Shanghai Jae Tong University, Shanghai, China. Y. Zhu, L. Song, R. Xie, and W. Zhang, "SJTU 4K video subjective quality dataset for content adaptive bit rate estimation without encoding," in Proc. IEEE Int. Symp. Broadband Multimedia Syst. Broadcast., 2016, pp. 1–4.

TS.42. Dataset of standard 512512 grayscale test images," http://decsai.ugr.es/cvg/CG/base.htm

TS.43. NASA image and video library, https://images.nasa.gov/n#/

TS.44. G. B. Huang et al., "Labeled faces in the wild: A database for studying face recognition in unconstrained environments," Dept. Comput. Sci., Univ. of Massachusetts, Amherst, MA, USA, Tech. Rep. 07–49, 2007.

TS.45. J. Deng et al., "ImageNet: A large-scale hierarchical image database," in Proc. IEEE Conf. Comput. Vision and Pattern Recognition (CVPR), pp. 248–255, June 2009.

TS.46. See [E488] the data of this presented database, including images, the ground truth depth maps and their associated DMOS, is publicly accessible (https://vaader-data.insa-rennes.fr/data/stian/ieee-tom/IETR DIBR Database.zip).

IEEE Journal on Emerging and Selected Topics in Circuits and Systems (JETCAS)

Special Issue on Screen Content Video Coding and Applications

Screen content video has evolved from a niche to a mainstream due to the rapid advances in mobile and cloud technologies. Real-time, low-latency transport of screen visuals between devices in the form of screen content video is becoming prevalent in many applications, e.g. wireless display, screen mirroring, mobile or external display interfacing, screen/desktop virtualization and cloud gaming. Today's commonly-used video coding methods, however, have been developed primarily with camera-captured content in mind. These new applications create an urgent need for efficient coding of screen content video, especially as the support of 4k or even 8k resolution begins to achieve mass market appeal.

Screen content video coding poses numerous challenges. Such content usually features a mix of computer generated graphics, text, and camera-captured images/video. With their distinct signal characteristics, content adaptive coding becomes necessary. This is without mentioning the varied level of the human's visual sensitivity to distortion in different types of content; visually or mathematically lossless quality may be required for all or part of the video.

Recognizing the demand for an industry standard for coding of screen content, the ISO/IEC Moving Picture Experts Group and ITU-T Video Coding Experts Group have since January 2014 been developing new extensions for HEVC. The Video Electronics Standards Association also recently completed a Display Stream Compression (DSC) standard for next-generation mobile or TV/Computer display interfaces. The development of these standards introduced many new ideas, which are expected to inspire more future innovations and benefit the varied usage of screen content coding.

Besides coding, there are many other challenging aspects related to screen content video. For instance, in applications like screen mirroring and screen/desktop virtualization, low-latency video processing and transmission are essential to ensure an immediate screen response. In addition to real-time streaming technologies, these applications need a parallel-friendly screen encoding algorithm that can be performed efficiently on modern mobile devices or remote servers in the data center, and require, in certain use cases, the harmony of their computing resources, to keep the processing time to a minimum. At the receiver side, best-effort decoding with consideration for transmission errors, along with visual quality enhancement, is expected. Addressing these constraints requires research from multiple disciplines as is the case for other applications.

The intent of this special issue is to present the latest developments in standards, algorithms, and system implementations related to the coding and processing of screen content video. Original and unpublished research results with topics in any of the following areas or beyond are hereby solicited.

- Screen content video coding techniques and standards, e.g. HEVC extensions and DSC
- Visually or mathematically lossless screen content video coding
- Application-specific screen content coding, e.g. display stream or frame memory compression
- Screen-content related pre/post-processing, e.g. resizing and post-filtering
- Visual quality assessment for screen content video
- Parallel-friendly, low-delay encoding optimization
- Robust decoding with error and power control
- Hardware/software/cloud-based screen codec implementations
- Real-time, adaptive screen content transport over Internet or wireless networks
- Design examples of novel screen content video applications, e.g. screen/desktop virtualization and cloud gaming
- System performance analysis and characterization

[SCC15] IEEE Journal on Emerging and Selected Topics in Circuits and Systems (JETCAS)

Special Issue on Screen Content Video Coding and Applications, vol. 6, issue 4, pp. 389–584, Dec. 2016.

See the overview paper: W.-H. Peng, et al., "Overview of screen content video coding: Technologies, standards and beyond," IEEE JETCAS, vol. 6, issue 4, pp. 393–408, Dec. 2016.

S.-H. Tsang, Y.-L. Chan and W.-C. Siu, "Fast and Efficient Intra Coding Techniques for Smooth Regions in Screen Content Coding Based on Boundary Prediction Samples," ICASSP2015, Brisbane, Australia, April. 2015.

J. Nam, D. Sim and I.V. Bajic, "HEVC-based Adaptive Quantization for Screen Content Videos," IEEE Int. Symposium on Broadband Multimedia Systems, pp. 1–4, Seoul, Korea, 2012.

HOW TO ACCESS JCT-VC DOCUMENTS - JCT-VC DOCUMENTS can be found in JCT-VC document management system http://phenix.int-evry.fr/jct

All JCT-VC documents can be accessed. [online].
http://phenix.int-evry.fr/jct/doc_end_user/current_meeting.php?id_meeting=154&type_order=&sql_type=document_number

Accessing JCT-VC Document Database

- In an internet browser, visit http://phenix.int-evry.fr/jct/.
- Create an account by clicking the link shown in the web page.
- Enter your details, and put THE UNIVERSITY OF TEXAS AT ARLINGTON in the "organization" field and also put in the appropriate motivation like STUDIES/ COURSEWWORK, etc.,
- Check for email suggesting the registration was successful.
- You are good to access the content like ALL MEETINGS and NEXT MEETING.

VCEG documents   http://wftp3.itu.int video site folder)

HM-16.9 Software-> https://hevc.hhi.fraunhofer.de/svn/svn_HEVC Software/tags/HM-16.9/

HM-16.9 Software Manual -> https://hevc.hhi.fraunhofer.de/svn/svn_HEVC Software/tags/HM-16.9/doc/software-manual.pdf

ElecardHEVCAnalyser: http://www.elecard.com/en/products/professional/analysis/hevc-analyzer.html

Scalable Extension of HEVC-> https://hevc.hhi.fraunhofer.de/svn/svn_HEVCSoftware

Encoding time evaluation: Intel VTune AmplIfier XE Software profiler, Available http://software.intel.com (accessed May 6, 2014)

See reference 31 in [E242]

Performance Comparison of Video Standards (PCVS)

IEEE PCS 2018, San Francisco, CA, June 2018.

PCVS1: L. Guo, J. D. Cock and A. Aaron, "Compression Performance Comparison of X264, X265, Libvpx and Aomenc for On-demand adaptive streaming applications" IEEE PCS 2018.

PCVS2: T. Nguyen and D. Marpe, "Future Video Coding Technologies: a performance evaluation of AV1, JEM, VP9, AND HM" IEEE PCS 2018.

PCVS3: T. Laude et al., "A Comparison of JEM and AV1 with HEVC: Coding Tools, Coding Efficiency and Complexity" IEEE PCS 2018.

PCVS4: Y. Chen, "An Overview of Core Coding Tools in the AV1 Video Codec", IEEE PCS 2018.

Subjective Evaluaton of Compression Algorithms and Standards

SE.1. P. Hanhart and T. Ebrahimi, "Calculation of average coding efficiency based on subjective quality scores," J. VCIR, vol. 25, pp. 555–564, April 2014. This is a very interesting and valuable paper on subjective quality and testing. The references listed at the end are highly useful. A MATLAB implementation of the proposed model can be downloaded from http://mmspg.epfl.ch/scenic. This paper can lead to several projects (EE5359 Multimedia Processing).

SE.2. H.R. Wu et al., "Perceptual visual signal compression and transmission," Proc. IEEE, vol. 101, pp. 2025–2043, Sept. 2013.

SE.3. C. Deng et al., (Editors), "Visual Signal Quality Assessment: Quality of Experience (QoE)," Springer, 2015.

SE.4. T.K. Tan et al., "Video Quality Evaluation Methodology and Verification Testing of HEVC Compression Performance," IEEE Trans. CSVT, vol. 26, pp. 76–90, Jan. 2016. Abstract of this paper is reproduced here for ready reference.

Abstract

The High Efficiency Video Coding (HEVC) standard (ITU-T H.265 and ISO/IEC 23008-2) has been developed with the main goal of providing significantly improved video compression compared with its predecessors. In order to evaluate this goal, verification tests were conducted by the Joint Collaborative Team on Video Coding of ITU-T SG 16 WP 3 and ISO/IEC JTC 1/SC 29. This paper presents the subjective and objective results of a verification test in which the performance of the new standard is compared with its highly successful predecessor, the Advanced Video Coding (AVC) video compression standard (ITU-T H.264 and ISO/IEC 14496-10). The test used video sequences with resolutions ranging from 480p up to ultra-high definition, encoded at various quality levels using the HEVC Main profile and the AVC High profile. In order to provide a clear evaluation, this paper also discusses various aspects for the analysis of the test results. The tests showed that bit rate savings of 59% on average can be achieved by HEVC for the same perceived video quality, which is higher than a bit rate saving of

44% demonstrated with the PSNR objective quality metric. However, it has been shown that the bit rates required to achieve good quality of compressed content, as well as the bit rate savings relative to AVC, are highly dependent on the characteristics of the tested content.

This paper has many valuable references including subjective quality assessment methods recommended by ITU-T.

SE5. J.-S. Lee and T. Ebrahimi, "Perceptual video compression: A survey," IEEE J. Selected Topics on Signal Process., vol. 6, no. 6, pp. 684–697, Oct. 2012.

SE6. P. Hanhart et al., "Subjective quality evaluation of the upcoming HEVC video compression standard," SPIE Applications of digital image processing XXXV, vol. 8499, paper 8499-30, Aug. 2012.

SE7. W. Lim and J.C. Kuo, "Perceptual video quality metric: a survey," J. VCIR, vol. 22, pp. 297–312, 2011.

SE8. F. Zhang and D.R. Bull, "A Perception-based Hybrid Model for Video Quality Assessment ", IEEE Trans. CSVT, vol. 26, pp. 1017–1028, June 2016.

SE9. Y. Li et al., "No-reference image quality assessment using statistical characterization in the shearlet domain." Signal Processing: Image Communication, vol. 29, pp. 748–759, July 2014.

SE10. Y. Li et al., "No reference image quality assessment with shearlet transform and deep neural networks," Neurocomputing, vol. 154, pp. 94–109, 2015.

SE11. Y. Li et al., "No Reference Video Quality Assessment with 3D shearlet Transform and Convolutional Neural Networks," IEEE Trans. CSVT, vol. 26, pp. 1044–1047, June 2016.

SE12. Y. Li, "No-reference image quality assessment using shearlet transform and stacked auto encoders," IEEE ISCAS, pp. 1594–1597, May 2015.

SE13. T. Daede, "Video codec testing and quality measurement," https://tools.ietf.org/html/ draft-daede-netvc-testing, 2015.

SE14. ITU-R BT2022 (2012) General viewing conditions for subjective assessment of quality of SDTV and HDTV television pictures on at panel displays. International Telecommunication Union.

SE15. ITU-R BT500-13 (2012) Methodology for the subjective assessment of the quality of television pictures. International Telecommunication Union.

SE16. ITU-R BT500-13 (2012) Methodology for the subjective assessment of the quality of television pictures. International Telecommunication Union.
SE17. W. Lim and C.C. Jay-Kuo, "Perceptual visual quality metrics: A survey," J. Visual Communication Image Representation, vol. 22, pp. 297–312, May 2011. (Same as SE7)
SE18.. C.-C. Jay Kuo, "Perceptual coding: Hype or Hope?," $8th$ International Conference on QoMEX, Keynote, Lisbon, Portugal, 6-8 June 2016.
SE19. S.-H. Bae and M. Kim, "DCT-QM; A DCT-Based Quality Degradation Metric for Image Quality Optimization Problems," IEEE Trans. IP, vol. 25, pp. 4916–4930, Oct. 2016. This has several references related to image quality assessment.
SE20. IEEE DATA COMPRESSION CONFERENCE (DCC) Snowbird, Utah, April 4–7, 2017. http://www.cs.brandeis.edu/~dcc, Keynote address
"Video Quality Metrics" by Scott Daly, Senior Member Technical Staff, Dolby Laboratories.
SE21. F.M. Moss et al., "On the Optimal Presentation Duration for Subjective Video Quality Assessment," IEEE Trans. CSVT, vol. 26, pp. 1977–1987, Nov. 2016.

Apart from the comprehensive list of references related to subjective VQA, based on the DSCQS tests on video sequences, the authors [SE21] conclude

1) "There is a small but significant increase in accuracy if sequences are increased from 1.5 to 5, 7, or 10 s.
2) This effect becomes stronger if the difference in distortion between the reference and test video is reduced.
3) The main effect remains consistent between different but temporally consistent source videos.
4) Observers feel just as confident assessing the quality of videos that are 5 s as ones that are 10 s.

The practical implications of these findings are significant.
Our results indicate that critical observations of video quality do not significantly change if 10-s sequences are exchanged for 7-, or indeed, 5-s sequences." Limitations on these findings are elaborated in the conclusions.

SE22. S. Hu et al., "Objective video quality assessment based on perceptually weighted mean square error," IEEE Trans. CSVT, vol. 27, pp. 1844–1855, Sept. 2017.

SE23. S. Wang et al., "Subjective and objective quality assessment of compressed screen content images," IEEE JETCAS, vol. 6, issue 4, pp. 532–543, Dec. 2016.

SE24. H. Yang et al., "Subjective quality assessment of screen content images" Proc. Int. Workshop Quality Multimedia Experience (QoMEX), pp. 257–262 Sept. 2014.

SE25. S. Shi et al., "Study on subjective quality assessment of screen content images" IEEE Proc. Picture Coding Symp. (PCS), pp. 75–79 June 2015.

SE26. New strategy for image and video quality assessment

J. Electron. Imaging, Vol. 19, 011019 (2010); doi:10.1117/1.3302129

Published 18 February 2010
ABSTRACT
REFERENCES (27)

Qi Ma, Liming Zhang, and Bin Wang
Fudan University, Department of Electronic Engineering, No. 220 Handan Road, Shanghai, China

Image and video quality assessment (QA) is a critical issue in image and video processing applications. General full-reference (FR) QA criteria such as peak signal-to-noise ratio (PSNR) and mean squared error (MSE) do not accord well with human subjective assessment. Some QA indices that consider human visual sensitivity, such as mean structural similarity (MSSIM) with structural sensitivity, visual information fidelity (VIF)with statistical sensitivity, etc., were proposed in view of the differences between reference and distortion frames on a pixel orlocal level. However, they ignore the role of human visual attention (HVA). Recently, some new strategies with HVA have been proposed, but the methods extracting the visual attention are too complex for real-time realization. We take advantage of the phase spectrum of quaternion Fourier transform (PQFT), a very fast algorithm we previously proposed, to extract saliency maps of colour images or videos. Then we propose saliency-based methods for both image QA (IQA) and video QA (VQA) by adding weights related to saliency features to these original IQA or VQA criteria.experimental results show that our saliency-based strategy can approach more closely to human subjective assessment compared with these original IQA or VQA methods and does not take more time because of the fast PQFT algorithm.

SE27. l. Xu et al., "Multi-task learning for image quality assessment," IEEE Trans. CSVT, vol. 27, pp. 1833–1843, Sept. 2017.

SE28. K. Ma, "Waterloo exploration database" New challenges for image quality assessment models," IEEE Trans. IP, (early access)
This paper has extensive list of references related to IQA and also on image data bases. The authors suggest four different directions in terms of future work.

SE29. The following papers are presented in IEEE DCC 2017, April, 2017.

SESSION 7, *Quality Metrics and Perceptual Compression, Part 1*

9:40am: Reduced Reference Image Quality Assessment Based on Entropy of Classified Primitives 231
Zhaolin Wan[1], Yutao Liu[1], Feng Qi[2], and Debin Zhao[1]
[1]Harbin Inst. of Tech, [2]Inst. of Computing Tech.

10:00am: Revisiting Perceptual Distortion for Natural Images: Mean Discrete Structural Similarity Index 241
Christopher Hillar[1] and Sarah Marzen[2]
[1]Univ. of California, Berkeley, [2]Massachusetts Inst. of Tech.

10:20am: Semantic Perceptual Image Compression Using Deep Convolution Networks.......... 250
Aaditya Prakash, Nick Moran, Solomon Garber, Antonella Dilillo, and James Storer
Brandeis University

SE30. R. Fang, R. A.- Bayaty and D. Wu, "BNB method for no reference image quality assessment," IEEE Trans. CSVT, vol. 27, pp. 1381–1391, July 2017. (BNB: blurriness, noisiness and blockiness)

SE31. Q. Huynh-Thu and M. Ghanbari, "Scope of validity of PSNR in image/video quality assessment," IET Electronics Letters, vol. 44, (13): 800, 2008.

SE32. L. Zhang and H. LI, "SR-SIM: A fast and high performance IQA index based on spectral residual," IEEE ICIP, pp. 1473–1476, Sept./Oct., 2012.

SE33. VQEG, "Final report from the video quality experts group on the validation of objective models of video quality assessment," http://www.its.bldrdoc.gov/vqeg/vqeg-home.aspx, August 2015.

SE34. S. Bosse et al., "Deep neural networks for no-reference and full reference image quality assessment," IEEE Trans. IP, vol. 27, pp. 206–219, Jan. 2018. Several interesting papers are cited.

In the conclusions, the authors state "—its relative simplicity suggests that neural networks used for image quality assessment have lots of potential and are here to stay with us."

SE35. T.-J. Liu and K.-H. Liu, "No reference image quality assessment by wide perceptual domain scorer ensemble method," IEEE Trans. IP, Vol. 27, pp. 1138–1151, March 2018.

SE36. K. Ma et al., "End-to-end blind image quality assessment using deep neural networks," IEEE Trans. IP, vol. 27, pp. 1202–1213, March 2018.

SE37. Q. Wu et al., "Blind image quality assessment using local consistency aware retriever and uncertainity aware evaluator," IEEE Trans. CSVT, vol. 28, pp. 2078–2089, Sept. 2018. This has several references related to image quality assessment.

SE38. M. Cheon and. J-S. Lee, "Subjective and objective quality assessment of compressed 4K UHD videos for immersive experience," IEEE Trans. CSVT, Early access.

SE39. K. Seshadrinathan et al., "Study of subjective and objective quality assessment of video," IEEE Trans. IP, vol. 19, pp. 1427–1441, June 2010.

SE40. X. Shang et al., "Color-Sensitivity-based Combined PSNR for Objective Video Quality Assessment," IEEE Trans. CSVT, Early access.

Conclusion section is repeated here.

In this paper, we propose a colour-sensitivity-based combined PSNR (CSPSNR) to evaluate video quality by presenting a novel weighted MSE. To the best of our knowledge, this is the first combined PSNR that is based on rigorous subjective tests and theoretical analysis. The advantage of this method is that visual sensitivity is taken into consideration to derive reasonable weighting factors. By utilizing the transform between the RGB and YCbCr colour spaces, the highest contrast is achieved and is used for the testing of YCbCr sensitivities. Three checkerboard patterns are then designed to obtain the spatial resolutions of the three components. From these subjective tests, we derive the perceptual sampling rate of the Y, Cb and Cr components in a pixel, which are used to derive the weighting coefficients of the MSE. Videos with different degrees of distortions on the Y, Cb and Cr components are utilized to verify the correlation between the subjective score and the objective metric. Extensive experimental results and statistical analysis demonstrate that the proposed metric correlates better with the subjective score than existing combined PSNRs. Future work will focus on improving the rate-distortion optimization of the encoder according to the proposed combined PSNR to reconstruct a video with quality confirmed with HVS. In addition, finding the optimal combined SSIM that considers the different significances of three components is also an interesting future research.

- SE41. Y. Fang and P.L. Callet, "Video quality assessment: Theories, methodology and applications," 26th EUSIPCO 2018, Rome, Italy, 3–7 Sept. 2018.
- SE42. W. Zhang, R.R. Martin and H. Liu, "A saliency dispersion measure for improving saliency-based image quality metrics," IEEE Trans. CSVT, vol. 28, pp. 1462–1466, June 2018.

  This has several valuable references related to saliency and also image quality metrics.

Conclusion section is repeated here.

This paper considered how to reliably measure saliency dispersion in natural scenes and how it can be used to adaptively incorporate computational saliency into image quality metrics. Results show that adaptive use of saliency according to saliency dispersion significantly outperforms the fixed use of saliency in improving IQMs. We also intend to investigate the dependence of gain on saliency model dependence to maximize the IQM's performance as future work.

- SE43. L. Krasula et al., "Quality assessment of sharpened images: Challenges, methodology, and objective metrics," IEEE Trans. IP, vol. 26, pp. 1496–1508, March 2017.
- SE44. S. Bosse et al., "Assessing Perceived Image Quality Using Steady-State Visual Evoked Potentials and Spatio-Spectral Decomposition," IEEE Trans. CSVT, vol. 28, pp. 1694–1706, Aug. 2018.
- SE45. Z. Duanmu, K. Ma and Z. Wang, "Quality of experience for adaptive streaming videos: An expectation confirmation theory motivated approach," IEEE Trans. IP, EA. Several valuable references related to QOE are listed at the end. Last paragraph in the conclusion section is repeated here.

Many challenging problems remain to be solved. First, due to the limited capacity of the subjective experiments, we only investigate the impact of a single quality adaptation event to the QoE. A comprehensive study consisting of more content types and adaptation patterns is desired to better understand the behaviors of human viewers and to examine the generalizability of the current findings. Second, incorporating more advanced cross-resolution and cross-frame rate video quality assessment models have great potentials to further boost the performance. Third, the optimization of the existing video streaming frameworks based on the current findings is another challenging problem that desires further investigation.

SE46. R. Yang et al., "Enhancing quality of HEVC compressed images," IEEE Trans. CSVT, EA.

SE47. S. Winkler, "Analysis of public image and video database for quality assessment," IEEE J. Sel. Topics Signal Process., vol. 6, no. 6, pp. 616–625, Oct. 2012.

SE48. Z. Sinno and A.C. Bovik, "Large-scale study of perceptual video quality," IEEE Trans. IP, vol. 28, pp. 612–627, Feb. 2019.

SE49. G. Yue et al., "Analysis of Structural Characteristics for Quality Assessment of Multiple Distorted Images," IEEE Trans. MM, vol. 20, pp. 2722–2732, Oct. 2018.

SE50. I.ITU, "Methodology for the subjective assessment of video quality in multimedia applications," Rapport technique, International telecommunication union, 2007.

SE-P1. Bae and Kim [DCT10] have developed the DCT-Q Based Quality Degradation Metric for Image Quality Optimization Problem that has both high consistency with perceived quality and mathematically desirable properties. See [SE19] for details. The authors conclude that the DCT-QM can effectively be used not only for objective IQA tasks with unknown distortion types but also for image quality optimization problems with known distortion types. They also state "We plan to apply our DCT-QM for image/video coding and processing applications for the improvement of coding efficiency and perceptual visual quality." Explore this. Note that DCT-QM serves for still images only.

SE-P2. See SE-P1. Note that the performance comparison of HEVC (DCT-QM based HM16.0 and conventional HM16.0 for low delay configuration) is shown in Figure 9 for test sequences BasketballDrill, BQ Mall, PartyScene and RaceHorse. See also Figures 10 and 11. Please extend this comparison to HM 16.0 random access configuration.

SE-P3. See [SE19] The authors state "Full application of DCT-QM for specific codecs require extending DCT-QM for specific transform block sizes, which will be part of our future work." They imply other than 4×4-sized DCT kernel i.e., 8×8, 16×16, 32×32 and even 64×64. Consider specifically how the different sized DCT kernels influence the DCT-QM. Consider also the implementation complexity.

304 Appendix

SE-P4. See [SE40]. Explore the future work and future research suggested in the conclusion section.

SE-P5. See [SE41]. Explore the future work suggested in this paper.

SE-P6. See [SE43] Abstract is repeated here:

"Steady-state visual evoked potentials (SSVEPs) are neural responses, measurable using electroencephalography (EEG), that are directly linked to sensory processing of visual stimuli. In this paper, SSVEP is used to assess the perceived quality of texture images. The EEG-based assessment method is compared with conventional methods, and recorded EEG data are correlated to obtained mean opinion scores (MOSs). A dimensionality reduction technique for EEG data called spatio-spectral decomposition (SSD) is adapted for the SSVEP framework and used to extract physiologically meaningful and plausible neural components from the EEG recordings. It is shown that the use of SSD not only increases the correlation between neural features and MOS to r = -0.93, but also solves the problem of channel selection in an EEG-based image-quality assessment." In the conclusion section several extensions to this research are suggested. Explore these. Please note that these extensions require extensive lab/testing facilities and group effort.

SE-P7. See [SE44]. Several challenging problems are listed at the end. Investigate these problems and explore how they can be solved.

Books on HEVC and Related Areas

Book1. V. Sze, M. Budagavi and G.J. Sullivan (Editors), "High efficiency video coding: Algorithms and architectures," Springer 2014.

Book2. M. Wien, "High Efficiency Video Coding: Coding Tools and Specification," Springer, 2014.

Book3. I.E. Richardson, "Coding video: A practical guide to HEVC and beyond," Wiley, 2018.

Book4. K.R. Rao, D.N. Kim and J.J. Hwang, "Video coding standards: AVS China, H.264/MPEG-4 Part10, HEVC, VP6, DIRAC and VC-1," Springer, 2014.
Translated into Spanish by Dr. Carlos Pantsios M, Professor Titular/Telecommunicaciones, USB/UCA/UCV, Dept. of Electronica & Circuits, Simon Bolivar University, Caracas, Venejuela. Also published in Chinese by China Machine press.

Book5. S. Wan and F. Yang, "New efficient video coding – H.265/HEVC – Principle, standard and application," in Chinese, Publishing house of electronic industry, http://www.phei.com.cn, 2014.

Book6. S. Okubo (Editor-in-Chief), "H.265/HEVC Textbook," in Japanese, Impress, Japan, 2013.

Book7. W. Gao and S. Ma, "Advanced Video Coding Systems," Springer, 2015.

Book8. M. Kavanagh, "H.265/HEVC Overview and Comparison with H.264/AVC," Sold by: Amazon Digital Services, Inc. August, 2015.

Book9. B. Bing, "Next-Generation Video Coding and Streaming," Wiley, Hoboken, NJ, 2015.

Book10. J.-R. Ohm, "Multimedia signal coding and transmission," Springer, 2015.

Book11. A.N. Netravali and B.G. Haskell, "Digital pictures: representation and compression," Plenum Press, 1988. II Edition, Springer 1995.

Book12. D. Grois et al., "High Efficiency Video Coding: A Guide to the H.265/HEVC standard," Cambridge University Press, UK, 2018.

Book13. M.E. Al-Mualla, C.N. Canagarajah and D.R. Bull, "Video Coding for Mobile Communications: Efficiency, Complexity and Resilience," Academic Press, Orlando, FL, 2002.

Book14. Y.Q. Shi and H. Sun, "Image and video compression for multimedia engineering: Fundamentals, algorithms and standards," CRC Press, 1999.

Book15. J.-B. Lee and H. Kalva, "The VC-1 and H.264 video compression standards for broadband video services," Springer 2008.

Book16. K.R. Rao, J.J. Hwang and D.N. Kim, "High efficiency video coding and other emerging standards," River Publishers, 2017.

Book17. B.G. Haskell, A. Puri and A.N. Netravali, "Digital video: An introduction to MPEG2," Chapman & Hall, Kluwer, New York, 1997.

Book18. H.R. Wu and K.R. Rao "Digital video image quality and perceptual coding," Taylor and Francis, 2006.

Book19. F. Dufaux et al., "High Dynamic Range Video: From Acquisition to Display and Applications," Academic Press, 2016.

Overview Papers

OP1. G.J. Sullivan and J.-R. Ohm, "Recent developments in standardization of high efficiency video coding (HEVC)," Proc. SPIE, Applications of Digital Image Processing XXXIII, vol. 7798, pp. 77980V-1 through V-7, San Diego, CA, 1–3 Aug. 2010.

OP2. G.J. Sullivan et al., "Overview of the high efficiency video coding (HEVC) standard," IEEE Trans. CSVT, vol. 22, pp. 1649–1668, Dec. 2012.

OP3. HM9 High efficiency video coding (HEVC) Test Model 9 (HM 9) encoder description JCTVC-K1002v2, Shanghai meeting, Oct. 2012.

OP4. M.T. Pourazad et al., "HEVC: The new gold standard for video compression," IEEE Consumer Electronics Magazine, vol. 1, issue 3, pp. 36–46, July 2012.

OP5. G.J. Sullivan et al., "Standardized extensions of high efficiency video coding (HEVC)," IEEE Journal of selected topics in signal processing, vol. 7, pp. 1001–1016, Dec. 2013.

OP6. J. M. Boyce et al., "Overview of SHVC: Scalable Extensions of the High Efficiency Video Coding (HEVC) Standard," IEEE Trans. CSVT. vol. 26, pp. 20–34, Jan. 2016. (SPATIAL, TEMPORAL, QUALITY, COLOR GAMUT and BIT DEPTH SCALABILITIES)

OP7. D. Flynn et al., "Overview of the Range Extensions for the HEVC Standard: Tools, Profiles and Performance," IEEE Trans. CSVT. vol. 26, pp. 4–19, Jan. 2016.

OP8. J. Xu, R. Joshi and R. A. Cohen, "Overview of the Emerging HEVC Screen Content Coding Extension," IEEE Trans. CSVT. vol. 26, pp. 50–62, Jan. 2016.

OP9. G. Tech et al., "Overview of the Multiview and 3D extensions of high efficiency video coding," IEEE Trans. CSVT. vol. 26, pp. 35–49, Jan. 2016.

OP10. Detailed overview of HEVC/H.265: https://app.box.com/s/rxxxzr1a1lnh7709yvih

OP11. W.-H. Peng et al., "Overview of screen content video coding: Technologies, Standards and Beyond," IEEE JETCAS, vol. 6, issue 4, pp. 393–408, Dec. 2016.

Abstract is reproduced below as a review.

Abstract

This paper presents recent advances in screen content video coding, with an emphasis on two state-of-the-art standards: HEVC/H.265 Screen Content Coding Extensions (HEVC-SCC) by ISO/IEC Moving Picture Experts Group and ITU-T Video Coding Experts Group, and Display Stream Compression (DSC) by Video Electronics Standards Association. The HEVC-SCC enhances the capabilities of HEVC in coding screen content, while DSC provides lightweight compression for display links. Although targeting different application domains, they share some design principles and are expected to become the leading formats in the marketplace in the coming years. This paper provides a brief account of their background, key elements, performance, and complexity characteristics, according to their final specifications. As we survey these standards, we also summarize prior arts in the last decade and explore future research opportunities and standards developments in order to give a comprehensive overview of this field.

The list of references [OP11] at the end is highly comprehensive covering not only SCC, but also related areas such as BTC, JPEG, JPEG XS, MCC, MRC, CIC, LZ, SE, VQ, SSIM among others.

OP12. A. Heindel, E. Wise and A. Kaup, "Low complexity enhancement layer compression for scalable lossless video coding based on HEVC," IEEE Trans. CSVT, Early access.

OP13. W. Lin and C.C. Jay Kuo, "Perceptual visual quality metrics: A survey," J. VCIR, vol. 22, no. 2, pp. 297–312, May 2011.

OP14. Md. Asikuzzaman and M.R. Pickering, "An overview of digital video watermarking," IEEE Trans. CSVT, vol. 28, pp. 2131–2153, Sept.2018. Same as [E452].

OP15. A. Khan et al., "A recent survey of reversible watermarking techniques," Information sciences, vol. 279, pp. 251–272, 2014.

OP16. J.-R. Ohm and G.J. Sullivan, "Versatile video coding-towards the next generation of video compression," Invited talk, IEEE PCS 2019, San Francisco, CA, June 2018.

OP17. A. Massoudi et al., "Overview on selective encryption of image and video: Challenges and perspectives," EURASIP J. Info. Security, vol. 2008, no.1, Art. #. 179290, 2008.

OP18. Y. Peng, X. Huang and Y. Zhao, "An overview of cross-media retrieval: Concepts, methodologies, benchmark and challenges," IEEE Trans. CSVT, vol. 28, pp. 2372–2385, Sept. 2018.

Conclusion section is repeated here.

Conclusion

"Cross-media retrieval is an important research topic which aims to deal with the "media gap" for performing retrieval across different media types. This paper has reviewed more than 100 references to present an overview of cross-media retrieval, for building up the evaluation benchmarks, as well as facilitating the relevant research. Existing methods have been introduced mainly including the common space learning and cross-media similarity measurement methods. Common space learning methods explicitly learn a common space for different media types to perform retrieval, while cross-media similarity measurement methods directly measure cross-media similarities without a common space. The widely-used crossmedia retrieval datasets have also been introduced, including Wikipedia, XMedia, NUS-WIDE, Pascal VOC 2007 and Clickture Datasets. Among these datasets, XMedia which we have constructed is the first dataset with five media types for comprehensive and fair evaluation. We are further constructing a new dataset XMediaNet with five media types and more than 100,000 instances. The cross-media benchmarks, such as datasets, compared methods, evaluation metrics and experimental results have been given, and we have established a continuously updated website to present them. Based on the discussed aspects, the main challenges and open issues have also been presented in the future works. We hope these could attract more researchers to focus on cross-media retrieval, and promote the relevant research and applications."

OP19. M. M. Hannuksela et al., "Overview of the multiview high efficiency video coding (MV-HEVC) standard," Proc. IEEE Int. Conf. Image Process., pp. 2154–2158, Quebec City, QC, Canada, Sept. 2015.

OP.P1. See [OP18] and the conclusion section. Explore the challenges and open issues in cross media retrieval described in this paper.

Tutorials:

Tut1. N. Ling, "High efficiency video coding and its 3D extension: A research perspective," Keynote Speech, ICIEA, Singapore, July 2012.

Tut2. X. Wang et al., "Paralleling variable block size motion estimation of HEVC on CPU plus GPU platform," IEEE ICME workshop, 2013.

Tut3. H.R. Tohidpour, M.T. Pourazad and P. Nasiopoulos, "Content adaptive complexity reduction scheme for quality/fidelity scalable HEVC," IEEE ICASSP 2013, pp. 1744–1748, June 2013.

Tut4. M. Wien, "HEVC – coding tools and specifications," Tutorial, IEEE ICME, San Jose, CA, July 2013.

Tut5. D. Grois, B. Bross and D. Marpe, "HEVC/H.265 Video Coding Standard (Version 2) including the Range Extensions, Scalable Extensions, and Multiview Extensions," (Tutorial) Monday 29 June 2015

11.30am–3:00pm , IEEE ICME 2015, Torino, Italy, 29 June–3 July, 2015.

Tut6. D. Grois, B. Bross and D. Marpe, "HEVC/H.265 Video Coding Standard including the Range Extensions, Scalable Extensions, and Multiview Extensions," (Tutorial), IEEE ICCE , Berlin, Germany, 6–9 Sept. 2015.

Tut7. D. Grois, B. Bross and D. Marpe, "HEVC/H.265 Video Coding Standard (Version 2) including the Range Extensions, Scalable Extensions, and Multiview Extensions," (Tutorial) Sunday 27 Sept 2015, 9:00 am to 12:30 pm), IEEE ICIP, Quebec City, Canada, 27–30 Sept. 2015.

This tutorial is for personal use only.
Password: a2FazmgNK
https://datacloud.hhi.fraunhofer.de/owncloud/public.php?service=files&t=8edc97d26d46d4458a9c1a17964bf881

Tut8. Please find the links to YouTube videos on the tutorial - HEVC/H.265 Video Coding Standard including the Range Extensions, Scalable Extensions and Multiview Extensions below:
https://www.youtube.com/watch?v=TLNkK5C1KN8

Tut9. HEVC tutorial by I.E.G. Richardson, www.vcodex.com/h265.html

Tut10. C. Diniz, E. Costa and S. Bampi, "High Efficient Video Coding (HEVC): From Applications to Low-power Hardware Design ", IEEE ISCAS, Montreal, Canada, 22–25 May 2016.

Abstract is reproduced below:

Digital video applications are widespread in every consumer electronic devices. Ultra-high resolutions videos (UHD, e.g. 8k × 4k and 4k × 2k resolutions) are gaining importance in the market. The development of an improved video coding standard with higher efficiency for higher resolution led to the High Efficient Video Coding (HEVC), H.265, published in 2013.This new video coding standard reaches up to approximately double compression efficiency of H.264/AVC standard for similar video quality, due to its sophisticated block partitioning schemes and novel coding algorithms. Its higher compression efficiency comes with a significant increase in computational effort in the video codecs. Real-time HEVC encoding of UHD resolution videos is a challenge, especially considering mobile video capable hardware devices that must consume lower energy to increase battery life. This energy efficiency requirement for future multimedia processors is requiring hardware architecture innovations to integrate multi-core processors with many on-chip hardware accelerators for compute-intensive tasks of the video encoder/decoder. This tutorial covers the algorithms and the dedicated hardware accelerators which are more energy efficient than general purpose processors in performing video tasks. Hardware accelerators will be shown in the tutorial, either as dedicated (ASIC) architectures or as configured/reconfigured FPGA designs. The tutorial starts with an overview the basic concepts on video representation and video coding, before moving to the details of the new HEVC algorithms, data structures, and features. We present a detailed analysis of HEVC video coding application to identify the most compute-intensive tasks of video codec. The second part of the tutorial covers the algorithmic optimization for dedicated hardware design and implementation. The state-of-the-art hardware architectures for video codec blocks are presented. In the end we point to significant future challenges to design low-power HEVC video codec systems. Keywords—Video Coding, HEVC, Compression Algorithms, Hardware Architecture Design.

Tut11. Introduction to Motion estimation and Motion compensation—> http://www.cmlab.csie.ntu.edu.tw/cml/dsp/training/coding/motion/me1.html

HEVC presentation: http://www.hardware.fr/news/12901/hevc-passe-ratifie.html

Detailed overview of HEVC/H.265: https://app.box.com/s/rxxxzr1a1lnh7709yvih

HEVC white paper-Ittiam Systems: http://www.ittiam.com/Downloads/en/documentation.aspx

HEVC white paper-Elemental Technologies: http://www.elementaltechnologies.com/lp/hevc-h265-demystified-white-paper

HEVC white paper: http://www.ateme.com/an-introduction-to-uhdtv-and-hevc

Tut12. J.-R. Ohm and M. Wien, (ohm@ient.rwth-aachen.de)

"Future Video Coding – Coding Tools and Developments beyond HEVC
IEEE ICIP 2017, Beijing, China, Sept. 2017.

Abstract

While HEVC is the state-of-the-art video compression standard with profiles addressing virtually all video-related products of today, recent developments suggest significant performance improvements relative to this established technology. At the same time, the target application space evolves further towards higher picture resolution, higher dynamic range, fast motion capture, or previously unaddressed formats such as 360° video. The signal properties of this content open the door for different designs of established coding tools as well as the introduction of new algorithmic concepts which have not been applied in the context of video coding before. Specifically, the required ultra-high picture resolutions and the projection operations in the context of processing 360° video provide exciting options for new developments. This type of content further modifies the way of video consumption (enabling the use of head-mounted displays) as well as the methods of video content creation and production.

This tutorial will provide a comprehensive overview on recent developments and perspectives in the area of video coding. As a central element, the work performed in the Joint Video Experts Team (JVET) of ITU-T SG16/Q6 (VCEG) and ISO/IEC JTC1 SC29WG11 (MPEG) is covered, as well as trends outside of the tracks of standardization bodies. The focus of the presentation is on algorithms, tools and concepts with potential for competitive future video compression technology. In this context, also the potential of methods related to perceptional models, synthesis of perceptional equivalent content, and deep learning-based approaches will be discussed.

Tut13. Z. Chen and J.-Z. Xu, "Future video coding: Beyond H.265/HEVC," IEEE ICME 2017, Hong Kong, July 2017. (includes VCEG-KTA, AVS2, AVS3D, IEEE1857, JVET and 360 degree video coding) zzchen@whu.edu.cn

Tut14. "Perceptual Metrics for Image and Video Quality in a Broader Context: From Perceptual Transparency to Structural Equivalence" IEEE ICIP 2018, 7-10 Oct. 2018.
Summary is repeated here.

Presenters: Thrasyvoulos N. Pappas (Northwestern University, USA), Sheila S. Hemami (Draper, USA)

Summary:

We will examine objective criteria for the evaluation of image quality that are based on models of visual perception. Our primary emphasis will be on image fidelity, i.e., how close an image is to a given original or reference image, but we will broaden the scope of image fidelity to include structural equivalence. We will also discuss no-reference and limited-reference metrics.

We will examine a variety of applications with special emphasis on image and video compression. We will examine near-threshold perceptual metrics, which explicitly account for human visual system (HVS) sensitivity to noise by estimating thresholds above which the distortion is just-noticeable, and supra-threshold metrics, which attempt to quantify visible distortions encountered in high compression applications or when there are losses due to channel conditions. We will also consider metrics for structural equivalence, whereby the original and the distorted image have visible differences but both look natural and are of equally high visual quality. We will also take a close look at procedures for evaluating the performance of quality metrics, including database design, models for generating realistic distortions for various applications, and subjective procedures for metric development and testing. Throughout the course we will discuss both the state of the art and directions for future research.

Tut15. Visual Quality Assessment: Theories, Methodology, and Applications, VCIP 2018, Taichung, Taiwan, 9-12 Dec. 2018.

Tutorial description is repeated here.

Yuming Fang, Jiangxi University of Finance and Economics, China Patrick Le Callet, Ecole polytechnique de l'Université de Nantes, France.

In this tutorial, we will overview the trend of visual content perception from the visual quality assessment. In the first section, we briefly introduce the achievements of visual quality assessment during the past decade, and provide the key advantages and disadvantages of existing visual quality assessment metrics. Then we will introduce some of new visual quality assessment methods proposed in our team, mainly including two types: general-purpose and application-specific approaches. Some open problems in visual quality assessment are also discussed for the potential research in the future. In particular, we will present ongoing challenges to deal with uncertainty of ground truth data when developing visual quality assessment models at the age of deep learning.

Tut16. Y. Chen, K. Wu, and Q. Zhang, "From QoS to QoE: A tutorial on video quality assessment," IEEE Comm. Surv. Tutorials, vol. 17, no. 2, pp. 1126–1165, 2015.

HEVC/H.265 Video Coding Standard including the Range Extensions Scalable Extensions and..

Share your videos with friends, family, and the world

Watch now...

https://www.youtube.com/watch?v=V6a1AW5xyAw

HEVC/H.265 Video Coding Standard including the Range Extensions Scalable Extensions and..

Share your videos with friends, family, and the world

Watch now...

This MSU Codec Comparison is released. It's extremely well done.

http://www.compression.ru/video/codec_comparison/codec_comparison_en.html

MSU Video Codecs Comparison (video codec measurement)

MSU Video Codecs Comparison MSU Graphics & Media Lab (Video Group) Project head: Dr. Dmitriy Vatolin Testing, charts, analysis: Sergey Grishin

Translating: Daria ...

Read more...

314 *Appendix*

Special Sessions:

SS1. G.J. Sullivan and Y. Ye, "Recent developments in video coding," Session Chairs, IEEE DCC, Snow Bird, Utah, 4–7 April 2017.

SS2. R. Wang, L. Song and L. Zhang, "Immersive video coding," IEEE VCIP, St. Petersburg, FL, Dec. 2017.

SS3. P. Liu et al., "Deep learning for image and video coding," IEEE VCIP, St. Petersburg, FL, Dec. 2017.

Special sessions: IEEE DCC 2018, April 2018. Snowbird, Utah,

SS4. "Toward a Next-Generation International Video Coding Standard"
Gary Sullivan, Microsoft Research
Yan Ye, Interdigital, Inc.
Session Coordinators

SS5. "The Emerging AV1 Video Codec"
Jingning Han, Google
Iraj Sodagar, Microsoft
Session Coordinators

SS6. "Computation over Compressed Data"
Travis Gagie, University of Helsinki
Gonzalo Navarro, University of Chile
Session Coordinators
SPECIAL SESSIONS: IEEE PCS 2018, San Francisco, CA, June 2018

SS7. Machine learning for image and video coding

Organizers:

Thomas Sikora, Technische Universität Berlin

Glenn Van Wallendael, Ghent University - imec

Description

Today's image and video coding techniques are the result of 30 years of evolution of the same block based hybrid video coding principles. Machine learning techniques are replacing custom engineered solutions in such fields as financial trading, marketing personalization, fraud detection, and self-driving cars. One starts to wonder how long image and video coding will still be engineered instead of trained. This special session will show a glimpse of what can be expected in the future from trained image and video compressors.

SS8. Image and Video Quality Assessment with Industry Applications

Organizer:

Zhi Li, Netflix

Description

There has been increasing interest from the industry to use image and video quality assessment tools, for various purposes ranging from passive monitoring to active optimization. Example applications include but not limited to big-data dashboards, A/B tests, perceptual-quality driven video coding, coding standard comparison, streaming client rate adaptation and artifact inspection. This special session is dedicated to create an opportunity for conversations and exchanging ideas among the image and video quality assessment community and the industry to explore existing and upcoming techniques and discovery of new industrial applications.

SS9. Content preparation and compression for virtual reality

Special sessions: IEEE DCC2019, 26–29 March 2019, Snowbird, Utah
[SS10] "Advances in Video Coding and its Applications"
Yuriy Reznik, Brightcove, Inc.
Gary Sullivan, Microsoft Research
Jiangtao Wen, Tsinghua University
Yan Ye, Interdigital, Inc.
Session Coordinators
[SS11] "Plenoptic Image Compression"
Thomas Richter, Fraunhofer IIS
Peter Schelkens, Vrije Universiteit Brussel
Session Coordinators

Keynote speeches

KS1. G. Sullivan, "The new horizon for video coding," IEEE VCIP, St. Petersburg FL, Dec. 2017.

Transcoders:

Tr1. E. de la Torre, R.R. Sanchez and J.L. Martinez, "Fast video transcoding from HEVC to VP9," IEEE Trans. Consumer Electronics, vol. 61, pp. 336–343, Aug. 2015. (This has several valuable references on H.264 to HEVC, mpeg-2 to HEVC, and H.264 to SVC transcoders.)

Tr2. E. Peixoto and E. Izquierdo, "A new complexity scalable transcoder from H.264/AVC to the new HEVC codec," IEEE ICIP 2012, session MP.P1.13 (Poster), Orlando, FL, Sept.–Oct. 2012.

Tr3. T. Shanableh, E. Peixoto and E. Izquierdo, "MPEG-2 to HEVC transcoding with content based modeling," IEEE Trans. CSVT, vol. 23, pp. 1191–1196, July 2013.

References [E97] through [E101] are from VCIP 2012. Please access http://www.vcip2012.org

There are also some papers on HEVC in the Demo session.

Tr4. E. Peixoto et al., "An H.264/AVC to HEVC video transcoder based on mode mapping," IEEE ICIP 2013, WP.P4.3 (poster), Melbourne, Australia, 15–18 Sept. 2013.

Tr5. E. Peixoto, T. Shanableh and E. Izquierdo, "H.264/AVC to HEVC video transcoder based on dynamic thresholding and content modeling," IEEE Trans. on CSVT, vol. 24, pp. 99–112, Jan. 2014.

Tr6. S. J. Kaware and S. K. Jagtap, "A Survey: Heterogeneous Video Transcoder for H.264/AVC to HEVC," IEEE International Conference on Pervasive Computing (ICPC), pp. 1–3, Pune, India, 8–10 Jan. 2015.

Tr7. E. Peixoto et al., "Fast H.264/AVC to HEVC Transcoding based on Machine Learning," IEEE International Telecommunications Symposium (ITS), pp. 1–4, 2014.

Tr8. Y. Chen et al., "Efficient Software H.264/AVC to HEVC Transcoding on Distributed Multicore Processors," IEEE Trans. CSVT, vol. 25, pp. 1423–1434, Aug. 2015.

Tr9. A. J. D. Honrubia et al., "Adaptive Fast Quadtree Level Decision Algorithm for H.264 to HEVC Video Transcoding," IEEE Trans. CSVT. vol. 26, pp. 154–168, Jan. 2016.

Tr10. I. Ahmad et al., "Video transcoding: an overview of various techniques and research issues," IEEE Trans. Multimedia, vol. 7, pp. 793–804, 2005.

Tr11. A. J. D.-Honrubia et al., "A fast splitting algorithm for an H.264/AVC to HEVC intra video transcoder," IEEE DCC, Snow Bird, Utah, March-April 2016.

Tr12. D. Zhang et al., "Fast Transcoding from H.264 AVC to High Efficiency Video Coding," in IEEE ICME 2012, pp. 651–656, July 2012.

Tr13. D. Guang et al., "Improved prediction estimation based h.264 to HEVC intra transcoding," in Advances in Multimedia Information Processing PCM 2014, ser. Lecture Notes in Computer Science, vol. 8879. Springer International Publishing, pp. 64–73, 2014.

Tr14. D. Hingole, "H.265 (HEVC) bitstream to H.264 (MPEG 4 AVC) bitstream transcoder," M.S. Thesis, EE Dept., University of Texas at Arlington, Arlington, Texas, Dec. 2015.
http://www.uta.edu/faculty/krrao/dip click on courses and then click on EE5359 Scroll down and go to Thesis/Project Title and click on D. Hingole.

Tr15. M. Sugano et al., "An efficient transcoding from MPEG-2 to MPEG-1," IEEE ICIP 2001.

Tr16. N. Feamster and S. Wee, "An MPEG-2 to H.263 transcoder," SPIE conference on multimedia systems and applications II, vol. 3845, pp. 164–175, Sept. 1999.

Tr17. X. lu et al., "Fast mode decision and motion estimation for H.264 with a focus on MPEG-2/H.264 transcoding," IEEE ISCAS, pp. 1246–1249, Kobe, Japan, May 2005.

Tr18. M. Pantoja, H. Kalva and J.B. Lee, "P-frame transcoding in VC-1 to H.264 transcoders," IEEE ICIP, pp. V–297 ∼ V-300, 2007. Also several papers on transcoders.

Tr19. H. Yuan et al., "Motion-homogeneous based fast transcoding method from H.264/AVC to HEVC," IEEE Trans. Multimedia, Early access. Several papers on H.264/AVC to HEVC transcoders.

Tr20. T. Shen et al., "Ultra fast h. 264/avc to hevc transcoder," in IEEE Data Compression Conference (DCC), 2013.

Tr21. J.F. Franche and S. Coulombe, "Efficient H.264-to-HEVC transcoding based on motion propagation and post-order traversal of coding tree units," IEEE Trans. CSVT, Early access. In the references there are several papers related to H.264/AVC to HEVC transcoder. The last section conclusion is repeated here.

Conclusion

In this paper, we proposed a fast H.264-to-HEVC transcoding approach combining a motion propagation algorithm and a fast mode decision framework. The motion propagation algorithm requires no motion refinement, eliminates motion estimation computational redundancy, and has a very low complexity at the PU level. The mode decision framework is based on a post-order traversal of the CTU. It includes some candidate mode reduction techniques and a criterion to early terminate the RD cost computation. Finally, the structural split decision developed in our previous work was advantageously replaced by a motion-based split decision method. Our experiments show that the proposed transcoder is much faster than related works, and achieves a coding efficiency similar to a CPDT transcoder. Moreover, our novel split decision method outperforms our previous method. In our future research, we intend to adapt the proposed transcoder to other coding applications, such as multi-rate HEVC video coding. Furthermore, we intend to improve the transcoding performance of intra modes by replacing the RMD method with a faster method, such as the method proposed by Jamali et al., [32]. Finally, we plan to improve the proposed approach by supporting B-frames

Tr22. J. Bialkowski, M. Barkowsky and A. Kaup, "Overview of low – complexity video transcoding from H.263 to H.264," IEEE ICME, pp. 49–52, 2006.

Tr23. R. Luo. R. Xie and L. Zhang, "Fast AVS to HEVC transcoding based on ROI detection using visual characteristics," IEEE International Symp. on broadband multimedia systems and broadcasting (BMSB), pp. 1–6, 2014.

Tr24. Y. Zhang, J. Zha and H. Chao, "Fast H.264/AVC to HEVC transcoding based on compressed domain information," IEEE DCC 2018, Sun Bird, Utah, March 2018.

Tr25. T. Shen et al., "Ultra fast h.265/avc to hevc transcoder," IEEE DCC 2013, pp. 241–250, 2013.

Tr25. W. Kim et al., "Exploiting pseudo-quadtree structure for accelerating HEVC spatial resolution downscaling transcoder," IEEE Trans. Multimedia, vol. 20, pp. 2262–2275, Sept. 2018.

Tr26. G. Correa, L. Agostini and L. A. da Sliva Cruz, "Fast H.264/AVC to HEVC transcoder based on data mining and decision trees," Proc. IEEE ISCAS, pp. 2539–2542, May 2016.

Tr27. G. Gao, Y. Wen and C. Westphal, "Dynamic priority-based resource provisioning for video transcoding with heterogeneous QOS," IEEE Trans., EA. Coclusion an dfuture work is repeated here.

Conclusion and Future Work

"This paper designed and implemented a video transcoding system that can dynamically provision resources for video transcoding with heterogeneous QoS requirements. We adopted the preemptive resume priority discipline to design a multiple-priority transcoding mechanism that can improve resource utilization without affecting the QoS for delaysensitive videos. We designed an online algorithm based on the MPC framework to intelligently provision the right amount of resources to accommodate time-varying workloads while keeping robust to prediction noise. The experiment results showed that our method can guarantee the heterogeneous QoS requirements for transcoding while significantly reducing resource consumption. In the future work, we will consider the multi-core/multi-mode scheduling for video transcoding and the resource provisioning for hardware-accelerated video transcoding. Another possible extension of this paper is to divide the transcoding workloads into many classes, and each class of workload is specified with a deadline according to the QoS requirement of the class of video content."

Tr-P1. See [Tr19]. Review this paper and also the references listed at the end related to H.264 to HEVC transcoders. Explore improved transcoders and implement and compare their performances with the proposed one in [Tr19].

Tr-P2. See [Tr21] Implement the H.263-to H.264 transcoder.

Tr-P3. See Tr-P2 Investigate, explore and implement H.264 to H.263 transcoder.

Tr-P4. See [Tr21] By thoroughly researching this and also several techniques implemented for the H.264/AVC to HEVC transcoders described in the references, can an efficient low complexity H.264 to HEVC transcoder be developed for all profiles. This is a challenging research.

Tr-P5. See the conclusion section in [Tr21].
"In our future research, we intend to adapt the proposed transcoder to other coding applications, such as multi-rate HEVC video coding." Investigate this.

Tr-P6. See [Tr21] and Tr-P5. The authors state
"Furthermore, we intend to improve the transcoding performance of intra modes by replacing the RMD method with a faster method, such as the method proposed by Jamali et al., [32]." Explore this.

Tr-P7. See [Tr21], [Tr-P5], and [Tr-P6] "Finally, we plan to improve the proposed approach by supporting B-frames." Carry out this research.
Tr-P8. See [Tr27] Explore the future work suggested in the conclusion section.
Tr-P9. See [Tr27] Explore the possible extension suggested in the conclusion section.

Encryption of HEVC Bit Streams

EH1. B. Boyadjis et al., "Extended selective encryption of H.264/AVC (CABAC) and HEVC encoded video streams," IEEE Trans. CSVT, (early access)
EH-P1. B. Boyadjis et al., [EH1], by incorporating selective encryption of CABAC in regular mode, have improved the encryption efficiency over the by-pass mode (see Table VI and Figure 11). In Section V (conclusion and future work), the authors suggest some further studies. Explore and implement future work.

File Format

FF1. M. M. Hannuksela, J. Lainema and V. K. M. Vadakital, "The high efficiency image file format standards," IEEE Signal Processing Magazine, vol. 32, pp. 150–156, IEEE Trans. CSVT. Vol. (Early Access).
FF2. A draft HEIF standard (03/2015) is available at http://mpeg.chiariglione.org/standards/mpeg-h/image-file-format/draft-text-isoiec-fdis-23008-12-carriagestill-image-and-image and the final HEIF standard (ISO/IEC 23008-12) is likely to appear among public ISO standards at http://standards.iso.org/ittf/PubliclyAvailableStandards/.
FF3. The JPEG, Exif, PNG, and GIF specifications are available at http://www.w3.org/Graphics/JPEG/itu-t81.pdf, http://www.cipa.jp/std/documents/e/DC-008-2012_E.pdf, http://www.w3.org/TR/PNG/, and http://www.w3.org/Graphics/GIF/spec-gif89a.txt, respectively.

Online Courses (OLC):

OLC1. Video Lecture on Digital Voice and Picture Communication by Prof. S. Sengupta, Department of Electronics and Electrical Communication Engineering, IIT, Kharagpur, India -> https://www.youtube.com/watch?v=Tm4C2ZFd3zE

Open Source Software:

OSS1. HEVC open source software (encoder/decoder):
https://hevc.hhi.fraunhofer.de/svn/svn_HEVCSoftware/tags/HM-16.9 JCT-VC (2014) Subversion repository for the HEVC test model reference software:
https://hevc.hhi.fraunhofer.de/svn/svn_HEVCSoftware/
http://hevc.kw.bbc.co.uk/svn/jctvc-al24 (mirror site)
Latest 8/23/2017
https://hevc.hhi.fraunhofer.de/svn/svn_HEVCSoftware/tags/HM-16.16/

OSS2. SHM – 7.0 SHVC reference software:
https://hevc.hhi.fraunhofer.de/svn/svn_SHVCSoftware/tags/SHM-7.0 (See V. Seregin and Y. He, "Common SHM test conditions and software reference configuration," document JCT-VC-P1019, San Jose, CA, Jan. 2014.

OSS3. JSVM 9.19.15 (H.264/AVC)
http://ftp3.itu.ch/av-arch/jvt-site/2005_07_Poznan/JVT-P001.doc

OSS4. JCt-3V. (2015, May) MV – and 3D-HEVC reference software, HTM – 14.1 [Online]. Available: https://hevc.hhi.fraunhofer.de/svn/svn_3DVCsoftware/tags/HTM-14.1/

OSS5. JVT KTA reference software
http://iphome.hhi.de/suehring/tml/download/KTA/

OSS6. JM11KTA2.3 (http://www.h265.net/2009/04/kta-software-jm11kta23.html).

OSS7. JSVM – joint scalable video model reference software for scalable video coding. (online) http://ip.hhi.de/imagecom_GI/savce/downloads/SVC-Reference-software.htm

OSS8. T. Ritchter : "JPEG XT Reference Codec 1.31 (ISO License)," available online at http:// www.jpeg.org/jpegxt/software.html (retrieved July 2015)

JPEG XT: A backwards-compatible extension of JPEG for intermediate and high-dynamic range image compression.

Abstract

JPEG XT is an extension of the well-known JPEG standard that allows lossy or lossless coding of intermediate (8-16 bit integer) or high-dynamic range (floating point) images such that any legacy JPEG decoder is still able to reconstruct an 8-bit/component image from it.

In this talk, the JPEG XT coding technology will be introduced, parts and profiles of the standard will be discussed and a short discussion of the coding performance of the JPEG XT profiles in comparison with other HDR coding technologies will be given.

OSS9. Dirac software: http://sourceforge.net/projects/dirac/

OSS10. AVS-china software download: ftp://159.226.42.57/public/avs_doc/avs_software

OSS11. HEVC SCC Extension Reference software https://hevc.hhi.fraunhofer.de/svn/svn_HEVCSoftware/tags/HM-15.0+RExt-8.0+SCM-2.0/ (same as [E256])

OSS12. Kakadu Software 6.0. [Online]. Available: http://www.kakadusoftware.com/ (JPEG 2000)

OSS13. FFmpeg: Open Source and Cross-Platform Multimedia Library. [Online]. Available: http://www.ffmpeg.org, accessed July 2014. Also https://ffmpeg.org

OSS14. SHVC Reference Software Model (SHM). [Online]. Available: https:// hevc.hhi.fraunhofer.de/svn/svn_SHVCSoftware/tags/SHM-7.0, accessed July 2014.

OSS15. 3H HEVC Reference Software Model (HTM). [Online]. Available: https://hevc.hhi.fraunhofer.de/svn/svn_3DVCSoftware/, accessed Jul. 2014.

OSS16. Open Source HEVC Decoder (Open HEVC). [Online]. Available: https://github.com/OpenHEVC, accessed July 2014.
OSS16 Kvazaar ultravideo.cs.tut.fi

OSS17. F265 f265.org

OSS18. CES265 ces.itec.kit.edu/ces265.php

OSS19. M. Masera, Adaptive Approximated DCT Architectures for HEVC, 2015 Dec. [Online]. Available: http://personal.det.polito.it/maurizio.masera/material/HEVCSoftware.zip This software implements the HEVC using these DCT architectures.

OSS20. T.835: Information technology - JPEG XR image coding system -Reference software," *ITU [Online]*, Available: https://www.itu.int/rec/TREC-T.835-201201-I/.

OSS21. libde open source www.libde265.org

OSS22. Kvazaar HEVC encoder https://github.com/ultravideo/kvzaar open source code supported platforms: x86 and x64 on Windows and Linux

OSS23. cclxv https://bitbucket.org/prunedtree/cclxv OSS HEVC decoder

OSS24. libde265 https://github.com/strukturag/libde265 OSS HEVC decoder

HM SOFTWARE MANUAL

F. Bossen, D. Flynn, K. Sharman, and K. Sühring. (2015). HM Software Manual, accessed on Jan. 18, 2016. [Online]. Available: https://hevc.hhi.fraunhofer.de/svn/svn_HEVCSoftware/tags/HM16.5/doc/software-manual.pdf

http://x265.ru/en/ - HEVC software (open source)

http://www.3gpp.org/DynaReport/26410.htm - HE ACC V2 software High efficiency advanced audio codec

https://ffmpeg.org/download.htmlhttps://ffmpeg.org/download.html - ffmpeg software

OSS25. MPEG software simulation group software version 1.2 (July 1996) online available: http://www.mpeg.org/MPEG/video/mssg-free-mpeg-software.html (MPEG-2/H.262)

OSS26. P. Dollár. *MATLAB Evaluation Code for Standardized Pedestrian Detection Datasets.* [Online]. Available: http://www.vision.caltech.edu/Image_Datasets/CaltechPedestrians/

OSS 27. H.264/AVC reference software
http://iphome.hhi.de/suehring/tml

OSS28. jvet JEM software.

J. V. E. T. J. on Future Video Coding, "JVET JEM software," https://jvet.hhi.fraunhofer.de/svn/svnHMJEMSoftware/.

http://mkvtoolnix.en.softonic.com/ -mkvmerge software

Winzip and GZIP open source software

Commercial or open source lossless compression software, such as WinZIP and gzip, that can reduce the file sizes of JPEG coded images.

Winzip 19.0 Pro(11294) 64 Bit. Available: http://www.winzip.com/win/en/downwz.html

GZIP.

http://www.gzip.org

OSS28 SCM Test Model (HM-16.4+SCM-4.0), accessed on Apr. 2017. [online] Available: https://hevc.hhi.fraunhofer.de/svn/svn_HEVCSoftware/tags/ HM-16.4+SCM-4.0/

MKVToolnix supports the following formats

Video: 264, AVC, H264, AVI, DRC, MP4, MPG, MPEG, M1V, M2V, EVO, MKV, OGG, MOV, RA, RAM, RMVB, VC1
Audio: AC3, AAC, M4A, MP4, DTS, FLAC, OGG, MP2, MP3, MKA, TTA, WAV, WV
Untertitel: SRT, ASS, SSA, USF, XML, IDX
OSS28 CES265 (C++) ces.itec.kit.edu/ces265.php

Software for VTM, HEVC, JEM, and 360 Video (publicly accessible)

- https://jvet.hhi.fraunhofer.de/svn/svn_VVCSoftware_
- https://jvet.hhi.fraunhofer.de/trac/vvc/browser/vtm
- https://hevc.hhi.fraunhofer.de/svn/svn_HEVCSoftware/
- https://jvet.hhi.fraunhofer.de/svn/svn_HMJEMSoftware/
- https://jvet.hhi.fraunhofer.de/svn/svn_360Lib/

OSS29. CrypTool. (2017). An Open-Source Windows Program for Cryptography and Cryptanalysis. Accessed: Aug. 28, 2018. [Online]. Available: https://www.cryptool.org/en/

OSS30. HEVC Test Model (HM) Reference Software 16.6. Accessed: Nov. 19, 2016. [Online]. Available: https://hevc.hhi.fraunhofer.de/svn/svn_HEVCSoftware/tags/HM-16.6

OSS31. HDR-VQM software. [Online]. Available: https://sites.google.com/site/narwariam/hdr-vqm

This MSU Codec Comparison is released. October 2015.

http://www.compression.ru/video/codec_comparison/codec_comparison_en.html

MSU Video Codecs Comparison (video codec measurement)

MSU Video Codecs Comparison MSU Graphics & Media Lab (Video Group) Project head: Dr. Dmitriy Vatolin Testing, charts, analysis: Sergey Grishin Translating: Daria

X265 Source Code

x265 HEVC High Efficiency Video Coding H.265 Encoder. [Online] Available: http://x265.org/, accessed May 6, 2014.

http://appleinsider.com/articles/14/09/12/apples-iphone-6-iphone-6-plus-use-h265-codec-for-facetime-over-cellular

Apple's iPhone6, 6plus use H.265 CODEC for encode/decode FaceTime video over cellular. It seems they use A8SoC for this.

Mediatek (MT6595) developed mobile chip H.265 encoder.

VideoLAN x265 HEVC video encoder (online) http://hg.videolan.org/x265

vTune Amplifier by Intel

https://registrationcenter.intel.com/en/forms/?licensetype=2&productid=2489
– This is the direct link for vtune for students to register and download the product for free.

https://registrationcenter.intel.com/en/forms/?licensetype=2&programid=EDUCT&productid=2520
– This is the direct link for vtune for faculty

https://software.intel.com/en-us/qualify-for-free-software
– This is the home page link for all Intel software that are free for students, faculty and researchers.

Note: VTune is a bit hidden on this page. Look for Intel Parallel Studio, Cluster Edition. This is a bundle of many tools that includes VTune, compilers, libraries and cluster tools.

ITU-R DOCUMENTS

Itu.int/rec/R-REC-BT.709

Itu.int/rec/R-REC-BT.2020

General

General1. T. Ebrahimi and M. Kunt, "Visual data compression for multimedia applications," Proc. IEEE, vol. 86, pp. 1109–1125, June 1998.

General2. T. Wiegand and B. Girod, "Lagrange multiplier selection in hybrid video coder control," in Proc. IEEE ICIP, vol. 3, pp. 542–545, Oct. 2001.

General3. A.M. Patino et al., "2D-DCT on FPGA by polynomial transformation in two-dimensions," Proceedings of the IEEE Intnl. Symp.

on Circuits and Systems, ISCAS'04, vol. 3, pp. 365–368, May 2004.

General4. G.A. Davidson et al., "ATSC video and audio coding," *Proc. of the IEEE*, vol. 94, pp. 60–76, Jan. 2006.

General5. J. Golston and A. Rao, "Video compression: System trade-offs with H.264, VC-1 and other advanced CODECs," Texas Instruments, White Paper, Nov. 2006. This paper was written for and presented at the Embedded Systems Conference – Silicon Valley, San Jose, CA, April 2006.

General6. C.-H. Yu and S.-Y. Chen, "Universal colour quantisation for different colour spaces," IEE Proc. Vision, Image and Signal Processing, vol. 153, pp. 445–455, Aug. 2006.

General7. R.A. Burger et al., "A survey of digital TV standards China," IEEE Second Int'l. Conf. Commun. and Networking in China, pp. 687–696, Aug. 2007.

General8. T. Wiegand and G.J. Sullivan, "The picture phone is here. Really," IEEE Spectrum, vol. 48, pp. 50–54, Sept. 2011.

General9. N. Jayant, "Frontiers of audiovisual communications: New convergences of broadband communications, computing, and rich media," Proc. IEEE, Special Issue, Scanning the Issue, vol. 100, pp. 821–823, April 2012. Several papers on audio visual communications.

General10. H.R. Wu et al., "Perceptual visual signal compression and transmission," Proc. IEEE, vol. 101, pp. 2025–2043, Sept. 2013.

General11. M.A. Isnardi, "Historical Overview of video compression in consumer electronic devices," IEEE ICCE, pp. 1–2, Las Vegas, NV, Jan. 2007.

General12. G. J. Sullivan and T. Wiegand, "Video compression – from concepts to the H.264/AVC Standard," Proc. IEEE, vol. 93, pp. 18–31, Jan. 2005.

General13. G. J. Sullivan et al., "Future of Video Coding and Transmission," IEEE Signal Processing Magazine, Vol. 23, pp. 76–82, Nov. 2006.

General14. T. Ebrahimi and M. Kunt, "Visual Data Compression for Multimedia Applications," Proc. IEEE, vol. 86, pp. 1109–1125, June 1998.

General15. Proc. IEEE, Special Issue on Global Digital Television: Technology and Emerging Services, vol. 94, Jan. 2006.

General16. N. Jayant, J. Johnston, and R. Safranek, "Signal compression based on models of human perception," Proc. IEEE, vol. 81, no. 10, pp. 1385–1422, Oct. 1993.

General17. R. Schafer and T. Sikora, "Digital video coding standards and their role in video communications," Proc. IEEE, vol. 83, pp. 907–924, June 1995.

General18. A.K. Jain, "Image data compression: A review," Proc. IEEE, vol. 69, pp. 349–384, March 1981.

JVT REFLECTOR. Queries/questions/clarifications etc regarding H.264/H.265

jvt-experts-bounces@lists.rwth-aachen.de; on behalf of; Karsten Suehring [karsten.suehring@hhi.fraunhofer.de]

The JCT experts reflector is used for discussions related to H.264/AVC and HEVC. You can subscribe yourself here: http://mailman.rwth-aachen.de/mailman/listinfo/jvt-experts

For discussions related to HEVC you should use the JCT-VC email reflector that can be found here:

http://mailman.rwth-aachen.de/mailman/listinfo/jct-vc

The draft standard: High Efficiency Video Coding Draft 10, Document JCTVC-L1003, available at: http://phenix.it-sudparis.eu/jct/doc_end_user/current_document.php?id=7243

JVT KTA reference software: http://iphome.hhi.de/suehring/tml/download/KTA/

HEVC Quality Evaluation

http://www.slideshare.net/touradj_ebrahimi/subjective-quality-evaluation-of-the-upcoming-hevcvideo-compression-standard

Fraunhofer HHI Institute, Germany: Image processing research group
http://www.hhi.fraunhofer.de/en/fields-of-competence/image-processing/research-groups/image-video-coding/team.html

JCT-VC Documents

JCT-VC documents can be found in the JCT-VC document management system at http://phenix.int-evry.fr/jct
Also publicly available (also downloadable) at http://wftp3.itu.int/av-arch in the video site

Video Test Sequences
[28] Repository for freely-redistributable test sequences. [Online] Available: www.media.xiph.org.
[29] "Video Test Sequences." [Online] Available: ftp://hvc:US88Hula@ftp.tnt.uni-hannover.de/testsequences

References on SSIM

SSIM1. Z. Wang, et al., "Image quality assessment: From error visibility to structural similarity," IEEE Trans. Image Processing, vol. 13, pp. 600–612, Apr. 2004.

SSIM2. Z. Wang, L. Lu and A. C. Bovik, "Video quality assessment using structural distortion measurement," IEEE International Conference on Image Processing, vol. 3, pp. 65–68, Rochester, NY, Sep. 22–25, 2002.

SSIM3. The SSIM Index for Image Quality Assessment http://www.ece.uwaterloo.ca/~z70wang/research/ssim/

SSIM4. What is 3-SSIM? 2/8/11 Alexander Parshin [videocodec-testing@graphics.cs.msu.ru]
It is our implementation of
http://live.ece.utexas.edu/publications/2010/li_jei_jan10.pdf

SSIM5. W. Malpica and A. Bovik, "Range image quality assessment by structural similarity," IEEE ICASSP 2009, pp. 1149–1152, 19–24 April 2009.

SSIM6. J. Zujovic, T.N. Pappas and D.L. Neuhoff, "Structural similarity metrics for texture analysis and retrieval," IEEE ICIP 2009, Cairo, Egypt, Nov. 2009 (paper MP.L1.5)

SSIM7. X. Shang, "Structural similarity based image quality assessment: pooling strategies and applications to image compression and digit recognition," M.S. Thesis, EE Department, The University of Texas at Arlington, Aug. 2006.

SSIM8. Z. Wang and Q. Li, "Video quality assessment using a statistical model of human visual speed perception," *Journal of the Optical Society of America A,* vol. 24, no. 12, pp. B61–B69, Dec. 2007.

SSIM9. Z. Wang and X. Shang, "Spatial pooling strategies for perceptual image quality assessment," *IEEE International Conference on Image Processing,* pp. 2945–2948, Atlanta, GA, Oct. 8–11, 2006.

SSIM10. Z. Wang and E.P. Simoncelli, "Translation insensitive image similarity in complex wavelet domain," *IEEE International Conference on Acoustics, Speech and Signal Processing,* vol. II, pp. 573–576, Philadelphia, PA, March 2005.

SSIM11. Z. Wang, L. Lu, and A.C. Bovik, "Video quality assessment based on structural distortion measurement," *Signal Processing: Image Communication*, special issue on "Objective video quality metrics," vol. 19, no. 2, pp. 121–132, Feb. 2004.

SSIM12. Z. Wang, E. P. Simoncelli and A. C. Bovik, "Multi-scale structural similarity for image quality assessment," Invited Paper, *IEEE 37th Asilomar Conference on Signals, Systems and Computers, vol. 2, pp. 1398–1402*, Nov. 2003.
Dr. Zhou Wang's web site
http://www.ece.uwaterloo.ca/~z70wang/temp/DIP.zip
Anush K. Moorthy and Alan C. Bovik
Efficient motion weighted spatio-temporal video SSIM index
Proc. SPIE, Vol. 7527, 75271I (2010); doi:10.1117/12.844198
Conference Date: Monday 18 January 2010 Conference Location: San Jose, California, USA.
Conference Title: Human Vision and Electronic Imaging XV

SSIM13. S. Wang, S. Ma and W. Gao, "SSIM based perceptual distortion rate optimization coding," SPIE, VCIP, vol. 7744–91, Huangshan, China, July 2010.
Is & T Electronic Imaging, Science and Technology, SPIE, Image quality and system performance, vol. 7867, San Francisco, CA, Jan. 2011 (plenary speech by A. C. Bovik).

SSIM14. A. Bhat, I. Richardson and S. Kanangara, "A new perceptual quality metric for compressed video based on mean squared error," SP:IC, pp. 1–4, 30 July, 2010.

SSIM15. Y.-T. Chen et al., "Evaluation of video quality by CWSSIM method," SPIE, Mathematics of Data/Image Coding, Compression, and Encryption with Applications XII, vol. 7799, pp. 7790T–1 through 7, Aug. 2010.

SSIM16. J. Wang et al., "Fractal image coding using SSIM," IEEE ICIP 2011.

SSIM17. A. Rehman and Z. Wang, "SSIM based non-local means image denoising," IEEE ICIP 2011.

SSIM18. W. Lin and C.-C.J. Kuo, "Perceptual visual quality metrics: A survey," J. VCIR, vol. 22, pp. 297–312, May 2011.

SSIM19. C. Vu and S. Deshpande, "ViMSSIM: from image to video quality assessment," ACMMoVid 12, Proc. 4*th* workshop on mobile video, Chapel Hill, N.C., Feb. 2012.

SSIM20. A. Horne and D. Ziou, "Image quality metrics: PSNR vs. SSIM," IEEE ICPR, pp. 2366–2369, 2010.

SSIM21. C. Yeo, H.L. Tan and Y.H. Tan, "SSIM-based adaptive quantization in HEVC," IEEE ICASSP, pp. 1690–1694, Vancouver, Canada, 2013 (IVMSP-P3.2: Video coding II Poster).
"SSIM-based Error-resilient Rate Distortion Optimized H.264/AVC Video Coding for Wireless Streaming" Signal Processing: Image Communication, pp. 2332–2335, June 2013.

SSIM22. T.-S. Ou and Y.-H. Huang and H.H. Chen, "SSIM-based perceptual rate control for video coding," IEEE Trans. CSVT, vol. 21, pp. 682–691, May 2011.

SSIM23. A. Rehman and Z. Wang, "SSIM-inspired perceptual video coding for HEVC," IEEE ICME 2012, pp. 497–502, July 2012. (**Same as SSIM29**)

SSIM23. S. Wang et al., "SSIM-motivated rate distortion optimization for video coding," IEEE Trans. CSVT, vol. 22, pp. 516–529, April 2012.

SSIM24. C. Yeo, H.L. Tan and Y.H. Tan, "On rate distortion optimization using SSIM," IEEE Trans. CSVT, vol. 23, pp. 1170–1181, July 2013.

SSIM25. T. Zhou and Z. Wang, "On the use of SSIM in HEVC," IEEE Asilomar conf. on circuits, systems and computers, pp. 1107–1111, Nov. 2013.

SSIM26. M. Hassan and C. Bhagavati, "Structural similarity measure for color images," Int. J. Comput. Appl., vol. 43, pp. 7–12, 2012.

SSIM27. K. Naser, V. Ricordel and P. L. Callet, "Experimenting texture similarity metric STSIM for intra prediction mode selection and block partitioning in HEVC," DSP, pp. 882–887, 2014.

SSIM28. S. Wang et al., "Perceptual Video Coding Based on SSIM – Inspired Divisive Normalization," IEEE Trans. on Image Processing, vol. 22, no. 4, pp. 1418–1429, Apr. 2013.

SSIM29. A. Rehman and Z. Wang, "SSIM – Inspired Perceptual Video Coding for HEVC," Proc. Int. Conf. Multimedia and Expo, pp. 497–502, Melbourne, Australia, July. 2012.

SSIM30. C.L. Yang, "An SSIM-Optimal H.264/AVC Inter Frame Encoder," Proceedings of 8th IEEE/ACIS International Conference on Computer and Information Science (ICIS 2009), pp. 291–295, Shanghai, China, June 2009.

SSIM31. G.H. Chen, "Edge-based Structural Similarity for Image Quality Assessment," IEEE ICASSP, vol. 2, pp. 933–936, Toulouse, France, May 2006.

SSIM32. M.-J. Chen and A. C. Bovik, "Fast structural similarity index algorithm," Journal of Real-Time Image Processing, vol. 6, no. 4, pp. 281–287, Dec. 2011.

SSIM33. W.-T. Loh and D.B.L. Bong, "Temporal video quality assessment method involving structural similarity," IEEE- ICCE Taiwan, May 2016.

SSIM34. S. Wang et al., "SSIM-motivated two-pass VBR video coding for HEVC," IEEE Trans. CSVT, vol. 27, pp. 2189–2203, Oct. 2017. This has several references related to SSIM.

Emmy award for SSIM (Oct. 2015)

http://www.emmys.com/news/awards-news/sthash.c1DcwJf7.dpuf

Zhou Wang, Alan Bovik, Hamid Shiekh and Eero Simoncelli for Structural Similarity (SSIM) Video Quality Measurement Model. Structural Similarity (SSIM) is an algorithm for estimating the perceived quality of an image or video. Its computational simplicity and ability to accurately predict human assessment of visual quality has made it a standard tool in broadcast and post-production houses throughout the television industry.

SSIM uses powerful neuroscience-based models of the human visual system to achieve breakthrough quality prediction performance. Unlike previous complex error models that required special hardware, it can be easily applied in real time on common processor software. SSIM is now the most widely-used perceptual video quality measure, used to test and refine video quality throughout the global cable and satellite TV industry. It directly affects the viewing experiences of tens of millions of viewers daily.

SSIM by Zhou wang on Youtube
Objective image quality assessment, what's beyond
https://www.youtube.com/watch?v=ibiCs_NJgCQ

SSIMPLUS Index for Video Quality-of-Experience Assessment

SSIM35. Z. Wang "SSIMplus," https://ece.waterloo.ca/~z70wang/research/ssimplus (A. Rehman, K. Zeng and Z. Wang, "Display device-adapted video quality-of-experience assessment," IS&T-SPIE Electronic Imaging, Human vision and electronic imaging XX, Feb. 2015.

SSIMPLUS describes the features, applications and advantages in terms of visual quality assessment. Please access:
https://ece.uwaterloo.ca/~z70wang/research/ssimplus/

SSIM36. S. Channappayya, A.C. Bovik and J.R.W. Heath, "Rate bounds on SSIM index of quantized images," IEEE Trans. CSVT, Early Access.

Bjontegaard Metric

BD1. G. Bjontegaard, "Calculation of average PSNR differences between RD-Curves," ITU-T SG16, Doc. VCEG-M33, 13th VCEG meeting, Austin, TX, April 2001. http://wfpt3.itu.int/av-arch/video-site/0104_Aus/VCEG-M33.doc
A graphical explanation of BD-PSNR and BS-bitrate is shown in [E202].

BD2. G. Bjontegaard, "Improvements of the BD-PSNR model," ITU-T SG16 Q.6, Doc. VCEG-AI11, Berlin, Germany, July 2008.

BD3. K. Anderson, R. Sjobetg and A. Norkin, "BD measurements based on MOS," (online), ITU-T Q6/SG16, document VCEG-AL23, Geneva, Switzerland, July 2009.
Available: http://wfpt3.itu.int/av-arch/video-site/0906_LG/VCEG-AL23.zip

BD4. S. Pateux and J. Jung, "An Excel add-in for computing Bjontegaard metric and its evolution," VCEG Meeting, Document VCEG-AE07, Marrakech, Morocco, Jan. 2007.

BD5. F. Bossen, "Excel template for BD-rate calculation based on piece-wise cubic Interpolation," JCT-VC Reflector, 2011.
BD Metrics computation using MATLAB source code [Online] Available: http://www.mathworks.com/matlabcentral/fileexchange/41749-bjontegaard-metric-calculation–bd-psnr-/content/bjontegaard2.m BD Metrics: VCEG-M34. [Online] Available: http://wftp3.itu.int/av-arch/video-site/0104_Aus/VCEG-M34.xls

BD6. M. Wien, "High Efficiency Video Coding: Coding Tools and Specification," Springer, 2014. Section 3.6.1.3, pages 96–97
Figure below explains clearly the BD-bit rate and BD-PSNR [E202

BD7. K. Senzaki et al., "BD-PSNR/Rate Computation tool for five data points," document JCTVC-B055, Geneva, Switzerland, July 2010.

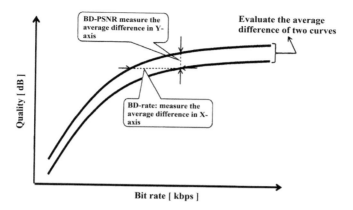

VP8, VP9:

VP.1. D. Grois, et al., "Performance comparison of H.265/MPEG-HEVC, VP9, and H.264/MPEGAVC encoders," in Proc. 30th IEEE Picture Coding Symposium, pp. 394–397, Dec. 2013.

VP.2. A. Grange. *Overview of VP-Next*. [Online]. Available: http://www.ietf.org/proceedings/85/slides/slides-85-videocodec-4.pdf, accessed Jan. 30, 2013.

VP.3. J. Bankoski, P. Wilkins, and Y. Xu, "Technical overview of VP8, an open source video codec for the web," in Proc. IEEE Int. Conf. Multimedia and *Expo*, pp. 1–6 , Jul. 2011.

VP.4. M. Rerabek and T. Ebrahimi, "Comparison of compression efficiency between HEVC/H.265 and VP9 based on subjective assessments," SPIE Optical Engineering + Applications, vol. 9217, San Diego, CA, Aug. 2014.

VP.5. D. Mukherjee et al., "An overview of new video coding tools under consideration for VP10: the successor to VP9," [9599–50], SPIE. Optics + photonics, Applications of DIGITAL IMAGE PROCESSING, XXXVIII, San Diego, California, USA, 9–13, Aug. 2015. Debargha Mukherjee; email: debargha@google.com

Please see the last slide below:
  Conclusions

- Modest progress towards a next generation codec
- A long way to go still
- Need a few big ideas
- Welcome to join the VP10 effort!

- All development in the open.
- Join the discussion with fellow developers
- The mailing list and group information can be found at

  http://www.webmproject.org/about/discuss/

VP.6. D. Mukherjee et al., "The latest open – source video codec VP9 – an overview and preliminary results," Proc. IEEE picture coding Symposium, pp. 390–393, San Jose, CA, Dec. 2013.

VP.7. D. Mukherjee et al., "A Technical overview of VP9 - the latest open – source video codec," SMPTE Motion Imaging Journal, Jan/Feb 2015.

VP.8. E. Ohwovoriole, and Y. Andreopoulos, "Rate-distortion performance of contemporary video codecs: comparison of Google/WebM VP8, AVC/H.264 and HEVC TMuC" Proc. London Communications Symposium (LCS), pp. 1–4, Sept. 2010.

VP.9. J. Bankoski, P. Wilkins and X. Yaowu, "Technical overview of VP8, an open source video codec for the web," IEEE International Conference on Multimedia and Expo (ICME), pp. 1–6, 11–15 July 2011.

VP.10. Chromium® open-source browser project, VP9 source code, Online: http://git.chromium.org/gitweb/?p=webm/libvpx.git;a=tree;f=vp9;hb= aaf61dfbcab414bfacc3171 501be17d191ff8506.

VP.11. J. Bankoski, et al., "Towards a next generation open-source video codec," Proc. SPIE 8666, Visual Information Processing and Communication IV, pp. 1–13, Feb. 21, 2013,

VP.12. E. de la Torre, R.R. Sanchez and J.L. Martinez, "Fast video transcoding from HEVC to VP9," IEEE Trans. Consumer Electronics, vol. 61, pp. 336–343, Aug. 2015. (This has several valuable references on H.264 to HEVC, MPEG-2 to HEVC, and H.264 to SVC transcoders.) also "HEVC to VP9 transcoder," IEEE VCIP 2015.

VP.13. See J. Padia, "Complexity reduction for VP6 to H.264 transcoder using motion vector reuse," M.S. Thesis, EE Dept., University of Texas at Arlington, Arlington, Texas, May 2010. http://www.uta.edu/faculty/krrao/dip click on courses and then click on EE5359 scroll down and go to Thesis/Project title and click on Jay Padia. Develop similar technique for VP9 to HEVC transcoder.

VP.14. "VP9 video codec," https://www.webmproject.org/vp9
Google's VP CODECS http://tinyurl.com/gto33oz

VP.15. G. Paim et al., "An efficient sub-sample interpolator hardware for VP9-10 standards," IEEE ICIP, Phoenix, Arizona, Sept. 2016.

VP.16. D. Mukherjee, et al., "A technical overview of VP9 – the latest open-source video codec," *SMPTE*, vol. 2013, no. 10, pp. 1–17, Annual technical conference and exhibition, 22–24 Oct. 2013.

VP9 is royalty free. Used for all Ultra HD content on YouTube and also embedded into Chrome Firefox and Opera browsers. Support from chip makers – Broadcom, MediaTek, Nvidia, ST Micro and Sigma Systems. Device manufacturers: Galaxy's S6, Sony Playstation etc.

VP.17. C.-B. Wiu, L.-H. Wang and Y.-L. Chou, "Hardware-and-memory-sharing architecture of deblocking filter for VP8 and H.264/AVC," IEEE Trans. CE, vol. 63, #3, pp. 2162–2224, Aug. 2017.

VP-P1. By cleverly using the information from the decoding process (HEVC) to accelerate the encoding process (VP9), the authors [VP12] have achieved significant reduction in encoding complexity (VP9) with negligible loss in PSNR. However BD bit rate BD PSNR and SSIM metrics have not been used. Include these metrics in evaluating the transcoder.

VP-P2. See VP.5. VP10 being developed as next generation codec by Google and successor to VP9 is an open source royalty free codec. VP9 is currently being served extensively by YouTube resulting in billions of views daily. This paper describes several tools that are proposed (beyond VP9 with the objective of improving the compression ratio (reduced bit rate) while preserving the same visual quality as VP9. (This is similar to development of H.265/HEVC in comparison with H.264/AVC). VP10 project is in still early stages and Google welcomes proposals in its development from researchers and developers. The tools (listed below) being evaluated in [VP-1] are subject to change including the exploration of new tools. Implement each of these tools separately and evaluate their performance compared with baseline codec VP9. Use the three test sets as described under Section 3 coding results. Please confirm the improvements caused by these tools over VP9 as described in Tables 1 through 5. Implementation complexity caused by these tools needs also to be considered as another performance metric. Evaluate this in comparison with VP9. Explore/innovate additional tools that can further improve the coding efficiency of VP10. The authors state that VP10 development is an open – source project, and they invite the rest of the video coding community to join the effort to create tomorrow's royalty-free codec.

- New coding tools, new ways of combining existing/new methods
- Several experiments and investigative threads underway
- High bit-depth internal
- Prediction tools
- Transform coding tools
- Screen content coding
- Miscellaneous

VP-P3. See VP.4 and VP.8 Performance comparison of HEVC, VP9 and H.264/MPEG 4 AVC in terms of subjective evaluations showing actual differences between encoding algorithms in terms of perceived quality is implemented. The authors indicate a dominance of HEVC based encoding algorithms compared to both VP9 and H.264/MPEG 4 AVC over wide range of bit rates. Extend this comparison in terms of PSNR, BD-bit rates, BD-PSNR, SSIM and implementation complexity. Consider various test sequences at different spatial and temporal resolutions.

VP-P4. Performance comparison of SDCT with DCT and DDCT is investigated in [E392]. In VP10 INTDCT is used. Replace INTDCT with INTSDCT and implement the performance comparison based on various test sequences including 4K and 8K and different block sizes. See the tables and figures in [E392].

Integrating inside VP10 might require a significant amount of work, as the transform has to be inserted in the rate-distortion optimization loop, and auxiliary information may have to be signaled. What can be easily done is to take the VP10 integer transform and rotate the transform using the technique developed in [E392], thus obtaining a rotated transform.

VP-P5 See [E393]. Apply the FCDR to VP10 codec both as a post processing operation and as an in loop (embedded) operation. Implement the projects for VP10 similar to those described in P.5.268 and P.5.269. The authors state that the merging of plenoptic imaging and correlation quantum imaging has thus the potential to open a totally new line of research. Explore this.

VP-P6 See [H78]. In the conclusions the authors state "we would like to extend this sketch attack framework to handle different video standards such as High Efficiency Video Coding (HEVC), Audio Video Standard (AVS) and Google VP9 as our future work." Extend this research to VP9.

JPEG 2000:

J2K1. D. S. Taubman and M. W. Marcellin, "JPEG2000: Image compression fundamentals, standards and practice," Springer, 2002.

J2K2. M. Rabbani, Review of the book: (D.S. Taubman and M.W. Marcellin, "JPEG2000: Image compression fundamentals, standards and practice," Springer, 2002), *J. Electron. Imaging,* vol. 11, no. 2, p. 286, Apr. 2002 (see 1 in JPEG 2000).

J2K3. Kakadu Software 6.0. [Online]. Available: http://www.kakadusoftware.com/

J2K4. H. Oh, A. Bilgin and M.W. Marcellin, "Visually lossless encoding for JPEG 2000," IEEE Trans. Image Processing, vol. 22, no.1, pp. 189–201, Jan. 2013.

J2K5. C. Christopoulus, A. Skodras and T. Ebrahimi, "The JPEG 2000 still image coding system: An overview," IEEE Trans. Consumer Electronics, vol. 46, pp. 1103–1127, Nov. 2000.

J2K6. J. Hunter and M. Wylie, "JPEG 2000 image compression: A real time processing challenge," Advanced Imaging, vol. 18, pp. 14–17, April 2003.

J2K7. D. Marpe, V. George and T. Wiegand, "Performance comparison of intra – only H.264/AVC HP and JPEG 2000 for a set of monochrome ISO/IEC test images," JVT – M014, pp. 18–22, Oct. 2004.

J2K8. D. Marpe et al., "Performance evaluation of motion JPEG 2000 in comparison with H.264/operated in intra – coding mode, "Proc. SPIE, vol. 5266, pp. 129–137, Feb 2004.

J2K9. JPEG2000 latest reference software (Jasper Version 1.900.0) Website: http://www.ece.uvic.ca/~frodo/jasper/

J2K10. M.D. Adams, "JasPer software reference manual (Version 1.900.0)," ISO/IEC JTC 1/SC 29/WG 1 N 2415, Dec. 2007.

J2K11. M.D. Adams and F. Kossentini, "Jasper: A software-based JPEG-2000 codec implementation," in *Proc. of IEEE Int. Conf. Image Processing,* vol. 2, pp. 53–56, Vancouver, BC, Canada, Oct. 2000.

J2K12. C. Christopoulos, J. Askelof and M. Larsson, "Efficient methods for encoding regions of interest in the upcoming JPEG2000 still image coding standard," IEEE Signal Processing Letters, vol. 7, pp. 247–249, Sept. 2000.

J2K13. F. Dufaux, "JPEG 2000 – Extensions," ITU-T VICA Workshop ITU Headquarters – Geneva, 22–23 July, 2005. (PPT slides)

J2K14. T. Fukuhara et al., "Motion-JPEG2000 standardization and target market," IEEE ICIP, vol. 2, pp. 57–60, 2000.

J2K15. M. Rabbani and R. Joshi, "An overview of the JPEG 2000 still image compression standard," Signal Processing: Image Communication, vol. 17, pp. 3–48, Jan. 2002.

J2K16. J. Hunter and M. Wylie, "JPEG2000 Image Compression: A real time processing challenge," Advanced Imaging, vol. 18, pp. 14–17, and 43, April 2003.

J2K17. "The JPEG 2000 Suite" P. Schelkens, editor; A. Skodras and T. Ebrahimi, co-editors —an addition to the IS&T/Wiley Series, 2010.

J2K18. P. Topiwala, "Comparative Study of JPEG2000 and H.264/AVC FRExt I– Frame Coding on High-Definition Video Sequences," SPIE, vol. 5909-31, Sept. 2005.

J2K19. T.D. Tran, L. Liu and P. Topiwala, "Performance Comparison of leading image codecs: H.264/AVC Intra, JPEG2000, and Microsoft HD Photo," http://www.uta.edu/faculty/krrao/dip/Courses/EE5351/AIC7.pdf. SPIE Intl. Symp., Digital image processing, vol. 6696, 66960B, San Diego, CA, Sept. 2007.

J2K20. D.T. Lee, "JPEG2000: Retrospective and new developments," Proc. IEEE, vol. 93, pp. 32–41, Jan. 2005.

J2K21. Special issue on JPEG 2000, Signal Processing and Image Communication, vol. 17, Jan. 2002.

J2K22. C. Li, C. Deng and B. Zhao, "Textural and gradient feature extraction from JPEG2000 code stream for airfield detection," IEEE DCC, Snow Bird, Utah, March-April 2016.

J2K23. D. Barina, O. Klima and P. Zemcik, "Single-loop software architecture for JPEG2000," IEEE DCC, Snow Bird, Utah, March-April 2016.

J2K24. T. Richter and S. Simon, "Towards high-speed, low-complexity image coding: variants and modification of JPEG 2000," in Proc. SPIE, vol. 8499, 2012.

J2K25. F. De Simone et al., "A comparative study of JPEG 2000, AVC/H.264, and HD photo," in SPIE Optics and Photonics, Applications of Digital Image Processing, vol. 6696, 2007.

J2K26. F. Liu et al., "Visibility thresholds in reversible JPEG2000 compression," IEEE DCC, Snow Bird, Utah, April 2017.

J2K27. C. Bi and J. Liang, "Joint source-channel coding of JPEG2000 image transmission over two-way multi-relay networks," IEEE Trans. IP, Early access. Section VII conclusion is repeated here.

Conclusion

In this paper, a two-way multi-relay transmission system is proposed, which uses an adaptive TDBC protocol with optimal mode selection, relay selection and power allocation under the data rate fairness constraint. Further, a low-complexity JSCC scheme is presented to optimize the transmission of progressive error-resilient JPEG 2000 bit streams over our proposed system. We then combine the proposed JSCC and the proposed cooperative transmission system, and simulation results show that it outperforms other similar works. Our future work includes extending the proposed two-way multi-relay framework to cognitive radio, as cognitive radio is designed to fully use the available spectrum, which is suitable for multimedia applications. Another future work is to study 3D video transmission over the proposed framework.

J2K28. Z. Wu, A. Bilgin and M.W. Marcellin, "Error resilient decoding of JPEG2000," IEEE Trans. IP, Early access.

J2K29. S. Jallouli et al., "A preprocessing technique for improving compression performance of JPEG2000 for images with sparse or locally sparse histograms," EUSIPCO2017, KOS Island, Greece, Aug. 2017.

J2K30. A. Skodras, C. Christopoulos, and T. Ebrahimi, "The JPEG 2000 still image compression standard," *IEEE Signal Processing Mag.,* vol. 18, no. 5, pp. 36–58, 2001.

J2K31. International standard ISO/IEC 15444-1, "Information technology JPEG 2000 image coding system: Core coding system," *second edition,* Sept. 2004.

J2K32. ITU-T T.809, "Information technology JPEG 2000 image coding system: Extensions for three-dimensional data," *SERIES T: Terminals for telematic services still-image compression JPEG 2000,* May, 2011.

J2K33. P. Schelkens, A. Munteanu, A. Tzannes, and C. Brislawn, "JPEG2000. Part 10. Volumetric data encoding," in *Proc. IEEE International Symposium on Circuits and Systems, (ISCAS),* pp. 3874–3877, 2006.

J2K34. C.-J. Liau et al., "Lifting based discrete wavelet transform architecture for JPEG2000," IEEE ISCAS, vol. 2, pp. 445–448, 2001.

J2K35. B.A. Banister, B. Belzer and T.R. Fischer, "Robust image transmission using JPEG2000 and turbo codes," IEEE Signal Processing Letters, vol. 9, pp. 117–119, April 2002.

J2K36. M. Charrier, D.S. Cruz and M. Larsson, "JPEG 2000, the next millennium compression standard for still images," IEEE ICMCS 99, pp. 131–132, Florence, Italy, June 2001.

J2K37. J. Zhang et al., "Evaluation of JP3D for lossy and lossless compression of hyperspectral imagery," Proc. int. Geoscience and Remote sensing symposium, vol. 4, pp. 474–477, Cape Town, S. Africa, 2009.

J2K38. T. Ebrahimi, M. Chamik and S. Winkler, "JPEG vs. JPEG 2000: An objective comparison of image encoding quality," Proc. SPIE Applications of Digital Image Processing, Vol. 5558, pp. 300–308, Denver, CO, Aug 2–6, 2004.

J2K39. G.A.D. Punchihewa, D.G. Bailey, and R.M. Hodgson, "Objective evaluation of edge blur and ringing artefacts: Application to JPEG and JPEG 2000 image codecs." Image and Vision Computing New Zealand, pp. 61–66, Dunedin, New Zealand, Nov. 28–29, 2005.

J2K40. J.C. Maturana-Espinosa et al., "Rate allocation for motion compensated JPEG2000," IEEE DCC 2018, Snow Bird, UTAH, March 2018.

J2K41. Y. Lin et al., "Task-based JPEG 2000 image compression: An information theoretic approach," IEEE DCC 2018, Snow Bird, UTAH, March 2018.

J2K42. F. Liu et al., "S visual discrimination model for JPEG2000 compression," IEEE DCC 2018, Snow Bird, UTAH, March 2018.

J2K43. P. Schelkens et al., "JPEG 2000 Part 10, volumetric data encoding," IEEE ISCAS, pp. 3874–3877, 2006.

J2K44. J. Zhang et al., "Evaluation of JP3D for lossy and lossless compression of hyperspectral imagery" Intl. Geoscience and remote sensing symposium, vol. 4, pp. 474–477, Cape Town, S.A., 2009.

J2K45. R. Biswas, S.R. Malreddy and S. Banerjee, "A high precision-low area unified architecture for lossy and lossless 3D multi-level discrete wavelet transform," IEEE Trans. CSVT, EA.

J2K46. ITU-T T.809, "Information technology JPEG 2000 image coding system: Extensions for three-dimensional data," SERIES T: Terminals for telematic services still-image compression JPEG 2000, May, 2011.

J2K47. C. Bland and J. Liang, "Joint source-channel coding of JPEG2000 image transmission over two-way multi-relay networks," IEEE Trans. IP, vol. 26, pp. 3594–3608, July 2017.

J2K48. Y. Lin et al., "Task-based JPEG 2000 image compression: An information theoretic approach," IEEE DCC, Sun Bird, Utah, March 2018.

J2K49. F. Liu et al., "A visual discrimination model for JPEG2000 compression," IEEE DCC, Sun Bird, Utah, March 2018.

J2K50. M.A. Martinez-del-Amor, V. Bruns and H. Sparenberg, "Parallel efficient rate control methods for JPEG 2000," SPIE, Applications of digital image processing XL, vol. 10396, San Diego, CA, Aug. 2017.

J2K51. D. Taubman, A. Naman and R. Mathew, "A fast block coding option for JPEG 2000," SPIE, Applications of digital image processing XL, vol. 10396, San Diego, CA, Aug. 2017.

J2K52. B.-F Wu, and C.-F Lin, "A High-Performance and Memory-Efficient Pipeline Architecture for the 5/3 and 9/7 Discrete Wavelet Transform of JPEG2000 Codec," IEEE Trans. CSVT, vol. 15, pp. 1615–1625, Dec. 2005.

J2K53. T. Tillo and G. Olmo, "A novel multiple description coding scheme compatible with JPEG 2000 decoder," IEEE Signal Processing Letters, vol. 11, no. 11, pp. 908–911, 2004.

J2K54. C. Deng, W. Lin, B.-S. Lee, C. Lau, and M. Paul, "Comparison between H.264/AVC and Motion JPEG2000 for super-high definition video coding," Proc. 17th IEEE Int. Conf. Image Process. (ICIP), pp. 2037–2040, Sept. 2010.

Digital Cinema

The Digital Cinema (DC) Ad Hoc Group within the JPEG Committee has been successful in seeing their work adopted by the industry. The Digital Cinema Initiatives (www.dcimovies.com) organization has adopted JPEG 2000 for the distribution of digital movies to theatres. The successful roll-out of this solution continues unabated with over 5000 theatres supporting digital cinema, including 1000 stereoscopic theatres. The first live-action stereoscopic feature encoded with JPEG 2000, Beowulf, is being released on November 16, 2007. The Digital Cinema Ad Hoc Group has initiated work on studio broadcast applications. In addition, the group has begun work on the archival of motion pictures and related contents.

JPEG 2000 Related Work

At the Kobe meeting, the JPEG Committee confirmed the endorsement of Intellectual Resource Initiative (IRI) as JTC 1 Registration Authority (RA) for use in JPSEC. IRI is a Non Profit Organization, based in Japan, established to create policy proposals on the importance of intellectual information in society.

JPEG 2000 Part 9 known as JPIP, allows powerful and efficient network access to JPEG 2000 images and their metadata in a way that exploits the best features of the JPEG 2000 standard. Interoperability testing among several JPIP implementations continued. Participation from additional organizations is solicited, and testing will continue over the Internet between meetings. For more information please contact jpip@jpeg.org.

JPEG 2000 Part 10 Ad Hoc Group has been working on the extension of JPEG 2000 to three-dimensional images such as Computer Tomography (CT) scans and scientific simulations. JP3D is currently in balloting phase for International Standard (IS) status. Due to an increased interest for compression technologies for floating-point data, the JPEG Committee has issued a call for information on applications and compression technology for floating-point data. Responses will be reviewed at the 44th WG1 San Francisco Meeting, March 31–April 4, 2008.

JPEG 2000 Part 11 Wireless, also known as JPWL, has become an International Standard (ISO/IEC 15444-11). JPWL has standardized tools and methods to achieve the efficient transmission of JPEG 2000 imagery over an error-prone wireless system.

JPEG 2000 Part 13 standardizes an entry level JPEG 2000 encoder with widespread applications, intended to be implemented on a license and royalty fee free basis. It is published as ISO/IEC 15444-13.

JPSearch

ISO/IEC 24800, Still Image Search, known as JPSearch, is a project that aims to develop a standard framework for searching large collections of images. This project is divided in five parts.

- Part 1 – Framework and System Components, is a Technical Report that introduces the JPSearch architecture and outlines the organization of the JPSearch specifications.
- Part 2 – Schema and Ontology Registration and Identification, standardizes a format for the import, export and exchange of ontology.
- Part 3 – JPSearch Query Format, which is developed jointly with MPEG, allows for the expression of search criteria, the aggregation of return results and the management of query process.
- Part 4 – Metadata Embedded in Image Data (JPEG and JPEG 2000) file format, standardizes image data exchange format with associated metadata.
- Part 5 – Data Interchange Format between Image Repositories, standardizes a format for the exchange of image collections and respective metadata between JPSearch compliant repositories.

At the meeting, requirements for each part have been reviewed and updated, and Working Drafts for Part 2 and Part 4 and Committee Draft of Part 3 have been produced.

The new JasPer (JPEG-2000) release is available from the JasPer Project Home Page (i.e., http://www.ece.uvic.ca/~mdadams/jasper)

JasPer 1.900.21 is available from the JasPer home page: (Nov. 2016) http://www.ece.uvic.ca/~mdadams/jasper

(i.e., http://www.jpeg.org/software) JasPer is the software in C for JPEG2000. Also http://jj2000.epfl.com

http://www.kakadusoftware.com (kakadu software by David Taubman) JPEG-2000 JasPer software http://www.ece.uvic.ca/~mdadams/jasper JPEG-2000 tutorial http://www.ece.uvic.ca/~mdadams

Dear JasPer Users, 12/30/2016

I am pleased to announce the release of version 2.0.10 of theJasPer software. This new release can be obtained from the"Downloads" section of the JasPer web site:

http://www.ece.uvic.ca/~frodo/jasper/#download
If you encounter any problems with the software, please report them to the issue-tracker for JasPer on GitHub:

https://github.com/mdadams/jasper/issues

- - -
Michael Adams, Associate Professor, Ph.D., P.Eng., IEEE Senior Member
Dept. of Electrical and Computer Engineering, University of Victoria
PO Box 1700 STN CSC, Victoria, BC, V8W 2Y2, CANADA
E-mail: mdadams@ece.uvic.ca, Web: http://www.ece.uvic.ca/~mdadams

Sweden Records Missile Tests Using JPEG2000 Technology April 2009

The Swedish Armed Forces deploys a sophisticated, medium-range ground-to-air defense missile system, called RBS 23 BAMSE, manufactured by Swedish company Saab Bofors Dynamics (SBD).

To facilitate the test system, SBD utilizes RGB Spectrum's advanced DGy ultra high resolution recording systems. The DGy^{TM} recorders were selected based upon their superior ability to record and reproduce the most intricate details using advanced JPEG2000 compression. Previously, SBD used scan converters and conventional VCRs which rendered "soft" imagery with difficult to discern text and graphics. The DGy recorders preserve the system's complex symbologies and provide sharp graphics and alpha-numerics for after-action-review.

RGB Spectrum Real Time Newsletter 6/25/09
Advanced JPEG2000 Recording Deployed For Helicopter Mission Simulators

Rockwell Collins, the prime contractor for the U.K. Army's Aviation Command and Tactics Trainer (ACTT), was contracted to execute the sixth phase upgrade of this simulator. The upgrade was designed to provide advanced command and tactical training capability for the Army's helicopter pilots for operation in high threat environments.

The upgrade required a system to record the intricate symbology used in the simulated Integrated Helmet and Display Sight System (IHADSS) unit. Rockwell Collins selected RGB Spectrum's DGY^{TM} digital recording system to meet this demanding requirement. The DGy system was chosen for its superior ability to reproduce intricate detail, achieved through advanced wavelet-based JPEG2000 near lossless compression.

Project:

J2K-P1. See references [36] through [39] in [LL1]. These references relate to parallel processing methods and advanced graphic processing units for significant speed up in execution time of JPEG 2000. Review these references and evaluate the encoding and decoding speeds of JPEG 2000. Consider the encoder/decoder complexity reduction using various test images.

J2K-P2. See J2K24. By replacing the EBCOT coding with a simple Huffman-run length scheme, the JPEG 2000 implementation complexity is significantly reduced. Implement JPEG 2000 Huffman coding and evaluate how the simplified VLC reduces the encoder/decoder complexity.

J2K-P3. See J2K27 The authors state that
"Our future work includes extending the proposed two-way multi-relay framework to cognitive radio, as cognitive radio is designed to fully use the available spectrum, which is suitable for multimedia applications." Explore this.

J2K-P4. See J2K-P3. The authors further state that
"Another future work is to study 3D video transmission over the proposed framework." Explore this.

J2K-P5. See [J2K19 and J2K25] Extend the performance comparison of leading video codecs to HEVC intra coding.

J2K-P6. See [J2K19]. For future work, we are planning a much more comprehensive comparison of the three codecs with a more diverse set of test images and video sequences across a wider range of bit rates, data resolution, bit depth, and color format. We are particularly interested in the high-quality 4:4:4 setting with 10-bit and 12-bit video data for high-definition applications. Lossless coding performance of H.264, scalability performance of JPEG2000 versus HD Photo, and complexity as well as memory requirement. Implement this comprehensive comparison and state the conclusions.

J2K-P7. In [J2K45] the applications of 3D DWT are outlined. Part of the conclusion section is repeated here.

The design described in [J2K45] can be an integral part of the JPEG 2000 part 10 [J2K43 and J2K46]. The speed of operation of JPEG 2000 is limited by the high computing EBCOT block, so the present architecture can be modified with a lesser hardware utilization to accommodate with JPEG 2000 system. Modify the architecture and implement in JPEG-Part 10.

JPEG XR (Extended Range)

JPEG XR

JXR1. F. Dufaux, G. J. Sullivan, and T. Ebrahimi, "The JPEG XR image coding standard [Standards in Nutshell]," IEEE Signal Processing Magazine, vol. 26, no. 6, pp. 195–199 and 204, Nov. 2009.

JXR2. Kodak Lossless True Color Image Suite. [Online] Available: http://r0k.us/graphics/kodak/

JXR3. JPEG Core Experiment for the Evaluation of JPEG XR Image Coding. [Online]. Available: http://jahia-prod.epfl.ch/site/mmspl/op/edit/page-58334.html

JXR4. Microsoft HD photo specification: http://www.microsoft.com/whdc/xps/wmphotoeula.mspx

JXR5. ITU. (2012) JPEG XR image coding system reference software. [online]. Available: http://www.itu.int/rec/T-REC-T.835-201201-I/en

JXR6. C. Tu et al., "Low-complexity hierarchical lapped transform for lossy-to-lossless image coding in JPEG XR/HD Photo", Proc. SPIE,

Vol. 7073, pp. 70730C-1 through 70730C-12, San Diego, CA, Aug. 2008.

JXR7. T. Suziki and T. Yoshida, "Lower complexity lifting structures for hierarchical lapped transforms highly compatible with JPEG XR standard," IEEE Trans. CSVT. vol. 27, pp. 2652–2660, Dec. 2017. The authors state, "JPEG XR (eXtended Range) is a newer lossy –to – lossless image coding standard. JPEG XR has half the complexity of JPEG 2000 while preserving image quality."

JXR8. T.835: Information technology – JPEG XR image coding system – Reference software," *ITU [Online]*, Available: https://www.itu.int/rec/TREC-T.835-201201-I/.

JXR9. "JPEG core experiment for the evaluation of JPEG XR image coding," *EPFL, Multimedia Signal Processing Group [Online]*, Available: http://mmspg.epfl.ch/iqa.

JXR10. "The New Test Images - Image Compression Benchmark," *Rawzor-Lossless Compression Software for Camera RAW Images [Online]*, Available: http://imagecompression.info/test images/.

JPEG XR is a proposed new part (Part 2) of the recently established new work item in JPEG, known as JPEG Digital Imaging System Integration (ISO/IEC 29199 – JPEG DI). JPEG DI aims to provide harmonization and integration between a wide range of existing and new image coding schemes, in order to enable the design and delivery of the widest range of imaging applications, across many platforms and technologies. JPEG DI aims to leverage the rich array of tools developed in and around JPEG and JPEG 2000 to support new image compression methods such as JPEG XR. JPEG XR is designed explicitly for the next generation of digital cameras, based extensively on the technology introduced by Microsoft in its Windows Media Format proposals, at present known as HD Photo. At the Kobe meeting, The JPEG XR specification (Working Draft) was reviewed and will be balloted for promotion to Committee Draft (CD) status before the 44th WG1 San Francisco meeting, March 31 to April 4, 2008. In addition to JPEG XR itself, the creation of two other new parts of the standard on compliance testing and reference software was approved at the meeting. Now adopted as a standard (Nov. 2009).

AIC Advanced Image Coding (latest activity in still frame image coding) please access the web site below.

http://www.bilsen.com/index.htm?http://www.bilsen.com/aic/

Please access the web site below about ELYSIUM activity in JPEG/ JPEG2000/JPEG XR and AIC www.elysium.ltd.uk *(Richard Clark JPEG webmaster* richard@elysium.ltd.uk*)*

HD PHOTO

(JPEG Convener Daniel Lee daniel.t.lee@ebay.com convener@jpeg.org*)*
Microsoft HD photo specification
http://www.microsoft.com/whdc/xps/wmphotoeula.mspx

JXR-P1. The authors in [JXR7] state, "that they have presented lifting structures for hierarchical lapped transforms (HLTs) highly compatible with the JPEG XR standard [JXR1] and that are highly compatible with JPEG XR standard and are lower in complexity in terms of the number of operations and lifting steps compared with JPEG XR. These lifting structures will have any signal processing and communication applications because they are based on commonly used rotation matrices." Explore these applications and see how these lifting structures for HLTs can be used.

JPEG-LS:

JLS1. M. J. Weinberger, G. Seroussi and G. Sapiro, "The LOCO-I Lossless Image Compression Algorithm: Principles and Standardization into JPEG-LS," IEEE Trans. on Image Processing, vol. 9, pp. 1309–1324, Aug. 2000. Website: http://www.hpl.hp.com/loco/HPL-98-193RI.pdf

JLS2. J. Weinberger, G. Seroussi, and G. Sapiro, "LOCO-I: A low complexity, context-based, lossless image compression algorithm," Hewlett-Packard Laboratories, Palo Alto, CA.

JLS3. Ibid, "LOCO-I A low Complexity Context-based, lossless image compression algorithm," Proc. IEEE DCC, pp. 140–149, Snowbird, Utah, Mar. 1996.

JLS4. See Z. Zhang, R Veerla and K.R. Rao, "A modified advanced image coding" CANS University Press, pp. 110–116, 2010.

JLS5. UBC JPEG-LS codec implementation [online]. Available: http://www.stat.columbia.edu/~jakulin/jpeg-ls/mirror.htm

JLS6. http://www.ece.ubc.ca/spmg/research/jpeg/jpeg_ls/jpegls.html

JLS7. LOCO-I Software (public domain) (JPEG-LS) http://spmg.ece.ubc.ca/research/jpeg/jpeg_ls/jpegls.html

JLS8. LOCO_I technical report www.hpl.hp.com/loco (HP Labs)

JLS9. JPEG-LS (LOCO-I) http://www.hpl.hp.com/loco/HPL-98-193R1.pdf

JLS10. T. Richter et al., "Entropy coding and entropy decoding improvements of JPEG-XS," IEEE DCC 2018, Snow Bird, Utah March–April 2018.

Project:

JLS-P1. This paper involves comparison of various image coding standards such as JPEG, JPEG-LS, JPEG2000, JPEG XR, advanced image coding (AIC) and modified AIC (MAIC) (Also H.264/MPEG4 AVC intra mode only). Extend this comparison to HEVC intra mode only. Consider test images at various spatial resolutions and at different bit rates. Include BD-bit rate and BD-PSNR as metrics besides PSNR and SSIM. Consider also implementation complexity.

JPEG:

JPEG1. G. K. Wallace, "The JPEG still picture compression standard," Commun. ACM, vol. 34, no. 4, pp. 30–44, Apr. 1991. Also "The JPEG still picture compression standard," IEEE Trans. CE, vol. 38, pp. 18–34, Feb. 1992.

JPEG2. Independent JPEG Group. [Online]. Available: http://www.ijg.org/ www.jpeg.org

JPEG3. J. Aas. Mozilla Advances JPEG Encoding With Mozjpeg 2.0. [Online]. Available: https://blog.mozilla.org/research/2014/07/15/mozilla-advances- jpeg-encoding-with-mozjpeg-2-0/, accessed July 2014.

JPEG4. W. B. Pennebaker and J. L. Mitchell, "JPEG Still Image Data Compression Standard," Van Nostrand Reinhold, New York, 1992.

JPEG5. JPEG reference software Website: ftp://ftp.simtel.net/pub/simtelnet/msdos/graphics/jpegsr6.zip

JPEG6. F. Huang et al., "Reversible data hiding in JPEG images," IEEE Trans. CSVT, vol. 26, pp. 1610–1621, Sept. 2016. Several papers related to reversible and other data hiding schemes are listed as references at the end.

JPEG7. X. Zhang et al., "Lossless and reversible data hiding in encrypted images with public key cryptography," IEEE Trans. CSVT, vol. 26, pp. 1622–1631, Sept. 2016.

JPEG8. P.A.M. Oliveira et al., "Low-complexity image and video coding based on an approximate discrete Tchebichef transform," IEEE Trans. CSVT, vol. 27, pp. 1066–1076, May 2017. (This is same as DCT34).

JPEG9. T. Richter, "JPEG on steroids: Common optimization techniques for JPEG image compression," IEEE ICIP2016, Phoenix, AZ, Sept. 2016.

JPEG10. H. Lee et al., "A novel scheme for extracting a JPEG image from an HEVC compressed data set" IEEE ICCE, Las Vegas, Jan. 2017.

JPEG11. C. Liu et al., "Random walk graph Laplacian-based smoothness prior for soft decoding of JPEG images," IEEE Trans. IP, vol. 26, pp. 509–524, Feb. 2017.

JPEG12. M.M. Siddeq and M.A. Rodrigues, "A novel high-frequency encoding algorithm for image compression," EURASIP J. on advances in signal processing," 2017:2017:26.

JPEG13. A. O. Zaid and A. Houimli, "HDR image compression with optimized JPEG coding," EUSIPCO 2017, KOS Island, Greece, Aug. 2017.

JPEG14. C. Zhao et al., "Reducing Image Compression Artifacts by Structural Sparse Representation and Quantization Constraint Prior," IEEE Trans. CSVT, vol. 27, pp. 2057–2071, Oct. 2017.

JPEG15. J. Qian et al., "New framework of reversible data hiding in encrypted JPEG bitstreams," IEEE Trans. CSVT, Early access. Conclusion and discussion are repeated here.

Conclusion and Discussion

This paper proposes a new framework of reversible data hiding in encrypted JPEG bitstreams. A JPEG encryption and decryption algorithm is developed to hide the content of the original image. The encryption is format-compliant to the popular JPEG decoders. On the cloud side, the server can embed a large amount of additional bits into the encrypted bitstream. We propose to do the embedding using a combination of code mapping and ordered embedding. Once an authorized user requires a downloading operation, the server extracts the additional message and recovers the original encrypted bitstream. Since the recovery has been done before downloading, the user obtains an image exactly the same as the original. The proposed framework outperforms the previous JPEG RDH-EI frameworks on three aspects. First, the proposed embedding method provides a much larger payload than the other methods. Second, the proposed framework frees the computation burdens on both the owner and the user sides. While pre-processing or post-processing is required in the traditional works, the propose framework requires the owner or the user to no extra tasks except encryption or decryption. Finally, the proposed JPEG encryption algorithm is also secure against the ciphertext-only attack. Access the following paper on RDH:

Z. Ni, Y. Shi, N. Ansari, and S. Wei, "Reversible Data Hiding," IEEE Trans. CSVT, vol. 16, pp. 354–362, March 2006.

JPEG16. A. Prakash et al., "Protecting JPEG images against adversarial attacks," IEEE DCC 2018, Snow Bird, Utah, March 2018.

JPEG17. M. Hopkins, M. Mitzenmacher and S. Wagner-Carena, "Simulated annealing for JPEG quantization," IEEE DCC 2018, Snow Bird, UTAH, March 2018.

JPEG18. X.F. Zhang et al., "Just noticeable distortion based perceptual quantization for JPEG compression," IEEE Signal Processing Letters, vol. 24, pp. 96–100, Jan. 2017.

JPEG19. A.S. Al-Fohoum and A.M. Reza, "Combined edge crispening and statistical differencing for deblocking JPEG compressed images," IEEE Trans. IP, vol. 10, pp. 1288–1298, Sept. 2011.

JPEG20. X. Liu et al., "Prior-based quantization bin matching for cloud storage of JPEG images," IEEE Trans. IP, vol. 27, pp. 3222–3235, July 2018. The section conclusion is repeated here.

"To reduce the cloud storage cost of images uploaded from social media sites, in this paper we propose a JPEG requantization/restoration framework at the expense of a controlled increase in computation mostly during download. Specifically, selective blocks of an uploaded JEPG image are coarsely quantized before storage in a central cloud, and when it is requested, a reverse mapping operation is performed to restore fine quantization bin indices from coarse quantization bin indices (quantization bin matching (QBM)) using a small over-complete dictionary and a sparse graph. Our framework operates in both lossless and near-lossless modes, which deterministically or statistically guarantees recovery of the original JPEG image with different coding overhead. Experimental results show significant compression gain over competing schemes for a large range of test images. The simplified quantization bin matching is actually a JPEG soft decoding problem, which can be addressed by a deep learning technique. As future work, we will explore this direction as an alternative implementation."

JPEG21. F. Huang, J. Huang and Y.Q. Shi, "New channel selection rule for JPEG stegonagraphy," IEEE Trans. Inf. Forensics and Security, vol. 7, no. 4, pp. 1181–1191, Apr. 2012.

JPEG22. O. Watanabe, H. Kobayashi and H. Kiya, "Two-layer lossless HDR codIng considering histogram sparseness with backward compatibility to JPEG," IEEE PCS2018, San Francisco, CA, June 2018.

JPEG23. F. Jiang et al., "An end-to-end compression framework based on convolutional neural networks," IEEE Trans. CSVT, EA.

Conclusion section is repeated here.

In this paper, we propose an effective end-to-end compression framework based on two CNNs, one of which is used to produce compact intermediate representation for encoding using an image encoder. The other CNN is used to reconstruct the decoded image with high quality. These two CNNs collaborate each other and are trained using a unified optimization method. Experimental results demonstrate that the proposed compression framework achieves state-of-the-art performance and is much faster than most post-processing algorithms. Our work indicates that the performance of the proposed compression framework can be significantly improved by applying the proposed framework, which will inspire other researchers to design better deep neural networks for image compression along this orientation.

JPEG24. Z. Wang et al., "D3: Deep dual-domain based fast restoration of jpeg-compressed images," in Proceedings of the IEEE Conference on CVPR, 2016.

JPEG25. H. Chang, M. K. Ng, and T. Zeng, "Reducing artifacts in jpeg decompression via a learned dictionary," IEEE Trans. SP, vol. 62, no. 3, pp. 718–728, 2014.

JPEG26. X.F. Zhang et al., "Just-noticeable difference-based perceptual optimization for JPEG compression," IEEE SP Letters, vol. 24, pp. 96–100, Jan. 2017.

JPEG27. G. Lakhani, "Modifying JPEG binary arithmetic codec for exploiting inter/intra-block and DCT coefficient sign redundancies," IEEE Trans. IP, vol. 22, pp. 1326–1339, April 2013.

JPEG28. A. Sengupta et al., "Low-cost obfuscated JPEG CODEC IP core for secure CE hardware," IEEE Trans. CE, EA.

Conclusion and future work is repeated here.

"This paper proposed a low-cost obfuscation for JPEG CODECIP core. Future works aims to amplify the SoO of proposed approach by including stronger structural obfuscation in the form of multi-level transformation in different abstraction levels. We also aim to integrate functional obfuscation with structural obfuscation during DSP core synthesis."

PEG29. P. Li and K.-T. Lo, "A content-adaptive joint image compression and encryption scheme," IEEE Trans. multimedia, vol. 20, pp. 1960–1972, Aug. 2018. Conclusion section is repetaed here.

"In this paper, we propose a new joint image compression and encryption scheme, in which the encryption key is dependent on the plain-image. The whole scheme has good diffusion property, and is compression friendly to JPEG. To produce the image content-sensitive key, we adopt BLAKE2 hash function to produce the 256-bit encryption key. Our encryption operations include three parts: new orthogonal 8×8 transforms transformation, DC coefficients encryption, and AC encryption. We generate new transforms by introducing rotation angles into order-8 DCT's flow-graph structure, and apply them alternatively for 8×8 blocks' transformation, controlled by the encryption key. DC coefficients are encrypted after quantization by 8×8 blocks' permutation and XOR operation. For AC coefficients encryption, we embed the 256-bit encryption key into the bitstream of some certain AC coefficients, and the data embedding/extracting operation is controlled by another secret key, called embedding key. This data embedding strategy can reduce the cost of sending different 256-bit encryption keys to the decoder when different plain-images are compressed and encrypted. Extensive experiments and security analysis are conducted to show the good compression and encryption performance of our proposed encryption scheme. There are few extensions we believe that can be considered in our following work. First, currently the encryption scheme cannot achieve perfect correlation removal effect and diffusion property, because the whole encryption work is realized in 8×8 block unit. Hence, in the next step, we may try to do transformation in different block sizes and to execute permutation in some smaller units, perhaps 4×4 blocks permutation or image pixels' bit-plane level permutation. Additionally, our encryption scheme is mainly realized in the transformation stage and entropy coding stage of JPEG, other existing JPEG encryption methods, such as zig-zag scanning order change or quantization table change, can be combined with our current proposed encryption scheme, in order to enlarge the encryption space by one dimension. Finally, we will try to extend our current encryption operations for other image/video compression standards, such as JPSEC standard and MPEG-4."

Related reference: B. Schneider, "Applied cryptography; Protocols, algorithms and source code in C," Wiley, New York, NY, 1996.

JPEG30. The JPEG Committee, "ISO/IEC JTC 1/SC 29/WG 1, JPEG Privacy and Security Call for Proposals," 2017. [Online]. Available:

https://jpeg.org/downloads/privacy and security/wg1n74015 Draft CfP.pdf
Objective is to develop various security functions in lossy JPEG in order to realize secure image information sharing.

JPEG31. Z. Qian, X. Zhang and S. Wang, "Reversible data hiding in encrypted JPEG bitstream," IEEE Trans. Multimedia, vol. 16, no. 5, pp. 1486–1491, Aug. 2014.

JPEG32. S. Ong, K. Wong, X. Qi, and K. Tanaka, "Beyond format-compliant encryption for JPEG image," Signal Process.: Image Commun., vol. 31, pp. 47–60, 2015.

JPEG33. Z. Qian et al., "Separable reversible data hiding in encrypted JPEG bitstreams," IEEE Trans. Depend. Security Comput., to be published.

JPEG34. Q. Liu, "An improved approach to exposing JPEG seam carving under recompression," IEEE Trans. CSVT, EA. Several valuable references related to seam carvig, image forgery/tampering detection, inpainting forgery and steganography are cited at the end.

Conclusion section is repeated here.
"To improve the detection of JPEG image seam carving under the recompression at the same or at a lower image quality, in this paper, we propose a hybrid large feature mining-based approach that contains directional derivative and minimal neighboring difference-based joint density in the spatial domain, high order joint density features in the DCT domain, and Gabor residual-based marginal density and neighboring joint density features. Ensemble learning is used to deal with the high feature dimensionality. Compared to other popular low-dimensionaldetection methods, proposed approach significantly improves the detection accuracy. The proposed approach is also superior to our previously proposed hybrid large feature mining-based detection method, and it is superior to rich modeling-based approach on large image sizes and comparable on small image sizes. In our approach, directional derivative-based feature set and Gabor residual-based feature set generally perform the best. Our study also indicates that feature selection may greatly reduce the feature number while maintaining a better or comparable detection accuracy. It also shows that the detection is more challenging while the image size is small. Our future study includes but not limited to optimizing the parameters in feature extraction, investigating judicious feature selection methods and feature fusion techniques, and further expanding our seam-carving forgery database. We will also integrate our approach with rich

modeling-based approach, and explore blind detection methods without the aid of a large external image database."

JPEG35. S.I. Young, A.T. Naman and D. Taubman, "COGL: Coeffficient graph Laplacians for optimized JPEG image decoding," IEEE Trans. IP, vol. 28, pp. 34–355, Jan. 2019. Conclusion section is repeated here.

Conclusion

In this paper, we presented a novel yet minimalistic method for optimized decoding of JPEG images based on a coefficient graph Laplacian regularizer. Our method regularizes variations in the DCT coefficients, rather than in the pixel intensity values themselves. We also related our graph to the nonlocal means de-noising algorithm of Buades, showing that fast Gaussian filtering can be utilized within the proximal gradientdescent method to solve our problem efficiently. Our method achieves an average of 1.4 dB gain compared to the standard decoding by a JPEG decoder, and is on par with other wellperforming JPEG decoding methods while being significantly faster. Further, we applied our Laplacian regularizer to the problem of decoding M-JPEG video by extending our graph Laplacian in the temporal direction. The spatio-temporal graph Laplacian produces a further gain of 0.45 dB on average compared to the spatial regularizer. Another possible extension of our work could be to introduce a dictionary-based regularization term as in [10], which we defer to future work.

JPEG36. J. He et al., "JPEG Image Encryption with Improved Format Compatibility and File Size Preservation," IEEETrans. MM, vol. 20, pp. 2645–2658, Oct. 2018. Several valuable papers related to JPEG encryption are cited.

Conclusion section is repeated here.

Conclusion

"JPEG is generally not only the default format of the images produced by many image capture devices but also the most commonly used image format in our daily lives. Image encryption techniques can be utilized to ensure the security and privacy of valuable JPEG images. In this paper, we present a novel bitstream-based JPEG image encryption method. The experimental

results and security analysis demonstrated the improved performance of our scheme. The main contributions of this paper are as follows: 1) Novel encryption techniques toward DC and AC codes are proposed, which contribute to the new JPEG image encryption scheme presented in this paper. The file size of an encrypted JPEG image generated with the proposed method can be preserved except for slight variations because of byte alignment. Moreover, better format compatibility with the JPEG standard is obtained. 2) All the encryption operations are conducted directly on the bitstream without the need of entropy encoding once again to generate an encrypted JPEG image. Additionally, the number of iterations of grouping and swapping DC codes can be adjusted to reach a compromise between visual security and computational efforts. 3) An image-content-related encryption key is employed to provide better security against known plaintext attacks. The proposed scheme can provide a large key space enough to defend against brute-force attacks. And the chaotic-based permutations help our method to resist cryptanalysis."
MATLAB JPEG TOOLBOX

Phil Sallee's MATLAB Jpeg Toolbox. http://dde.binghamton.edu/download/stego algorithms.

JPEG37. J. Kodovsky and J. Fridrich, "Steganalysis of JPEG images using rich models," in Proc. SPIE, Electronic imaging, Media Watermarking, Security, and Forensics XIV, vol. 8303, pp. 0A 1–13, 2012.

JPEG38. W. Su et al, "A new distortion function design for JPEG steganography using the generalized uniform embedded strategy, IEEE CSVT, EA.

JPEG39. l. Guo, J. Ni and Y.Q. Shi, "Uniform embedding for efficient JPEG steganography," IEEE Trans. information forensics and security, vol. 9, pp. 814–825, May 2014.

JPEG40. L. Guo et al, "Using statistical image model for JPEG steganography; Uniform embedding revisited," IEEE Trans. information forensics and security, vol. 10, pp. 2669–2680, Dec. 2015.

JPEG41. C. Wang and J. Ni, "An efficient JPEG steganographic scheme based on the block-entropy of DCT coefficients," IEEE ICASSP, pp. 1785–1788, March 2012.

JPEG42. V. Holub and J. Fridrich, "Low complexity features for JPEG steganalysis using undecimated DCT," IEEE Trans. information forensics and security, vol. 10, pp. 219–228, 2015.

356 *Appendix*

JPEG43. V. Holub and J. Fridrich, "Phase-aware projection model for steganalysis of JPEG images," SPIE electronic imaging, media watermarking, security and forensics, vol. 9409, March 2019.

JPEG44. X. Song et al, "Steganalysis of adaptive JPEG steganography using 2D Gabor filters," 3rd ACM workshop on information hiding and multimedia security, pp. 15–23, 2015.

JPEG45. X. Liu et al, "Graph-based joint dequantization and contrast enhancement of poorly lit JPEG images," IEEE Trans. IP, EA.

Conclusion section is repeated here.

Conclusion

To restore images captured in poorly lit environments that suffer from both JPEG compression artifacts and low contrast, we propose a graph-based joint dequantization/contrast enhancement scheme. The key idea is to decompose the target image into a product of illumination and reflectance components via Retinex theory, where each component can be mathematically described as a signal smoothness prior on an appropriately defined graph. Specifically, we use signed graphs where positive edges connect similar pixels for averaging to combat noise, and negative edges connect dissimilar pixels to accentuate their differences for enhancement. Experimental results show cleaner and higher-contrast images compared to state-of-the-art methods. There are some limitations to our current work. First, the determination of positive and negative edge weights is tuned manually and thus not optimal in any sense. Second, the connectivity of the graph used (4-connected graph, where each pixel is connected to its four nearest neighbors) is not optimized. While our hand-tuned parameters and connectivity led to good visual performance in our experiments, we plan to study these issues more theoretically as future work.

JPEG46 Z. Wang al, "D3: Deep dual-domain based fast restoration of JPEG-compressed images," arXiv preprint arXiv:1601.04149, 2016.

Winzip and GZIP open source software

Commercial or open source lossless compression software, such as WinZIP and gzip, which can reduce the file sizes of JPEG-coded images.

Winzip 19.0 Pro(11294) 64 Bit. Available: http://www.winzip.com/win/en/downwz.html [35] GZIP.

http://www.gzip.org

BOSSbase image data set http://www.agents.cz/boss/BOSSFinal IMAGE DATA SET

JPEG-P1. Oliveira et al., [JPEG8] have developed a new low complexity approximation the discrete Tchebichef transform (DTT) and embedded it in JPEG and H.64/AVC. Extend this to 16-point DTT and develop a fast algorithm. Draw a flow graph similar to that shown in Figure 1.

JPEG-P2. Develop hardware aspects of the 16-point DTT as described in VI Hardware section. This implies that the 16- point DTT needs to be implemented on Xilinx FPGA board.

JPEG-P3. See JPEG-P1. Embed the new DTT in HEVC Main profile and compare its performance with anchor HEVC. (See Figure 4)

JPEG-P4. Huang et al., [JPEG6] have proposed a new histogram shifting based RDH (reversible data hiding) scheme in JPEG images that realize high embedding capacity and good visual quality while preserving the JPEG file size. Implement this scheme and confirm the results as described in the figures.

JPEG-P5. See JPEG-P4. In Section 4 Conclusions the authors state that the proposed novel block selection strategy can result in better visual quality and less JPEG file size. They also state that this technique may be applied to other RDH schemes to improve their performance. Several references related to RDH in images are listed at the end in [JPEG6]. Apply this strategy in the RDH schemes and evaluate their performances. Use different test images at various quality factors. (See also the references listed at the end in [JPEG7]).

JPEG-P6. See JPEG10. Implement JPEG extraction from the HEVC bit stream in the compressed domain. Confirm the results shown in Table I for the test sequences.

JPEG-P7. See [JPEG11]. In conclusion the authors state "In future work, we will work on speeding up the proposed algorithm to make it more practical." Explore this.

JPEG-P8. See [JPEG14] Conclusion section is repeated here.

To reduce the blocking artifacts and obtain high-quality images, this paper proposed a novel image-deblocking algorithm by utilizing SSR prior and QC prior simultaneously under the MAP framework. To make the proposed optimization problem tractable, a new SBI-based method with an adaptively adjusted regularization parameter is developed. This parameter-adaptive advantage makes the entire algorithm more attractive and practical. Note that, although the proposed SSRQC algorithm outperforms the current state-of-the-art algorithms in terms of both PSNR and visual perception, it still usually suffers from the introduction of distracting contours in smooth backgrounds for compressed images at low bit rates. The main reason for this may be that the quantization noise model adopted in this paper is a naive Gaussian assumption for interpreting overall quantization noises all over the entire image, which is not consistent with region characteristics and thus not accurate. Therefore, more accurate quantization noise models that are adaptive to region characteristics are expected to achieve better results. Our future work will focus on region-adaptive quantization noise modelling for image deblocking.

Go through the SSRQC-based image de-blocking algorithm and confirm all the simulation results described in the Tables and graphs. Explore the future work suggested in this paper.

JPEG-P9. See [JPEG15]. Implement this new framework of reversible data hiding in encrypted JPEG bit streams and confirm the conclusions.

JPEG-P10. See [JPEG20]. Explore the future work suggested in the conclusion section.

JPEG-P11. See [JPEG23]. Explore how you can implement end-to-end compression framework (See Figure 2) to other image codecs such as JPEG-LS, JPEG-XR, JPEG-XT and JPEG-XS. Compare the coded image quality metrics (PSNR/SSIM) of the latter with those listed in [JPEG23] using the same test images. Consider complexity also.

JPEG-P12. See [JPEG28]. Explore the future work suggested in the conclusion section.

JPEG-P13. See [JPEG28]. The authors state that the proposed design paradigm can be undertaken for audio and video.
Explore this for all the video coding standards such as H.264, H.265 (HEVC), AV1 by AOM, DIRAC etc. Please note each one of these is quite complex and requires group effort and also hardware facilities.

JPEG-P14. See [JPEG29] and the conclusion section. This is an excellent paper on joint image compression and encryption scheme applied to JPEG. Several extensions to this approach are suggested. Explore these.

JPEGP15. See [JPEG34] Explore the future studies suggested in the conclusion section.

JPEGP16. See [JPEG34]. The last sentence before the conclusion section is repeated here.
"We would also note that by integrating rich-modeling based approach [9], the detection accuracy may be further improved." Explore this.
JPEG-P17. See [JPEG35] and the conclusion section. Explore the future work suggested in this section.
JPEG-P18. See [JPEG36]. Implement the encryption scheme-applied to JPEG images described in this paper and verify the results.
JPEG-P19. Review the papers on JPEG steganography [JPEG21, JPEG37 thru JPEG44]. Evaluate the various JPEG steganography schemes and draw some conclusions.
JPEG-P20. See JPEG-P19. After reviewing the the various JPEG steganography schemes develop an efficient and robust JPEG steganography technique.
JPEG-P21. See JPEG45 and the conclusion section. Explore the future work suggested by the authors.

Commercial or open source lossless compression software, such as WinZIP and gzip, that can reduce the file sizes of JPEG coded images.

JPEG XT

JXT1. T. Richter et al., "The JPEG XT suite of standards: status and future plans," [9599 – 30], SPIE. Optics + Photonics, San Diego, California, USA, 9–13, Aug. 2015.
JXT2. M. Fairchild: "The HDR Photographic survey," available online at http://rit-mcsl.org/fairchild/HDR.html (retrieved July 2015).
JXT3. R. Mantiuk, "pfstools: High Dynamic Range Images and Video," available online at http://pfstools.sourceforge.net/(retrieved July 2015).
JXT4. T. Ricther, "JPEG XT Reference Codec 1.31 (ISO License)," available online at http://www.jpeg.org/jpegxt/software.html (retrieved July 2015).
JXT5. T. Richter, "Lossless Coding Extensions for JPEG," IEEE Data Compression Conference, pp. 143–152, Mar. 2015.

(See JXT1) JPEG XT is a standardization effort targeting the extension of the JPEG features by enabling support for high dynamic range imaging, lossless and near lossless coding and alpha channel coding, while also guaranteeing backward and forward compatibility with the JPEG legacy format. JPEG XT has nine parts described in detail in Figure 2. Further extensions relate to JPEG Privacy and Security and others such as JP Search and JPEG systems and are listed in the conclusion. JPEG XT is forward and backward compatible with legacy JPEG unlike JPEG-LS and JPEG2000.

JXT6 A. Artusi et al., "Overview and Evaluation of the JPEG XT HDR Image Compression Standard," J. of real time image processing, vol. 10, Dec. 2015.

Swaminathan s <swaminathan1986@gmail.com>
For ready reference, the abstract is reproduced below.

Abstract

Standards play an important role in providing a common set of specifications and allowing interoperability between devices and systems. Until recently, no standard for High Dynamic Range (HDR) image coding had been adopted by the market, and HDR imaging relies on proprietary and vendor specific formats which are unsuitable for storage or exchange of such images. To resolve this situation, the JPEG Committee is developing a new coding standard called JPEG XT that is backwards compatible to the popular JPEG compression, allowing it to be implemented using standard 8-bit JPEG coding hardware or software. In this paper, we present design principles and technical details of JPEG XT. It is based on a two-layer design, a base layer containing a Low Dynamic Range (LDR) image accessible to legacy implementations, and an extension layer providing the full dynamic range. The paper introduces three of currently defined profiles in JPEG XT, each constraining the common decoder architecture to a subset of allowable configurations. We assess the coding efficiency of each profile extensively through subjective assessments, using 24 naive subjects to evaluate 20 images, and objective evaluations, using 106 images with five different tone-mapping operators and at 100 different bit rates. The objective results (based on benchmarking with subjective scores) demonstrate that JPEG XT can encode HDR images at bit rates varying from 1:1 to 1:9 bit/pixel for estimated mean opinion score (MOS) values above 4:5 out of 5, which is considered as fully transparent in many applications. This corresponds to 23-times bit stream reduction compared to lossless OpenEXR PIZ compression.

In the conclusions the authors [JXT6] state "The benchmarking showed that, in terms of predicting quality loss due to coding artifacts, simple metrics, such as PSNR, SNR, and MRSE computed in linear space are unsuitable for measuring perceptual quality of images compressed with JPEG XT, however, the prediction of these metrics improve when applied to the pixels converted in the perceptually uniform space. Also, HDR-VDP-2 provides the best performance as compared to other tested metrics." This again confirms that that these objective metrics do not correlate with the subjective (perceptual) quality.

JXT7. M. Shah, "Future of JPEG XT: Scrambling to enhance privacy and security," M.S. Thesis, EE Dept., UTA, Arlington, TX, USA. http://www.uta.edu/faculty/krrao/dip click on courses and then click on EE5359 Scroll down and go to recent Theses/Project Title and click on Maitri Shah.

JXT8. R. K. Mantiuk, T. Richter and A. Artusi, "Fine-tuning JPEG-XT compression performance using large-scale objective quality testing," IEEE ICIP, Phoenix, Arizona, Sept. 2016.

Two publicly available (image datasets) used in [JXT6].
Fairchild's HDR photographic survey. http://rit-mcsl.org/fairchild/HDR.html
HDR-eye dataset of HDR images. http://mmspg.ep.ch/hdr-eye
The dataset contained scenes with architecture, landscapes, portraits, frames extracted from HDR video, as well as computer generated images.

JXT9. P. Korshunov et al., "EPFL's dataset of HDR images'" 2015 (online), Available: http://mmsp.epfl.ch/hdr-eye

JXT10. T. Richter, "JPEG on steroids: Common optimization techniques for JPEG image compression," IEEE ICIP, Phoenix, AZ, Sept. 2016.

JPEG SOFTWARE:

JXT11. J. Aas: "Mozilla JPEG encoder project," available online at https://github.com/mozilla/mozjpeg (See [JXT10]).

JXT12. T. Richter, "Error bounds for HDR image coding with JPEGXT," IEEE DCC, pp. 22–130, April 2017.

Mozilla JPEG Encoder Project

This project's goal is to reduce the size of JPEG files without reducing the quality or compatibility with the vast majority of the world's deployed decoders.

The idea is to reduce transfer times for JPEGs on the Web, thus reducing page load times.

'mozjpeg' is not intended to be a general JPEG library replacement. It makes tradeoffs that are intended to benefit Web use cases and focuses solely on improving encoding. It is best used as part of a Web encoding workflow. For a general JPEG library (e.g. your system libjpeg), especially if you care about decoding, we recommend libjpeg-turb libjpeg.

A complete implementation of 10918-1 (JPEG) coming from jpeg.org (the ISO group) with extensions for HDR currently discussed for standardization.

JXT13. Jpegxt demo software T. Richter: "libjpeg: A complete implementation of 10918-1 (JPEG)," available online at https://github.com/thorfdbg/libjpeg (See [JXT10]).

JXT14. T. Richter, "On the standardization of the JPEG XT image compression," IEEE PCS, pp. 37–40, Dec. 2013.

JXT15. H. Kobayashi and H. Kiya, "An extension of JPEG XT with JPEG2000," IEEE ICCE Taiwan, pp. 373–374, June 2017.

JXT16. T. Richter, "On the integer coding profile of JPEG XT," Proc. SPIE, vol. 9217, pp. 9217∼1 through, 9217∼19, 2013.

JXT17. A. Artusi et al., "JPEG XT: A compression standard for HDR and WCG images [standards in a nutshell]," IEEE Signal Process. Mag., vol. 33, no. 2, pp. 118–124, March 2016.

JXT18. T. Richter, A. Artusi and T. Ebrahimi, "JPEG XT: A new family of JPEG backward-compatible standards," IEEE Multimedia, vol. 2, no. 3, pp. 80–88, July 2016.

JPEG XT Projects:

The JPEG XT suite of standards described in the paper [JXT1] can lead to several projects. Review this paper in detail and implement these standards. Please access the reference software.

JXT-P1. JPEG Privacy & Security: With social media having a huge impact on every individual, it disrupts their privacy. Hence, protecting the privacy and security is becoming very important, not only because of the social media, but also because images/meta-data use cloud for storage purposes over private repositories. Encrypting the images and providing access to only the authorized person can help in maintaining the privacy of the image/meta-data.

JXT-P2. Recoding Image-editing Operations: While editing any image or text, only the original data and the final output after editing is usually saved. But instead, if every intermediate step is also recorded, it gives a possibility to revert the data to the original image or any of the intermediate steps. Saving every step while developing a special effect (sharpen, emboss, smooth) can help to develop new, faster and easier effects.

JXT-P3. In the conclusion section, the authors [JXT6] state: "The encoding or decoding hardware can be in fact designed based on a pair of existing JPEG coding chips, as shown in Figure 2, resulting in a minimal hardware change in the existing hardware infrastructure without influencing its real-time performances." Implement the encoding and decoding hardware.

JXT-P4. See [JXT7] As part of JPEG XT future extensions, JPEG privacy and security, Shah has developed an algorithm to encrypt and decrypt the images using C++ and OpenCV libraries. This algorithm performs well for protecting an individual's identity by blurring the image. By resolving the compatibility issues between the OpenCV libraries and JPEG XT code this algorithm can be incorporated in the JPEG XT code tom provide the security feature automatically. Explore this.

JXT-P5 See JXT-P4. Shah suggests that the amount of blurring can be changed according to the user preference by changing the kernel size. This gives the user full control over how the image should be seen by others. Explore this in detail and draw some conclusions.

JXT-P6 See JXT-P4. Several other tasks also are proposed as future work in Chapter of this thesis. Investigate and implement these tasks.

JXT-P7 See [JXT10], [JXT11] and [JXT12]. Richter has proposed some optimization techniques to improve the legacy JPEG in terms of rate distortion performance close to JPEG-XR (See Figure 5 in JXT10). Using the software described in [JXT11] and [JXT12] confirm the enhanced JPEG rate distortion performance shown in [JXT10].

JXT-P8 JPEG-XT standard defines a normative decoding procedure to reconstruct an HDR image from two JPEG regular code streams named the base layer (visible to legacy decoders) and an extension layer and how to merge them together to form one single image. See also JXT1.The standard does not, however, define the encoding procedure and leaves large freedoms to the encoder for defining the necessary decoder configuration. Mantiuk, Richter and Artusi [JXT8], explored the whole space of possible configurations to achieve the best possible R-D performance. Review this paper in detail and verify the performance results described in the figures [JXT8].

JXT-P9   See [JXT1, JXT4 and JXT5]. The JPEGXT encoder (review the encoder/decoder block diagrams) is both forward and backward compatible with legacy JPEG encoder. See references on JPEG. Verify the forward compatible characteristic of JPEGXT. This implies that the JPEGXT decoder should be able to decode and display the images based on the JPEG coded bit streams.

JXT-P10 See JXT-P10. Verify the backward compatible characteristic of JPEGXT. This implies that legacy JPEG decoder should be able to extract part of the JPEGXT coded bit stream, decode, and display the JPEG images. (See the references on JPEG).

JXT-P11 See [JPEG22]. Watanabe, Kobayashi and H. Kiya have proposed a two-layer coding method using the histogram packing technique with the backward compatibility to the legacy JPEG. The authors claim that this method has a better lossless compression performance than that of the JPEG XT. Using the test images verify this claim?

JXT-P12 Review the papers related to JPEG XT. Implement JPEG XT and evaluate its performance.

JXT-P13 See [JPEG22]. The authors claim improved performance of lossless compression using histogram packing technique. Go through the various references listed in [JPEG22] and verify this claim.

JPEG PLENO

JP1. T. Ebrahimi, "Towards a new standard for plenoptic image compression," IEEE DCC, 29 March–1 April 2016. Abstract of this valuable paper is reproduced below:

Abstract

JPEG format is today a synonymous of modern digital imaging, and one of the most popular and widely used standards in recent history. Images created in JPEG format now exceeds one billion per day in their number, and most of us can count a couple, if not more JPEG codecs in devices we regularly use in our daily lives; in our mobile phones, in our computers, in our tablets, and of course in our cameras. JPEG ecosystem is strong and continues an exponential growth for the foreseeable future. A significant number of small and large successful companies created in the last two decades have been relying on JPEG format, and this trend will likely continue.

A question to ask ourselves is: will we continue to have the same relationship to flat snapshots in time (the so-called Kodak moments) we call pictures, or could there be a different and enhanced experience created when capturing and using images and video, that could go beyond the experience images have been providing us for the last 120 years? Several researchers, artists, professionals, and entrepreneurs have been asking this same question and attempting to find answers, with more or less success. Stereoscopic and multi-view photography, panoramic and 360-degree imaging, image fusion, point cloud, high dynamic range imaging, integral imaging, light field imaging, and holographic imaging are among the examples of solutions that have been proposed as future of imaging.

Recent progress in advanced visual sensing has made it feasible to capture visual content in richer modalities when compared to conventional image and video. Examples include Kinect by Microsoft, mobile sensors in Project Tango by Google and Intel, light-field image capture by Lytro, light-field video by Raytrix, and point cloud acquisition by LIDAR (Light Detection and Ranging). Likewise, image and video rendering solutions are increasingly relying on richer modalities offered by such new sensors. Examples include Head Mounted Displays by Oculus and Sony, 3D projector by Ostendo and 3D light field display solutions by Holografika. This promises a major change in the way visual information is captured, processed, stored, delivered, and displayed.

JPEG PLENO evolves around an approach called plenoptic representation, relying on a solid mathematical concept known as plenoptic function. This promises radically new ways of representing visual information when compared to the traditional image and video, offering richer and more holistic information. The plenoptic function describes the structure of the light information impinging on observers' eyes, directly measuring various underlying visual properties like light ray direction, multi-channel colours, etc.

The road-map for the JPEG PLENO follows a path that started in 2015 and will continue beyond 2020, with the objective of making the same type of impact that the original JPEG format has had on today's digital imaging starting from 20 years ago. Several milestones are in work to approach the ultimate image representation in well-thought, precise, and useful steps. Each step could potentially offer an enhanced experience when compared to the previous, immediately ready to be used in applications, with potentially backward compatibility. Backward compatibility could be either at the coding or at the file format level, allowing an old JPEG decoder of 20 years ago to still be able to decode an image, even if that image won't take full advantage of the intended experience, which will be only offered with a JPEG PLENO decoder.

This talk starts by providing various illustrations the example applications that can be enabled when extending conventional image and video models toward plenoptic representation. Doing so, we will discuss use cases and application requirements, as well as example of potential solutions that are or could be considered to fulfill them. We will then discuss the current status of development of JPEG PLENO standard and discuss the various milestones ahead. The talk will conclude with a list of technical challenges and other considerations that need to be overcome for a successful completion of JPEG PLENO.

JP2. M.D. Angelo et al., "Correlation plenoptic imaging," Physical Review Letters, vol. 116, no. 22, pp. 223602, June 2016.

Abstract is reproduced below.

Plenoptic imaging is a promising optical modality that simultaneously captures the location and the propagation direction of light in order to enable tridimensional imaging in a single shot. However, in the classical imaging systems, the maximum spatial and angular resolutions are fundamentally linked, thereby, the maximum achievable depth of field is inversely proportional to the spatial resolution. We propose to take advantage of the second-order correlation properties of light to overcome this fundamental limitation. In this paper, we demonstrate that the momentum/position correlation of chaotic light leads to the enhanced refocusing power of correlation plenoptic imaging with respect to the standard plenoptic imaging.

JP3. T. Ebrahimi et al., "JPEG Pleno: Toward an efficient representation of visual reality," IEEE Multimedia, vol. 23, no. 4, pp. 14–20, Oct./Dec. 2016.

JP4. X. Jin, H. Han and Q. Dai, "Plenoptic image coding using macropixel-based intra prediction," IEEE Trans. IP, vol. 25, pp. 3954–3968, Aug. 2018.

This is an excellent paper describing the characteristics of plenoptic imaging followed by a highly efficient compression technique. The vast list of references is a good resource and covers the state-of-the-art in this emerging field. The authors illustrate that their approach is superior to any other state-of-the-art methods on plenoptic image coding. The section Conclusion is repeated here.

This paper proposes a novel plenoptic image compression scheme, which can efficiently exploit the inherent correlation among macropixels. After applying the previously proposed invertible image reshaping method to the lenslet image, the reshaped image is compressed by adding the three proposed macropixel-based intra prediction modes as additional candidate modes. The proposed modes predict the current PU by the co-located blocks or their combinations in the spatially adjacent macropixels, which can bring significant compression performance improvement. A maximum of 80.5% bitrate reduction with an average of 47.0% bitrate reduction can be achieved relative to HEVC under the same reconstructed light field quality. Also, significant compression performance is demonstrated by outperforming state-of-the-art methods IBC/LLE/PVTA by an average of 27.7%/22.7/45.0% bitrate reduction. Moreover, a better tradeoff between the compression efficiency and computational complexity can be achieved by the combination of MWP and CSP if the computational resources are limited at the encoder. The performance of the proposed scheme can be further improved by designing a specific entropy-coding engine and a fast mode selection method, which are under investigating as future works.

JP5. JPEG
Pleno Dataset, EPFL. Light-Field Dataset, https://jpeg.org/plenodb/
/lf/epfl

JP6. JPEG Pleno Call for Proposals on Light Field Coding, document N74014, ISO/IEC JTC 1/SC29/WG1 JPEG, Geneva, Switzerland, Jan. 2017.

JP7. REFERENCES AND DOCUMENTATION ON JPEG PLENO
https://jpeg.org/jpegpleno/documentation.html JP8 Special session in IEEE DCC 2019, Sunbird, UT, March 2019.

"Plenoptic Image Compression"
Thomas Richter, Fraunhofer IIS
Peter Schelkens, Vrije Universiteit Brussel
Session Coordinators

JP8. I. Viola et al., "Grand challenge on light field image compression," IEEE ICME, Seattle, WA, 2016.

JP-P1. Please review the abstract in [JP1]. Focus on the research towards developing the JPEG PLENO standard with backward compatibility with legacy JPEG. Review the technical challenges and consider the application requirements. This is a challenging research.

JP-P2. Please review [JP2] in detail. In the conclusions and outlook, the authors state that the merging of plenoptic imaging and correlation quantum imaging has thus the potential to open a totally new line of research. Explore this.

JP-P3. See [JP4] Investigate the future work.

Considering that micro-lens array can be of any shape and any arrangement, the possible solutions in generalizing the proposed algorithm are discussed in this section, whose further improvement is also under investigation as our future work. Explore this future work.

JP-P4. See [JP4] On page 3965 (left column), the authors state "Considering that micro-lens array can be of any shape and any arrangement, possible solutions in generalizing the proposed algorithm are discussed in this section, whose further improvement is also under investigation as our future work. Explore this future work."

JP-P5. See [JP4] and go through this paper in detail. Using the test images (Figures 11 and 13), the authors have shown the superior performance of their technique. See the figures and tables. Implement their approach and validate their results. For computational complexity (Figure 12), execution time ratios will be based on your processing hardware.

JP-P6. See [JP6] and [JP8]. Take the grand challenge and submit proposals on light field coding.

JPEG XS

The overview paper on HEVC SCC extension [OP11] states that the JPEG has initiated the standardization of JPEG XS [JXS1, JXS3] a low latency lightweight image coding system with potential applications in video link compression, frame buffer compression, and real-time video storage. This paper compares the requirements and features of HEVC SCC extension, DSC and JPEG XS in Table 5 with the foot note that the requirements and features of JPEG XS are subject to change. The JPEG XS international standard is projected to be finalized in 2018. Willeme et al., [JXS2] conducted preliminary comparison of the several existing lightweight compression schemes.

UPDATES. See [JXS6]. At the 73rd JPEG meeting held in Chengdu, China (2017), JPEG decided as one of the top-performing contribution the proposal described in [JXS6] and the decision was taken to merge the two most promising proposals ([JXS6] and one from Intopix).

The information below on JPEG XS is from https://jpeg.org/jpegxs (April 2018).

The JPEG XS (ISO/IEC 21122) standard defines a compression algorithm with very low latency and very low complexity. By offering various degrees of parallelism, JPEG XS can be efficiently implemented on various platforms such as FPGAs, ASICs, CPUs, and GPUs and excels with high multi-generation robustness. It is particularly optimized for visual lossless compression as defined in ISO/IEC 29170-2 for both natural and synthetic images. The typical compression ratios are between 1:2 and 1:6 for both 4:4:4 and 4:2:2 images and image sequences with up to 16-bit component precision. Typical parameterizations address a maximum algorithmic latency between 1–32 video lines for a combined encoder-decoder suite.

Those properties allow JPEG XS to be applied in all applications that before operated with uncompressed image data. This includes among others professional video links (3G/6G/12G-SDI), IP transport (SMPTE 2022-5/6 & SMPTE 2110), real-time video storage, memory buffers, omni-directional video capture system, head-mounted displays for Virtual or Augmented Reality, and sensor compression for the Automotive Industry. By these means, systems can offer an increased resolution and frame rate, while offering a visually lossless quality.

JPEG XS is a multi-part specification being currently under development and including the following parts:

JXS1. JPEG Committee Initiates Standardization of JPEG XS a Low-Latency Lightweight Image Coding System, July 2015 [online] Available: http://jpeg.org/items/20150709_press.html.

JXS2. A. Willeme, et al., "Quality and error robustness assessment of low-latency lightweight intra-frame codecs" Proc. IEEE Data Compression Conf. (DCC), Mar. 2016.

JXS3. PEG XS—Call for Proposals for a Low-Latency Lightweight Image Coding System, Mar. 2016 [online] Available: https://jpeg.org/items/20160311_cfp_xs.html.

JXS4. T. Richter et al., "Entropy coding and entropy coding improvements of JPEG XS," IEEE DCC, Sun Bird, Utah, March 2018.

JXS5. A. Willeme et al., "JPEG XS-based frame buffer compression inside HEVC for power-aware video compression," SPIE Applications of

Appendix

Part 1: Core coding system	Part 2: Profiles and buffer models	Part 3: Transport and container
JPEG XS Part 1 (ISO/IEC 21122-1) normatively defines how a compressed JPEG XS codestream can be transformed into a decoded image in a bit exact manner. Moreover, it informatively explains the key algorithms enabling an encoder to generate a JPEG XS codestream.	JPEG XS Part 2 (ISO/IEC 21122-2) ensures interoperability between different implementations by specifying typical codestream parameterizations and properties. This allows deriving the hardware and software requirements for different purposes such as high compression ratios, low memory or low logic implementations. Moreover, implementation guidelines inform about how to achieve low latency implementations.	JPEG XS Part 3 (ISO/IEC 21122-3) defines how to embed a JPEG XS codestream into a more descriptive file format. Moreover, it contains all definitions that are necessary to transport a JPEG XS codestream by means of a transmission channel using existing transmission protocols defined by different standardization bodies.
Part 4, Conformance testing	Part 5, Reference software	
JPEG XS Part 4 defines conformance testing of JPEG XS.	JPEG XS Part 5 provides the JPEG XS reference software.	

- the digital image processing LX, vol. 10396, San Diego, CA, Oct. 2017.
- JXS6. T. Richter et al., "High-speed low-complexity video coding with EDiCTius: a DCT coding proposal for JPEG XS," SPIE Applications of the digital image processing LX, vol. 10396, San Diego, CA, Oct. 2017. Also listed as [DCT54].
- JXS7. T. Richter et al., "High speed, low-complexity image coding for IP-transport with JPEG XS," SPIE, vol. 9971, pp. 99711N10, 2016.
- JXS8. J. Keinert et al., "Introduction to JPEG XS- The new low complexity codec standard for professional video production," Proposal from Fraunhofer IIS, Germany and intoPIX SA, Belgium, One of the two proposals selected by JPEG.
- JXS9. JPEG XS DOCUMENTATION https://jpeg.org/jpegxs/documentation.html
- JXS10. V. Bruns et al., "Decoding JPEG XS on a GPU," IEEE PCS2018, San Francisco, CA, June 2018.

Abstract is reproduced here.

JPEG XS is an upcoming lightweight image compression standard that is especially developed to meet the requirements of compressed video-over-IP use cases. It is designed with not only CPU, FPGA or ASIC platforms in mind, but explicitly also targets GPUs. Though not yet finished, the codec is now sufficiently mature to present a first NVIDIA CUDA-based GPU decoder architecture and preliminary performance results. On a 2014 mid-range GPU with 640 cores a 12 bit UHD 4:2:2 (4:4:4) can be decoded with 54 (42) fps. The algorithm scales very well: on a 2017 high-end GPU with 2560 cores the throughput increases to 190 (150) fps. In contrast, an optimized GPU-accelerated JPEG 2000 decoder takes $2\times$ as long for high compression ratios that yield a PSNR of 40 dB and $3\times$ as long for lower compression ratios with a PSNR of over 50 dB.

JXS11. Info on JPEGXS SOFTWARE https://www.dropbox.com/s/dnz4hx2e 4b54vg4/jpeg-docs.zip?dl=0

JXS-P1. Review the web sites on JPEG XS [JXS1, JXS2] and [JXS2] and submit proposals to JPEG that can meet the requirements of JPEG XS. Note that this project requires group effort. This is outdated now as the final two proposals have been chosen by JPEG leading to the JPEG XS standard. Review the JPEG XS proposal described in [JXS6] and implement the same.

Two new evaluation metrics Objective single cycle quality evaluation and subjective multi cycle quality evaluation are described in [JXS8].

JXS-P2. See [JSX6] and [JXS7] and the updates. Access the other most outstanding proposal from Intopix for JPEG XS and merge the two resulting in the best proposal that meets the criteria stipulated by JPEG. Implement the merged proposal.

JXS-P3. Access [OP14]. Also listed as [E452] An overview of digital watermarking. Review this overview paper thoroughly and see how you can embed an imperceptible water mark in the JPEG XS standard.

JXS-P4. See [JXS8] In the conclusion and outlook, the authors state "Moreover HDR support will be investigated." Explore this.

JXS-P5. Now that the JPEG XS standard is (almost) finalized, it is appropriate to compare and evaluate the image coding standards in the JPEG series i.e., JPEG, JPEG LS, JPEG 2000, JPEG XT, JPEG XR, JPEG XS, JPEG AIC and JPEG PLENO. Using the standard image test data, thoroughly evaluate their capabilities, limitations and advantages.

JXS-P6. Video coding standards such as H. 264/AVC, H.265/HEVC, MPEG2, AVS China etc. also address image coding. Compare and evaluate the performance of JPEG XS with these standards. For example see [E78]. This describes intra coding of HEVC standard.

JPEG AIC

The following is reproduced from the JPEG web site:

Currently, the work of JPEG AIC resulted in two standards (ISO/IEC 29170-1 and ISO/IEC 29170-2). ISO/IEC 29170-1 provides a framework that allows evaluating the suitablilty of image compression algorithms in various application domains. To this end, this standard aims to provide a set of evaluation tools that allow multiple features of such codecs to be tested and compared. The second standard, ISO/IEC 29170-2, defines a set of evaluation procedures, similar to how the ITU-R BT.500-11 standard did for the subjective quality evaluation with television standards.

This activity is a work in progress and it is continuously updated to accommodate new coding architectures and evaluation methodologies.

JA1. Z. Zhang, R. Veerla and K.R. Rao, "A modified advanced image coding," CANS, Bucharest, Rumania, 2008.

JBIG

JBIG1. F. Ono et al., "JBIG2- the ultimate bi-level image coding standard," IEEE ICIP, pp. 140–143, Vancouver, B.C., Canada, Sept. 2000.

LAR-LLC

LL1. Y. Liu, O. Deforges and K. Samrouth, "LAR-LLC: A low complexity multiresolution lossless image codec," IEEE Trans. CSVT, vol. 26, pp. 1490–1501, Aug 2016.

Abstract

It compares the performance of LAR-LLC lossless codec with JPEGXR, JPEG2000, JPEGLS, and LJPEG (lossless JPEG) in terms of compression ratio, encoding, and decoding speeds (see Tables I–IV). LAR-LLC has much less encoding and decoding speeds (less complexity) compared to the JPEG series. It also supports spatial scalability similar to JPEG2000.

LL-P1. See [LL1]. In Table II, the bitrates (bpp) for lossless color image (24 bpp) codecs are shown, for various test images. Implement the JPEG lossless codecs and compare with LAR-LLC.

LL-P2. HEVC has lossless coding in intra (AI) profile. Implement this and compare with Table II.

LL-P3. Develop Tables similar to Tables V and VI for HEVC AI profile (lossless coding).

LL-P4. In the paragraph before VII conclusion, the authors state that the LAR-LLC currently runs in a mono thread configuration, but has potential parallel processing ability in multithreading and suggest how this can be implemented. Explore this and evaluate how the multithreading configuration can further reduce the LAR-LLC encoding/decoding speeds for lossless coding.

LL-P5. Extend the lossless image coding scheme to VP10. See Tables I to IV in [LL1].

LL-P6. See LL-P4. Extend the lossless image coding scheme to AVS2 – intra coding only.

PNG

PNG1. T. Boutell, "PNG (Portable Network Graphics) Specification version 1.0," Network Working group RFC 2083, available online at http.ietf.org/html/rfc2083 (retrieved November 2014).

WebP

Web.P1. *WebP—Google Developers*. [Online]. Available: http://code.google.com/speed/webp/

Web.P2. The WebP container format is specified at https://developers.google.com/ speed/webp/docs/riff_container, while the VP8L format is specified at https:// developers.google.com/speed/webp/docs/webp_lossless_bitstream_specification.

WebM

WebM1. The WebM Project. [Online]. Available: http://www.webmproject.org/

WebM2. WebM^{TM}: an open web media project, VP8 Encode Parameter Guide, 2-Pass Best Quality VBR Encoding, Online:

http://www.webmproject.org/docs/encoder-parameters/#2-pass-best-qualityvbr-encoding

WebM3. Paul Wilkins, Google® Groups "WebM Discussion," Online: https://groups.google.com/a/webmproject.org/forum/?fromgroups #topic/webmdiscuss/UzoX7owhwB0

WebM4. Ronald Bultje, Google® Groups "WebM Discussion," Online: https://groups.google.com/a/webmproject.org/forum/?fromgroups #topic/webmdiscuss/xopTll6KqGI

WebM5. John Koleszar, Google® Groups "Codec Developers," Online: https://groups.google.com/a/webmproject.org/forum/#!msg/codecdevel/yMLXzaohONU/m69TbYnEamQJ

VP8 HARDWARE

https://www.webmproject.org/hardware/vp8/#webm-video-decoder-hardware-ip

DIRAC (BBC)

Dirac website: www.diracvideo.org

[DR3] Dirac software: http://sourceforge.net/projects/dirac/ https://sourceforge.net/projects/dirac/files/dirac-codec/Dirac-1.0.2/

[DR4] A. Urs, "Multiplexing/demultiplexing Dirac video and AAC audio while maintaining lip sync," M.S. Thesis, EE Dept., University of Texas at Arlington, Arlington, Texas, May 2011.

[DR5] A. Ravi and K.R. Rao, "Performance analysis and comparison of the Dirac video codec with H.264/MPEG- 4, Part 10," Intrnl. J. of wavelets, multi resolution and information processing (IJWMIP), vol. 4, No. 4, pp. 635–654, 2011.

[DR6] See DR5. Implement performance analysis and comparison of the Dirac video codec with HEVC.

[DR7] BBC Research on Dirac: http://www.bbc.co.uk/rd/projects/dirac/index.shtml

[DR8] A. Mandalapu, "STUDY AND PERFORMANCE ANALYSIS OF HEVC, H.264/AVC AND DIRAC", project report, Spring 2016. http://www.uta.edu/faculty/krrao/dip click on courses and then click on EE5359 Scroll down and go to projects Spring 2016 and click on A. Mandalapu final report.

[DR9] S.S.K.K. Avasarala, "Objective Video quality assessment of Dirac and H.265" project report, Spring 2016. http://www.uta.edu/faculty/

krrao/dip click on courses and then click on EE5359 Scroll down and go to projects Spring 2016 and click on S.S.K.K. Avasarala final report.

[DR10] http://diracvideo.org/2012/01/schroedinger-1-0-11/ access to DIRAC reference software.

[DR11] https://en.wikipedia.org/wiki/Dirac (video compression format) – website on DIRAC.

[DR12] https://www.youtube.com/watch?v=HVFFq44UvLA a DETAILED TUTORIAL ABOUT dirac VIDEO CODEC by BBC.

[DR1] T. Borer and T. Davies, "Dirac – video compression using open technology," *EBU Technical Review*, pp. 1–9, July 2005.

DR-P1. See [DR8] Performance comparison of HEVC, DIRAC and H.264/AVC using CIF and QCIF test sequences shows, in general, superior performance of HEVC.

Extend this comparison to SD, HDTV, UHDTV, 4K and 8K test sequences. See also [DR9].

DR-P2. Fracastoro, Fossen and Magli [E392] developed steerable DCT (SDCT) and its integer approximation and applied them to image coding. They also compared this with wavelet coding (Table III). Confirm the results shown in this Table and extend this to higher resolution images.

DR-P3. See [E393]. Apply the FCDR to DIRAC codec both as a post processing operation and as an in loop (embedded) operation. Implement the projects for DIRAC similar to those described in P.5.268 and P.5.269.

DAALA

DAALA1. T.J. Daede et al., "A perceptually-driven next generation video codec," IEEE DCC, Snow Bird, Utah, March–April 2016. Abstract of this paper is repeated here:

The Daala project is a royalty-free video codec that attempts to compete with the best patent-encumbered codecs. Part of our strategy is to replace core tools of traditional video codecs with alternative approaches, many of them designed to take perceptual aspects into account, rather than optimizing for simple metrics like PSNR. This paper documents some of our experiences with these tools, which ones worked and which did not, and what we have learned from them. The result is a codec which compares favorably with HEVC on still images, and is on a path to do so for video as well.

DAALA2. Daala test media,"https://wiki.xiph.org/Daala Quickstart#Test Media.
https://xiph.org/daala
https://git.xiph.org/?p=daala.git
Daala Codec demo – http://people.xiph.org/~xiphmont/demo/daala/demo1.shtml

Daala video compression

Daala is the code-name for a new video compression technology. The effort is a collaboration between Mozilla Foundation, Xiph.Org Foundation and other contributors.

The goal of the project is to provide a free to implement, use and distribute digital media format and reference implementation with technical performance superior to h.265.

Technology demos

- Next generation video: Introducing Daala
- Introducing Daala part 2: Frequency Domain Intra Prediction
- Introducing Daala part 3: Time/Frequency Resolution Switching
- Introducing Daala part 4: Chroma from Luma
- Daala: Painting Images For Fun (and Profit?)
- Daala: Perceptual Vector Quantization (PVQ)

DAALA3. R.R. Etikala, "Performance comparison of DAALA and HEVC," project report, Spring 2016. http://www.uta.edu/faculty/krrao/dip click on courses and then click on EE5359 Scroll down and go to projects Spring 2016 and click on R.R. Etikala final report. This report implemented both HEVC and DALLA and compares their performances in terms of R-D plots for CIF and QCIF test sequences.

DAALA4. J.-M. Valin et al., "DAALA: A perceptually-driven still picture codec," IEEE ICIP, Session MA.L3: Image compression grand challenge, Phoenix, AZ, Sept. 2016.

DAALA5. Access https://en.wikipedia.org/wiki/AV1 Many references on DAALA transforms in AV1 by Nathan Egge.

DAALA-P1. Implement and improve Daala video codec using video test sequences and compare with HEVC/H.264/VP9 and other standard based codecs using implementation complexity, PSNR, SSIM, BD-bit rate and BD-PSNR as the metrics. See [DAALA1 and DAALA3].

DAALA-P2. Performance comparison of SDCT with DCT and DDCT is investigated in [E392]. In DAALA INTDCT is used. Replace INTDCT with INTSDCT and implement the performance comparison based on various test sequences including 4K and 8K and different block sizes. See the Tables and Figures in [E392].

DAALA-P3. See [E393]. Apply the FCDR to DAALA video codec both as a post processing operation and as an in loop (embedded) operation. Implement the projects for DAALA similar to those described in P.5.268 and P.5.269.

MPEG-DASH

MD1. I. Irondi, Q. wang and C. Grecos, "Subjective evaluation of H.265/HEVC based dynamic adaptive video streaming over HTTP (HEVC-DASH), SPIE-IS&T Electronic Imaging, vol. 9400, p. 94000B (session: real-time image and video processing), 2015.

MD.P1. See [MD1]. Irondi, Wang, and Grecos have presented a subjective evaluation of a HEVC-DASH system that has been implemented on a hardware test bed. In the abstract, the authors state "DASH standard is becoming increasingly popular for real-time adaptive HTTP streaming of internet video in response to unstable network conditions. Integration of DASH streaming technologies with the new H.265/HEVC video coding standard is a promising area of research." In the conclusions, they state "The results of this study can be combined with objective metrics to shed more light on the (QOE)–quality of experience- of DASH systems and correlation between the QOE results and the objective performance metrics will help designers in optimizing the performance of DASH systems." Papers listed as references in [MD1] relate to both subjective and objective performances (each separately) of DASH systems. Combine (integrate) both objective and subjective criteria with HEVC-DASH and evaluate the QOE. Based on this evaluation optimize the performance of HEVC-DASH system.

Bjontegaard Metric

http://www.mathworks.com/matlabcentral/fileexchange/27798-bjontegaardmetric/content/bjontegaard.m

BJONTEGAARD metric calculation: Bjontegaard's metric allows computing the average gain in PSNR or the average per cent saving in bitrate between two rate-distortions curves. See [E81, 82, E96, E198, E431]. This metric reflects the subjective visual quality better than MSE and PSNR.

Figure below explains clearly the BD-bit rate and BD-PSNR [E202].

AVS China:

[AVS1] AVS Video Expert Group, "Information Technology – Advanced coding of audio and video – Part 2: Video (AVS1-P2 JQP FCD 1.0)," Audio Video Coding Standard Group of China (AVS), Doc. AVS-N1538, Sep. 2008.

[AVS2] AVS Video Expert Group, "Information technology – Advanced coding of audio and video – Part 3: Audio," Audio Video Coding Standard Group of China (AVS), Doc. AVS-N1551, Sep. 2008.

[AVS3] L. Yu et al., "Overview of AVS-Video: Tools, performance and complexity," *SPIE VCIP*, vol. 5960, pp. 596021-1~596021-12, Beijing, China, July 2005.

[AVS4] L. Fan, S. Ma and F. Wu, "Overview of AVS video standard," *IEEE Int'l Conf. on Multimedia and Expo, ICME '04*, vol. 1, pp. 423–426, Taipei, Taiwan, June 2004.

[AVS5] W. Gao et al., "AVS – The Chinese next-generation video coding standard," National Association of Broadcasters, Las Vegas, 2004.

[AVS6] AVS-China official website: http://www.avs.org.cn

[AVS7] AVS-china software download: ftp://159.226.42.57/public/avs_doc/avs_software

[AVS8] IEEE Standards Activities Board, "AVS China - Draft Standard for Advanced Audio and Video Coding," adopted by IEEE as standard IEEE 1857. Website: stds.ipr@ieee.org

[AVS8] S. Ma, S. Wang and W. Gao, "Overview of IEEE 1857 video coding standard," IEEE ICIP, pp. 1500–1504, Sept. 2013.

[AVS9] D. Sahana and K.R. Rao, "A study on AVS-M standard," Calin ENACHESCU, Florin Gheorghe FILIP, Barna Iantovics (Eds.), Advanced Computational Technologies published by the Romanian Academy Publishing House, pp. 311–322, Bucharest, Rumania, 2012.

[AVS10] S. Sridhar, "Multiplexing/De-multiplexing AVS video and AAC audio while maintaining lip sync," M.S. Thesis, EE Dept., University of Texas at Arlington, Arlington, Texas, Dec. 2010. http://www.uta.edu/faculty/krrao/dip click on courses and then click on EE5359 Scroll down and go to Thesis/Project Title and click on S. Sridhar.

[AVS 11] W. Gao and S. Ma, "Advanced video coding systems," Springer, 2015. AVS2 Audio and video standard China adopted as IEEE 1857.4 https://en.wikipedia.org/wiki/Audio_Video_Standard has extensive coverage on AVS China.

[AVS 12] K. Fan et al., "iAVS2: A fast intra encoding platform for IEEE 1857.4," IEEE Trans. CSVT, vol. 27, pp. 2726–2738, Dec. 2017. Several papers related to fast intra coding in HEVC are listed at the end. See Table 4 for the list of test sequences used in AVS2 Standard.

[AVS 13] Z. He, L. Yu, X. Zheng, S. Ma and Y. He, "Framework of AVS2 Video Coding," IEEE Int. Conf. Image Process., pp. 1515–1519. Sept. 2013.

[AVS 14] S. Ma, T. Huang and C. Reader, "AVS2 making video coding smarter [standards in a nutshell]," IEEE Signal Process. Mag., vol. 32, no. 2, pp. 172–183, 2015.

[AVS 15] X. Zhang et al., "Adaptive loop filter for avs2," in Proc. 48th AVS Meeting, 2014.

[AVS16] Screen and mixed content coding working draft 1, Document AVS N2283, Apr. 2016.

[AVS17] Common test conditions for screen and mixed content coding, Document AVS N2282, Apr. 2016.

380   *Appendix*

AVS-P1. Review the references listed above on AVS China. In particular, review [AVS8]. Implement various profiles in AVS China and compare its performance with HEVC using various video test sequences. Comparison needs to be based on various metrics such as PSNR, SSIM, BD-bitrate, and BD-PSNR besides computational complexity. The profiles involve several projects.

AVS-P2. See [AVS 12] through [AVS 15] AVS2 is the second generation of the audio video coding standard to be issued as IEEE 1857.4 standard. It doubles the coding efficiency of AVS1 and H.264/AVC. Implement this standard and compare with HEVC based on standard performance metrics using various test sequences at different spatial/temporal resolutions.

AVS-P3. See AVS-P2. In the conclusions, the authors state [AVS 12] state "Owing to their similar frameworks, the proposed systematic solution and the fast algorithms can also be applied in HEVC intra encoder design." Explore this.

AVS-P4. Fan et al., [AVS12] proposed the fast intra encoding platform for AVS2 using numerous speedup methods. Using the test sequences shown in Table IV confirm the results shown in Tables V through X and Figures 13 through 16.

AVS-P5. Compare the performance of all intra image coding techniques including JPEG series, HEVC, MPEG-4/AVC, DIRAC, DAALA, THOR, AVS1, AVS2, VP8-10 and AV1 (developed by Alliance for Open Media – AOM). As always use various test sequences at different spatial resolutions.

AVS-P6. Performance comparison of SDCT with DCT and DDCT is investigated in [E392]. In AVS1 and AVS2, INTDCT is used. Replace INTDCT with INTSDCT and implement the performance comparison based on various test sequences including 4K and 8K and different block sizes. See the Tables and Figures in [E392].

AVS-P7. See [E393]. Apply the FCDR to AVS China codec (AVS2) both as a post processing operation and as an in loop (embedded) operation. Implement the projects for AVS2 similar to those described in P.5.268 and P.5.269.

AVS-P8. The AVS Workgroup of China has released the first working draft [AVS16] of the AVS SCC extension along with the reference software and common test conditions [AVS17] using the test sequences [TS27] developed for SCC. Implement the AVS China SCC extension and compare its performance with the HEVC SCC extension using the standard metrics. See [OP11]. There may be a restricted access to both [AVS16] and [AVS17].

AVS-P9. See [H78]. In the conclusions the authors state "we would like to extend this sketch attack framework to handle different video standards such as High Efficiency Video Coding (HEVC), Audio Video Standard (AVS) and Google VP9 as our future work." Extend this research to AVS China.

Thor Video Codec

Website: http://tinyurl.com/ismf77

TVC1 G. Bjontegaard et al., "The Thor video codec," IEEE DCC, pp. 476–485, Sun Bird, Utah, March–April 2016.

Thor video codec is being developed by Cisco Systems as a royalty free codec. Details regarding block structure, intra/inter (prediction, fractional sample interpolation, motion vector coding, transform, quantization, in-loop filtering, entropy coding, transform coefficient scanning and other functionalities are described in [TVC1]). Encoder follows the familiar motion compensated, transform predictive coding as in AVC/H.264 and HEVC/H.265. For performance comparison of TVC with VP9 and x265 see Table 3 (low delay and high delay) and Figures 5 and 6. For this comparison only HD sequences (classes B and E) and 1080p video conferencing sequences were used.

[TVC-P1] Extend this comparison using all classes of test sequences and develop graphs based on BD-bit rate and BD-PSNR versus encoded bit rate.

[TVC-P2] Consider implementation complexity as another metric and develop tables showing the complexity comparison of TVC with AVC/H.264 and HEVC/H.265 for all test sequences at various bit rates.

[TVC-P3] Carry out extensive research and explore how the various functionalities described in [TVC1] can be modified such that the TVC can be highly competitive with AVC/H.264 and HEVC/H.265. Keep in view that the TVC is royalty free.

[TVC-P4] Performance comparison of SDCT with DCT and DDCT is investigated in [E392]. In TVC, INTDCT is used. Replace INTDCT with INTSDCT and implement the performance comparison based on various test sequences including 4K and 8K and different block sizes. See tables and figures in [E392].

[TVC-P5] See [E393]. Apply the FCDR to Thor video codec both as a post processing operation and as an in loop (embedded) operation. Implement the projects for TVC similar to those described in P.5.268 and P.5.269.

References on Screen Content Coding

[SCC1] D. Flynn, J. Sole and T. Suzuki, "High efficiency video coding (HEVC) range extensions text specification," Draft 4, JCT-VC. Retrieved 2013-08-07.

[SCC2] M. Budagavi and D.-Y. Kwon, "Intra motion compensation and entropy coding improvements for HEVC screen content coding," IEEE PCS, pp. 365–368, San Jose, CA, Dec. 2013.

[SCC3] M. Naccari et al., "Improving inter prediction in HEVC with residual DPCM for lossless screen content coding," IEEE PCS, pp. 361–364, San Jose, CA, Dec. 2013.

[SCC4] T. Lin et al., "Mixed chroma sampling-rate high efficiency video coding for full-chroma screen content," IEEE Trans. CSVT, vol. 23, pp. 173–185, Jan. 2013.

[SCC5] Braeckman et al., "Visually lossless screen content coding using HEVC base-layer," IEEE VCIP 2013, pp. 1–6, Kuching, China, 17–20, Nov. 2013.

[SCC6] W. Zhu et al., "Screen content coding based on HEVC framework," IEEE Trans. Multimedia, vol. 16, pp. 1316–1326 Aug. 2014 (several papers related to MRC) MRC: mixed raster coding.

[SCC7] ITU-T Q6/16 Visual Coding and ISO/IEC JTC1/SC29/WG11 Coding of Moving Pictures and Audio.

[SCC8] Title: Joint Call for Proposals for Coding of Screen Content Status: Approved by ISO/IEC JTC1/SC29/WG11 and ITU-T SG16 Q6/16 (San Jose, 17 January 2014). Final draft standard expected: late 2015.

[SCC9] F. Zou, et al., "Hash Based Intra String Copy for HEVC Based Screen Content Coding," IEEE ICME (workshop), Torino, Italy, June 2015

[SCC10] T. Lin, et al., "Arbitrary Shape Matching for Screen Content Coding," in Picture Coding Symposium (PCS), pp. 369–372, Dec. 2013.

[SCC11] T. Lin, X. Chen and S. Wang, "Pseudo-2d-matching based dual – coder architecture for screen contents coding," IEEE International Conference in Multimedia and Expo Workshops (ICMEW), pp. 1– 4, July 2013.

[SCC12] B. Li, J. Xu and F. Wu, "1d dictionary mode for screen content coding," in Visual Communication and Image Processing Conference, pp. 189–192, Dec. 2014.

[SCC13] Y. Chen and J. Xu, "HEVC Screen Content Coding core experiment 4 (scce4) : String matching for sample coding," JCTVC – Q1124, Apr. 2014.

[SCC14] H. Yu, et al., "Common conditions for screen content coding tests," JCTVC – Q1015, vol. 24, no. 3, pp. 22–25, Mar. 2014.

[SCC15] IEEE Journal on Emerging and Selected Topics in Circuits and Systems (JETCAS) Special Issue on Screen Content Video Coding and Applications, vol. 6, issue 4, pp. 389–584, Dec. 2016. This has 21 papers on all aspects of screen content video coding including guest editorial, overview paper, future research opportunities and standards developments in this field.
See the overview paper: W.-H. Peng, et al., "Overview of screen content video coding: Technologies, standards and beyond," IEEE JETCAS, vol. 6, issue 4, pp. 393–408, Dec. 2016.

[SCC16] J. Nam, D. Sim and I.V. Bajic, "HEVC-based Adaptive Quantization for Screen Content Videos," IEEE Int. Symp. on Broadband Multimedia Systems, pp. 1–4, Seoul, Korea, 2012.

[SCC17] S.-H. Tsang, Y.-L. Chan and W.-C. Siu, "Fast and Efficient Intra Coding Techniques for Smooth Regions in Screen Content Coding Based on Boundary Prediction Samples," ICASSP2015, Brisbane, Australia, April. 2015.

[SCC18] HEVC SCC Extension Reference software: https://hevc.hhi.fraunhofer.de/svn/svn_HEVCSoftware/tags/HM-15.0+RExt-8.0+SCM-2.0/

[SCC19] HEVC SCC Software reference manual: https://hevc.hhi.fraunhofer.de/svn/svn_HEVCSoftware/branches/HM-SCC-extensions/doc/software-manual.pdf
Link to Screen content Test Sequences: http://pan.baidu.com/share/link?shareid=3128894651&uk=889443731

[SCC20] K. Rapaska et al., "Improved block copy and motion search methods for HEVC screen content coding," [9599-49], SPIE.

Optics + photonics, San Diego, California, USA, 9–13, Aug. 2015. Website: www.spie.org/op

[SCC21] B. Li, J. Xu and G. J. Sullivan, "Performance analysis of HEVC and its format range and screen content coding extensions," [9599-45], SPIE. Optics + photonics, San Diego, California, USA, 9–13, Aug. 2015. Website: www.spie.org/op

[SCC22] S. Hu et al., "Screen content coding for HEVC using edge modes," IEEE ICASSP, pp. 1714–1718, 26–31 May 2013.

[SCC23] H. Chen, A. Saxena and F. Fernandes, "Nearest-neighbor intra prediction screen content video coding," IEEE ICIP, pp. 3151–3155, Paris, 27–30, Oct. 2014.

[SCC24] X. Zhang, R. A. Cohen and A. Vetro, "Independent Uniform Prediction mode for screen content video coding," IEEE Visual Communications and Image Processing Conference, pp. 129–132, 7–10 Dec. 2014.

[SCC25] J. Nam, D. Sim and I. V. Bajic, "HEVC – based adaptive quantization for screen content videos," IEEE International Symposium on Broadband Multimedia Systems and Broadcasting (BMSB), pp. 1–4, June 2012.

[SCC26] L. Guo et al., "Color palette for screen content coding," IEEE ICIP, pp. 5556–5560, Oct. 2014.

[SCC27] D.-K. Kwon and M. Budagavi, "Fast intra block copy (intra BC) search for HEVC Screen Content Coding", IEEE ISCAS, pp. 9–12, June 2014.

[SCC28] J.-W. Kang, "Sample selective filter in HEVC intra-prediction for screen content video coding," IET Journals & Magazines, vol. 51, no. 3, pp. 236–237, Feb. 2015.

[SCC29] H. Zhang et al., "HEVC – based adaptive quantization for screen content by detecting low contrast edge regions," IEEE ISCAS, pp. 49–52, May 2013.

[SCC30] Z. Ma et al., "Advanced screen content coding using color table and index map," IEEE Trans. on Image Processing, vol. 23, no. 10, pp. 4399–4412, Oct. 2014.

[SCC31] R. Joshi et al., "Screen Content Coding Test Model 3 Encoder Description (SCM 3)," Joint Collaborative Team on Video Coding, JCTVC-S1014, Strasbourg, FR, 17–24 Oct. 2014.

[SCC32] S. wang et al., "Joint Chroma Downsampling and Upsampling for screen Content Image," IEEE Trans. on CSVT. (Early Access).

[SCC33] H. yang, Y. Fang and W. Lin, "Perceptual Quality Assessment of Screen Content Images," IEEE Trans. on Image processing. (Early Access).

[SCC34] I. J. S. W. R. Subgroup, "Requirements for an extension of HEVC for coding on screen content," in MPEG 109 meeting, 2014.

[SCC35] H. Yang et al., "Subjective quality assessment of screen content images," in International Workshop on Quality of Multimedia Experience (QoMEX), 2014.

[SCC36] "SIQAD," https://sites.google.com/site/subjectiveqa/.: Perceptual Quality Assessment of Screen Content Images.

[SCC37] S. Kodpadi, "Evaluation of coding tools for screen content in High Efficiency Video Coding" M.S. Thesis, EE Dept., University of Texas at Arlington, Arlington, Texas, Dec. 2015. http://www.uta.edu/faculty/krrao/dip click on courses and then click on EE5359 Scroll down and go to recent Theses/Project Title and click on Swetha Kodpadi.

[SCC38] N.N. Mundgemane, "Multi-stage prediction scheme for Screen Content based on HEVC," M.S. Thesis, EE Dept., University of Texas at Arlington, Arlington, Texas, Dec. 2015. http://www.uta.edu/faculty/krrao/dip click on courses and then click on EE5359 Scroll down and go to recent Theses/Project Title and click on N.N. Mundgemane.

[SCC39] T. Vernier et al., "Guided chroma reconstruction for screen content coding," IEEE Trans. CSVT, (EARLY ACCESS).

[SCC40] H. Yu et al., "Requirements for an extension of HEVC for coding of screen content," ISO/IEC JTC1/SC 29/WG 11 Requirements subgroup, San Jose, California, USA, document MPEG2014/N14174, Jan. 2014.

[SCC41] C. Lan et al., "Screen content coding," 2nd JCT-VC meeting, Geneva, Switzerland, document JCTVC-B084, July 2010.

[SCC42] B. Li and J. Xu, "Non-SCCE1: Unification of intra BC and inter modes," 18th JCT-VC meeting, Sapporo, Japan, document JCTVC-R0100, July 2014.

[SCC43] R.R. Barkur, "Production techniques for palette coding in screen content," project report, Spring 2016. http://www.uta.edu/faculty/krrao/dip click on courses and then click on EE5359 Scroll down and go to projects Spring 2016 and click on R.R. Barkur final report.

386 Appendix

[SCC44] C. Chen et al., "A STAIRCASE TRANSFORM CODING SCHEME FOR SCREEN CONTENT VIDEO CODING" IEEE ICIP, session MPA.P8.1, Phonex, AZ, Sept. 2016.

[SCC45] T. Vermeir et al., "Guided Chroma Reconstruction for Screen Content Coding," IEEE Trans. CSVT, vol. 26, pp. 1884–1892, Oct. 2016.

[SCC46] W. Xiao et al., "Weighted rate-distortion optimization for screen content coding," IEEE Trans. CSVT, vol. 28, pp. 499–512, Feb. 2018.

[SCC47] C.-C. Chen and W.-H. Peng, "Intra line copy for HEVC screen content coding," IEEE Trans. CSVT, (Early access).

[SCC48] W.-H. Peng et al., "Overview of screen content video coding: Technologies, Standards and Beyond," IEEE JETCAS, vol. 6, issue 4, pp. 393–408, Dec. 2016. This paper is part of the special issue on screen content video coding and applications. IEEE JETCAS, vol. 6, issue 4, pp. 389–584, Dec. 2016. This special issue can lead to several projects on SCC.

[SCC49] Z. Ni et al., "ESIM: Edge similarity for screen content image quality assessment," IEEE Trans. IP, Early access. This has several valuable references related to image/video quality assessment. Part of the introduction describing the screen content image/video is reproduced here.

In the era of multimedia communications, mobile and cloud computing, and the Internet of Things, the contents of digital images are no longer just limited to natural scenes. In fact, the contents of digital images nowadays can have a mixture of sources, such as natural scene, computer-generated graphics, sub titles, texts, animation, gaming, charts, maps, user's hand-writing and -drawing, and even some special symbols or patterns (e.g., logo, barcode, QR code) imposed and rendered by an electronic device or a photo editing software. Such kind of images is denoted as the screen content images (SCIs) and they are frequently encountered in various multimedia applications and services, such as online news and advertisement, online education, electronic brochures, remote computing, cloud gaming to name a few.

In the conclusions the authors state "that the proposed ESIM model delivers the highest performance compared to the other state-of-the-art IQA models on providing the more accurate and consistent assessment in accordance with what the HVS perceives and judges the SCIs."

The authors also have developed a large SCI database.

[SCC50] S. Wang et al., "Reduced-reference quality assessment of screen content images," IEEE Trans. CSVT, vol. 28, pp. 1–14, Jan. 2018.

The section Discussion is repeated here:

As one of the first attempts on this topic, the proposed method also has several limitations that should be improved in the future. Firstly, the current method is applicable to the screen content image. In practice, how to extend it to screen content video should be further investigated. In particular, the high correlation among screen video frames may further reduce the transmission and computational overhead of the extracted features by exploiting the inter-prediction like techniques for feature prediction. Secondly, it is worth mentioning that the design of the RR method does not make any assumption on the image distortion types, making it have the potential to be used for general-purpose applications. However, currently, the testing is based on eight distortion types. In the future, more distortion types will be involved for verification as well. Moreover, statistical features that exhibit more robust cross distortion type prediction will be studied in the future. Thirdly, the methodology of selecting the model parameters from the functionalities of HVS is worth further exploring. Though the investigation of viewing behaviors for the particular screen content is still at the starting stage, the fundamentally interesting differences between SCIs and natural images from the perspective of psychological studies may bring more inspirations to the model parameter selection in the future. Finally, how to design screen content enhancement algorithm using the statistical features from the proposed RRIQA algorithm is a topic worth further investigating. This poses new challenges to IQA and restoration research for SCIs and opens up new space for future exploration.

[SCC51] L. Pragnell and V. Sanchez, "JND-based perceptual video coding for 4:4:4 screen content data in HEVC," ICASSP 2018, Calgary, Canada, 2018.

[SCC52] Y. Zhang, D.M. Chandler and X. Mou, "Quality Assessment of Screen Content Images via Convolutional-Neural-Network-Based Synthetic/Natural Segmentation," IEEE Trans. IP, EA.

[SCC53] Y. Fu et al., "Screen content image quality assessment using multi-scale difference of Gaussian," IEEE Trans. CSVT, vol. 28, pp. 2428–2432, Sept. 2018.

SCC-P1. See [SCC50] Explore how the proposed method can be extended to screen content video.

SCC-P2. See [SCC-P1] and [SCC50] Explore the remaining future work suggested in the discussion above.

SCC-P3. See [SCC52] The authors claim that the CNN-based quality assessment of screen content images has the best performance compared to previous techniques. Go through this paper and other screen content image quality assessment approaches (see the references listed at the end) and verify this claim.

Access to JCT – VC resources:
Access to JCT-VC documents: http://phenix.it.-sudparis.eu/jct/
  Bug tracking: http://hevc.hhi.fraunhofer.de/trac/hevc
  Email list: http://mailman.rwth-aachen.de/mailman/listinfo/jct-vc

Beyond HEVC:

[BH1] J. Chen et al., "Coding tools investigation for next generation video coding based on HEVC," [9599 – 47], SPIE. Optics + photonics, San Diego, California, USA, 9–13, Aug. 2015.

[BH2] A. Alshin et al., "Coding efficiency improvements beyond HEVC with known tools," [9599-48], SPIE. Optics + photonics, San Diego, California, USA, 9–13, Aug. 2015. References [2] through [8] describe various proposals related to NGVC (beyond HEVC) as ISO/IEC JTC – VC documents presented in Warsaw, Poland, June 2015. Several projects can be implemented/explored based on [BH2] and these references.

[BH3] A. Alexander and A. Elina, "Bi-directional optical flow for future video codec," IEEE DCC, Sun Bird, Utah, March–April 2016.

[BH4] X. Zhao et al., "Enhanced multiple transform for video coding," IEEE DCC, Sun Bird, Utah, March–April 2016.

[BH5] Requirements for a Future Video Coding Standard v1, Standard, ISO/IEC JTC1/SC29/WG11, N15340, June 2015.

[BH6] Algorithm Description of Joint Exploration Test Model 1 (JEM 1), Standard ISO/IEC JTC1/SC29/WG11, N15790, Oct. 2015.

[BH17] S.-H. Park and E.S. Jang, "An efficient motion estimation method for QTBT structure in JVET," IEEE DCC, April 2017.

[BH18] M.J. Garrido et al., "A high performance FPGA-based architecture for the future video coding adaptive multiple core transform," IEEE Trans. CE, Early access.

Abstract is repeated here.

Abstract

Future Video Coding (FVC) will be the next generation video coding standard. Yet in the first stages of the standardization process, it is expected to replace HEVC beyond 2020. One of the enhancements in discussion is the Adaptive Multiple Core Transform, which uses five different types of 2D Discrete Sine/Cosine transforms (DCT-II, DCT-V, DCT-VIII, DST-I and DST-VII) and up to 64 × 64 Transform Unit sizes. This schema, increases the computational complexity of both, encoder and decoder. In this paper, a deeply pipelined high performance architecture to implement the five transforms for 4 × 4, 8 × 8, 16 × 16 and 32 × 32 sizes is proposed. The design has been described in VHDL and it has been synthesized for different FPGA chips, being able to process up 729 fps@3840 × 2160 for 4 × 4 transform sizes. Up the best of our knowledge, this is the first implementation of an architecture for the FVC Adaptive Multiple Core Transform supporting 4 × 4, 8 × 8, 16 × 16 and 32 × 32 sizes.

[BH19] H. Azgin et al., "Reconfigurable intra prediction hardware for future video coding," IEEE Trans. CE, vol. 63, pp. 419–425, Nov. 2017.
Abstract is repeated here:

In this paper, two Future Video Coding (FVC) reconfigurable intra prediction hardware are proposed. They are the first FVC intra prediction hardware in the literature. The first hardware implements multiplications with constants using adders and shifters instead of using multipliers. Therefore, it can be used in ASIC implementations of the FVC encoders. The second hardware implements multiplications with constants using DSP blocks in FPGA. Therefore, it can be used in FPGA implementations of the FVC encoders.

Projects on Beyond HEVC:

[BH-P1] In [BH2] several tools (some of these are straight forward extensions of those adopted in HEVC) are considered in NGVC (beyond HEVC). Some of these are increasing CU and TU sizes, up to 64 adaptive intra directional predictions, multi – hypothesis probability estimation for CABAC, bi-directional optical flow, secondary transform, rotational transform and multi-parameter intra prediction. Go through [BH2] in detail and evaluate the performance of each new tool, for all – intra, random access, low delay B, and low delay P (see Table 1). Compare the computational complexity of each tool with that of HEVC.

[BH-P2] See [BH-P1]. Evaluate performance impact of enlarging CU and TU sizes (see Table 2). Also consider computational complexity.

[BH-P3] See [BH-P1]. Evaluate performance impact of fine granularity Intra prediction vs HEVC with enlarged CU and TU sizes (see Table 3). Also consider computational complexity.

[BH-P4] See [BH-P1]. Evaluate performance impact of multi-hypothesis probability estimation vs HEVC with enlarged CU and TU sizes (see Table 4). Also consider computational complexity.

[BH-P5] See [BH-P1]. Evaluate performance impact of bi-directional optical flow vs HEVC with enlarged CU and TU sizes (see Table 6). Also consider computational complexity.

[BH-P6] See [BH-P1]. Evaluate performance impact of implicit secondary transform vs HEVC with enlarged CU and TU sizes (see Table 7). Also consider computational complexity.

[BH-P7] See [BH-P1]. Evaluate performance impact of explicit secondary transform vs HEVC with enlarged CU and TU sizes (see Table 9). Also consider computational complexity.

[BH-P8] See [BH-P1]. Evaluate performance impact of multi-parameter Intra prediction vs HEVC with enlarged CU and TU sizes (see Table 11). Also consider computational complexity.

[BH-P9] See [BH-P1]. Evaluate Joint performance impact of all tested tools on top of HEVC (see Table 6). Also consider computational complexity.

[BH-P10] Similar to [BH2], several coding tools (some of them were considered earlier during the initial development of HEVC) are investigated for NGVC based on HEVC [BH1]. These tools include large CTU and TU, adaptive loop filter, advanced temporal Motion Vector Prediction, cross component prediction, overlapped block Motion Compensation and adaptive multiple transform. The overall coding performance improvement resulting from these additional tools for classes A through F Video test sequences in terms of BD-rate are listed in Table 3. Class F sequences include synthetic (computer generated) video. Performance improvement of each tool for Random Access, All Intra and low-delay B is also listed in Table 4. Evaluate the performance improvement of these tools [BH1] over HM 16.4 and verify the results shown in Table 3.

[BH-P11] See [BH-P10]. Evaluate the performance improvement of each tool for All Intra, Random Access and Low-delay B described in Table 4. Test sequences for screen content coding (similar to class F in Table 3) can be accessed from http://pan.baidu.com/share/link?shareid=3128894651&uk=889443731.

[BH-P12] See [BH-P10] and [BH-P11]. Consider implementation complexity as another metric, as these additional tools invariably result in increased complexity. Evaluate this complexity for all classes (A through F) – see Table 3 and for All Intra, Random Access and low-delay B cases (see Table 4). See the papers below related to proposed requirements for Next Generation Video coding (NGVC).

1. ISO/IEC JTC1/SC29/WG11, "Proposed Revised Requirements for a Future Video coding Standard," MPEG doc.M36183, Warsaw, Poland, June. 2015.
2. M. Karczewicz and M. Budagavi, "Report of AHG on Coding Efficiency Improvements," VCEG-AZ01, Warsaw, Poland, June 2015.
3. J.-R. Ohm et al., "Report of AHG on Future Video Coding Standardization Challenges," MPEG Document M36782, Warsaw, Poland, June. 2015.

[BH-P13] In [BH1] for larger resolution video such as 4K and 8K, 64×64 INT DCT (integer approx.) is proposed to collaborate with the existing transforms (up to 32×32 INT DCT) and to further improve the coding efficiency. Madhukar and Sze developed unified forward and inverse transform architecture (2D-32 $\times$ 32 INT DCT) for HEVC (see IEEE ICIP 2012) resulting in simpler hardware compared to implementation separately i.e., forward and inverse. See also Chapter 6 HEVC transform and quantization by Budagavi, Fuldseth and Bjontegaard in [E202]. Implement similar uniform forward and inverse transform architecture for 2D-64 $\times$ 64 INT DCT and evaluate the resulting simpler hardware compared to implementation separately i.e., forward and inverse.

[BH-P14] Refer to Figures 6.7 and 6.8 of Chapter 6 cited in [BH-P13]. These QM are based on visual sensitivity of the transform coefficients. Similar QM have been proposed/adopted in JPEG, MPEG – 1, 2, 4, AVS china etc. Also see the book by K.R. Rao and P. Yip, "Discrete Cosine Transform," Academic press 1990, wherein the

392 *Appendix*

theory behind developing the QM is explained. See also related references at the end of Chapter 6 in [E202]. The QM matrices (Figure 6.8) for (16×16) and (32×32) transform block sizes are obtained by replicating the QM for (8×8) transform block size. These extensions are based on reducing the memory needed to store them. Develop QM for (16×16) and (32×32) transform block sizes independently based on their visual sensitivities (perception).

[BH-P15] See [BH-P14]. Develop QM for (64×64) INTDCT reflecting the visual perception of these transform coefficients. Again refer to the book by K.R. Rao and P. Yip, "Discrete Cosine Transform," Academic press 1990.

[BH-P16] See [BH-P13] and [BH-P14]. In Figure 6.2 (page 149) of Chapter 6 [E202] (32×32) INTDCT, (4×4), (8×8), (16×16) INTDCTs are embedded. Develop (64×64) INTDCT wherein the smaller size INTDCTs are embedded. Is it orthogonal? What are the norms of the 64 basis vectors?

[BH-P17] See "Y. Sugito et al., "A study on addition of 64×64 transform to HM 3.0.," Joint Collaborative Team on Video Coding (JCT-VC) of ITU-T SG16 WP3 and ISO/IEC JTC1/SC29/WG11, JCTVC-F192, Torino, Italy, July, 2011" which is reference [15] in [BH2]. Is the (64×64) INTDCT orthogonal? Embedding property?

[BH-P18] See [BH3]. Theoretical and implementation aspects of bi-directional optical flow which is part of JEM1 (Joint Exploration Model) See [2] in this paper are developed. Evaluate this MV refinement in terms of MC prediction compared to the ME process adopted in HEVC based on some standard test sequences. Sum of absolute MC prediction errors for various block sizes can be used as the comparison metric.

[BH-P19] In [BH4] enhanced multiple transform (EMT) is proposed and implemented on top of the reference software using various test sequences. See Table 4. This scheme is compared with HM-14.0 Main 10 for all intra (AI) and random access (RA) configurations. Verify these results and extend the simulation to low delay case. Use the latest HM software.

[BH-P20] See [BH19] and the abstract. Implement the two future video coding (FVC) reconfigurable IP hardware and confirm the results shown in Tables IV and V and Figure 5.

[BH-P21] See [BH-P20] In [BH19] the two IP hardware can implement 65 FVC intra angular prediction modes for 4 × 4, 8 × 8, 16×16 and 32 × 32 prediction units (Figure 1). In FVC, the angular prediction mode is extended to 62 × 62 PU. Design and implement the IP hardware for this large PU.

Post – HEVC Activity

Both MPEG and VCEG have established AHGs (ad hoc groups) for exploring the next-generation video coding.

Grois et al., (See item 7 under tutorials) have suggested as follows:

Focusing on perceptual models and perceptual quality, and perceptually optimized video compression provision: http://www.provision-itn.eu

PROVISION is a network of the leading academic and industrial organizations in Europe, including international researchers working on the problems with regard to the state-of-the-art video coding technologies.

The ultimate goal is to make noteworthy technical advances and further improvements to the existing state-of-the-art techniques of compression video material.

AV1 Codec (Alliance for Open Media – AOM)

https://en.wikipedia.org/wiki/AV1 This web site is very extensive coverage on AV1 codec.

http://tinyurl.com/zgwdo59

Website: http://www.streamingmedia.com/Articles/Editorial/What-Is-.../What-is-AV1-111497.aspx

AOM (chartered/founding members Amazon, Cisco, Google, Intel Corporation, Microsoft, Mozilla, IBM, and Netflix). In April 2016, ARM, AMD, Apple, and NVIDIA joined the alliance to help ensure that the codec is hardware friendly and to facilitate and accelerate AV1 hardware support. Hulu joined AOM as a member on 29 June 2017. NGCodec has joined as a member on 11 July 2017. Facebook has joined AOM as a founding member (NOV. 2017). There are now 25+ members (March 2018).

AOM has nearly finalized a codec (See the latest developments below); called AV1 based on CISCO's Thor, Google's VP10 and Mozilla's DAALA with the following goals:

- Interoperable and open
- Optimized for the web membership@aomedia.org

- Scalable to any modern device at any bandwidth
- Designed with a low computational footprint and optimized for hardware
- Capable of consistent, highest-quality, real-time video delivery
- Flexible for both commercial and non-commercial content, including user-generated content.

In terms of makeup, the Alliance members enjoy leading positions in the following markets:

- Codec development - Cisco (Thor) Google (VPX), Mozilla (Daala)
- Desktop and mobile browsers - Google (Chrome), Mozilla (Firefox), Microsoft (Edge)
- Content – Amazon (Prime), Google (YouTube), Netflix
- Hardware co-processing – AMD (CPUs, graphics), ARM (SoCs, other chips), Intel (CPUs), NVIDIA (SoC, GPUs)
- Mobile – Google (Android), Microsoft (Windows Phone)
- OTT – Amazon (Amazon Fire TV), Google (Chromecast, Android TV)

The Alliance is targeting an improvement of 50 percent over VP9/HEVC with reasonable increases in encoding and playback complexity. One focus is UHD video, including higher bitrate, wider colour gamut, and increased frame rates, with the group targeting the ability to play 4K 60fps in a browser on a reasonably fast computer. The base version of the codec will support 10-bit and 12-bit encoding, as well as the BT.2020 colour space. Another focus is providing a codec for WebRTC (Real-Time Communications), an initiative supported by Alliance members Google and Mozilla, and similar applications including Microsoft's Skype.

P.S: This material on AV1 codec is collected from Streaming Media Magazine, "What is AV1," Jan Ozer, Posted on June 3, 2016.

The AOMedia Video codec, AV1, the Alliances first video compression standard, strives to deliver improved compression efficiency over the current cutting-edge video compression standards HEVC/H.265 and VP9, in addition to providing the industry with an open source, royalty-free, interoperable video format. IEEE DCC 27–30 March 2018 has conducted a special session on "The emerging AV1 video codec" with J. Han and I. Sodegar as the session coordinators [E426].

For details on AOM, please access http://aomedia.org

The code base for AOM is here

https://aomedia.googlesource.com/

aomedia Git repositories - Git at Google

aomedia.googlesource.com

https://aomediacodec.github.io/av1-spec/
All-Projects Access inherited by all other projects. All-Users Individual user settings and preferences. AOM; Alliance for Open Media contributor-guide

The following link shows how to build and test AOM, and also provides test sequences.

http://aomedia.org/contributor-guide/

Alliance for Open Media - Contributor Guide

aomedia.org

Download Test Scripts. Test scripts (and this document) are available in the contributor-guide Git project: $ git clone

https://aomedia.googlesource.com/contributor...

Amsterdam- Sept. 6, 2018 – The Alliance for Open Media (AOMedia) today announced that it would lead a notable round of demo activities focusing on the AOMedia Video Codec (AV1) at the IBC Show Exhibition 2018, Sept. 14–18, 2018, at RAI Amsterdam. AOMedia member companies will present AV1 demonstrations throughout the IBC2018 event. AOMedia AV1 demos at IBC 2018:
Bitmovin, Google, Mozilla, NGCodec, Socionext and VideoLAN.

AOMedia Presents and Demos AV1 at the NAB Streaming Smmit, NYC, Oct. 17–18

Join AOMedia members Bitmovin, Google and Mozilla at the AV1 Happy Hour on Wednesday, Oct. 17, 2018, 5:00–6:00 p.m. at the CMIIP Theater

New York, USA – Oct. 16, 2018 - The Alliance for Open Media (AOMedia) today announced that members will be leading presentations and demos focusing on the AOMedia Video Codec (AV1) from Oct. 17–18, 2018 during the NAB Show's Streaming Summit in New York at the Jacob K. Javits Center. Steaming Media attendees are also invited to the AV1 Happy Hour on Wednesday, Oct. 17, 2018, 5:00–6:00 p.m. in the CMIIP Theater, sponsored by AOMedia, Bitmovin, Mozilla and Google.

Co-produced by streaming industry expert Dan Rayburn, the Streaming Summit at NAB will feature presentations, round-table discussions and fireside chats that will address trends and technologies that are having a direct impact on the continued development, proliferation and profitability of online video.

Track C: AV1, A Win for the Open Web: Demos by Google and Mozilla

When: Wednesday, Oct. 17, 2018, 3:15 p.m.– 4:00 p.m.

Where: Location: 3D10

What: Presenters from AOMedia members Google and Mozilla will share a look at AV1, the royalty-free video codec using Chrome and Firefox. AV1 offers a generational improvement in coding efficiency over any web codec and comes backed by nearly 40 companies, including some of the largest in the internet, tech, and video industries. This talk will cover why Google and Mozilla invest in web codecs, the benefits of AV1, where the codec stands going into 2019, and demos of AV1 already available in browsers today as well as on different platforms and for different use cases.

- PRESENTERS:
    - Jai Krishnan, Product Manager, Google
    - Nathan Egge, Video Codec Engineer, Mozilla

AV1 Happy Hour – Sponsored by AOMedia, Bitmovin, Mozilla, and Google

When: Wednesday, Oct. 17, 2018, 5:00–6:00 p.m.

Where: Location CMIIP Theater

What: Attendees are invited to the AV1 Happy Hour. AOMedia members Bitmovin, Mozilla, and Google will deliver a quick general interest update on AV1 and refreshments will be served.

Track B: AV1: How the Codec of the Future is Being Used Today

When: Thursday, Oct. 18, 2018, 10:15 a.m.–11:15 a.m.

Where: Location: 3D09

What: The next-generation video codec AV1 was standardized in the Alliance for Open Media earlier this year, and it is coming to the web in a hurry.

Content and encoding providers such as YouTube, Ateme, Harmonic, Netflix, and Bitmovin presented AV1 demos this fall, and both Chrome and Firefox have decode support in browser releases. Join panelists from Google, Mozilla, and Bitmovin to talk about why their companies have invested in AOM, the early gains shown by AV1, and what to expect in 2019 and beyond.

- MODERATOR:
    - John Luther, SVP, Technology, JW Player
- PANELISTS:
    - Nathan Egge, Video Codec Engineer, Mozilla
    - Steve Geiger, Director, Solutions, Bitmovin
    - Jai Krishnan, Product Manager, Google
    - Mike Coward, Director of Engineering, Facebook

Track C: AV1 Benchmarks, 6 Months Post-Launch

Launched in late March 2018 by the Alliance for Open Media (AOMedia), AV1 marks a significant shift in the balance of power in online entertainment. While HEVC and VP9 offer the best compatibility across devices and platforms, the newest standard promises a much higher compression efficiency and thereby also a better user experience. How does this new codec compare to the incumbents in compression and speed? This presentation will showcase the latest performance benchmarks and key encoding techniques that make AV1 an interesting choice for the future of your video delivery.

- PRESENTER
    - Steve Geiger, Director of Solutions, Bitmovin

About the Alliance for Open Media

Launched in 2015, the Alliance for Open Media (AOMedia) was formed to define and develop media technologies to address marketplace demand for an open standard for video compression and delivery over the web. Founding Members include Amazon, Apple, Arm, Cisco, Facebook, Google, IBM, Intel, Microsoft, Mozilla, Netflix, and NVIDIA. AOMedia is a Joint Development Foundation Projects, LLC series. The Joint Development Foundation is an independent nonprofit organization that provides the corporate and legal infrastructure to enable groups to establish and operate standards and source code development collaborations.

Alliance for Open Media, 401 Edgewater Place,
Suite 600, Wakefield, MA 01880

[AV1-1] IEEE DCC2018, Session 1. The emerging AV1 video codec Part 1, Snow Bird, Utah, March, 2018.

[AV1-2] IEEE DCC2018, Session 12. The emerging AV1 video codec Part 2, Snow Bird, Utah, March, 2018.

[AV1-3] S. Midtsokogen and J.-M. Valin, "The AV1 constrained directional enhancement filter," ICASSP 2018, Calgary, Canada, 2018.

[AV1-4] The Alliance for Open Media Kickstarts Video Innovation Era with "AV1" Release, 28 March 2018. AV1 Unleashes 4K and Higher UHD, Royalty-Free Video for All, AV1: AOMedia's royalty-free UHD video codec (jhennion@virtualmgmt.com). Alliance has announced the public release of the AOMedia Video Codec1.0 (AV1) specification. It is reported that AVI has 30% to 40% better quality for UHD 4K videos than HEVC/VP9. Encoding times at present are 100x times slower than VP9 and decoding times at present are $5 \times$ times slower than VP9. It is anticipated that these encoding/decoding times will drop down to $5 \times$ and $2 \times$ times respectively by year (2018) end.

Please click the links below:

http://www.streamingmedia.com/Articles/News/Online-Video-News/AV1-Is-Finally-Here-but-Intellectual-Property-Questions-Remain-124134.aspx
  https://bitmovin.com/apple-joins-av1-codec-consortium/

- http://aomedia.org/about-us/
- https://hacks.mozilla.org/2017/11/dash-playback-of-av1-video/
- https://youtu.be/thvSyJN1vsA?t=5452
- https://tools.ietf.org/id/draft-midtskogen-netvc-cdef-00.html
- https://arxiv.org/abs/1602.05975

Open this web site about AV1 codec
  https://bitmovin.com/cool-new-video-tools-five-encoding-advancements-coming-av1/

AV1 has three profiles Main, High, and Professional: (Levels being decided)

The profile specifies the bit depth and subsampling formats supported, while the level defines resolution and performance characteristics. A.1.1. Profiles There are three named profiles: "Main" compliant decoders must be able to decode streams with seq_profile equal to 0. "High" compliant decoders must be able to decode streams with seq_profile less than or equal to 1. "Professional" compliant decoders must be able to decode streams with seq_profile less than or equal to 2. A.1.2. Levels TODO: Still under discussion. Note: Main profile consists of YUV 4:2:0 or monochrome sequences with bit depth equal to 8 or 10. High profile consists of all main profile sequences plus YUV 4:4:4 sequences with bit depth equal to 8 or 10 (monochrome is not allowed when seq_profile = 1). Professional profile consists of all high and main profile sequences plus YUV 4:2:2 sequences with bit depth equal to 8 or 10 or 12, 4:2:0 sequences with bit depth equal to 12, 4:4:4 sequences with bit depth equal to 12, and monochrome sequences with bit depth equal to 12. DRAFT AV1 Bitstream & Decoding Process Specification Section:

[AV1-5] W.-T. Lin et al., "Efficient AV1 video using a multi-layer framework," IEEE DCC 2018, Sun Bird, Utah, March 2018.

[AV1-6] https://arewecompressedyet.com

AreWeCompressedYet

AreWeCompressedYet is an automatic service for compressing videos with AV1, VP9, Daala, Thor, and other codecs.

It generates graphs using different metrics, to enable comparisons between different versions of codec source code.

[AV1-7] K.-S. Lu et al., "Efficient rate-distortion approximation and transform type selection using laplacian operators." IEEE PCS 2018, San Francisco, CA, June 2018.

Abstract is repeated here.

Abstract

Rate-distortion (RD) optimization is an important tool in many video compression standards and can be used for transform selection. However, this is typically very computationally demanding because a full RD search involves the computation of transform coefficients for each candidate transform. In this paper, we propose an approach that uses sparse Laplacian operators to estimate the RD cost by computing a weighted squared sum of transform coefficients, without having to compute the actual transform coefficients. We demonstrate experimentally how our method can be applied for transform selection. Implemented in the AV1 encoder, our approach yields a significant speed-up in encoding time with a small increase in bitrate.

AV1-P1. As AV1 is designed to be a royalty free codec, gather all details about this codec (website, steaming media etc.) and implement using standard test sequences. Compare its performance with HEVC using the standard criteria such as PSNR, BD bitrate, BD PSNR, SSIM, implementation complexity etc.

AV1-P2. See AV1-P1 Based on the performance comparison, explore any changes in the functionalities (transform/quantization/ME-MC, entropy coding, intra-inter and other modes) of the codec that can result in further in improving AV1's compression performance with negligible increase in complexity.

AV1-P3. See AV1-P1 Can AV1 codec be applied to SCC? If so compare its performance with HEVC SCC extension.

AV1-P4. See AV1-P1 Explore how AV1 Codec can be modified for scalable video coding similar to HSVC.

AV1-P5. See AV1-P1 Explore how AV1 Codec can be modified for 3D-Multiview coding similar to HSVC.

AV1-P6. Performance comparison of SDCT with DCT and DDCT is investigated in [E392]. In AV1 INTDCT is used. Replace INTDCT with INTSDCT and implement the performance comparison based on various test sequences including 4K and 8K and different block sizes. See the Tables and Figures in [E392].

AV1-P7. See [E393]. Apply the FCDR to AV1 codec both as a post processing operation and as an in loop (embedded) operation. Implement the projects for AV1 similar to those described in P.5.268 and P.5.269.

AV1-P8. See [AV1-4]. Implement the royalty free AV1 codec. Compare its performance based on standard metrics with HEVC.

AV1-P9. It is stated that "AV1 is an open, royalty-free, next-generation video coding format from the Alliance of Open Media Video. It is designed to replace Google's VP9 and compete with H.265/HEVC. AV1 is targeting an expected improvement of about 30% over VP9/HEVC with only reasonable increases in encoding and playback complexity." Compare AV1's performance with VP9 based on standard metrics.

AV1-P10. Obtain the AV1 encoder/decoder block diagrams. See how you can implement the various functional blocks by FPGA, ZIlinx and other hardware.

AV1-P11. see the info on Opus 1.2 audio codec.

Opus 1.2 Codec Arrives on Your Phone: High Quality Audio at 32 kbps
The Xiph.Org Foundation just announced their latest improvement to the Opus audio codec with the release of their libopus 1.2 encoder. With this latest improvement, Xiph.Org has managed to make Opus usable for fullband stereo audio at just 32 kb/s, which will pair well with the upcoming royalty-free AV1 video format in the WebM container to bring higher quality audio and video on slower connections.

Implement multiplex/demultiplex AV1 coded bit stream with Opus audio codec bit stream while maintaining lip synch.

AV1-P12. See AV1-P11 Replace Opus audio codec with DTS codec.

AV1-P13. See AV1-P11 Replace Opus audio codec with Dolby audio codecs such as HE-AAC version 2.

AV1-P14. See [AV1-5] Conclusion and future work are repeated here.

Conclusion and Future Work

In this paper, we first introduce a new coding tool that extends the total number of reference frames in the AV1 baseline. We then propose a multi-layer framework for the encoder design, which leverages the new coding tool through the use of extended ALTREF FRAMEs and the virtual index manipulation. The multi-layer, multireference prediction framework substantially increases the overall coding efficiency over an abundant set of video clips of various content and motion characteristics with a wide range of resolutions, providing evidence for the effectiveness of the proposed framework. The computational complexity at the decoder side is negligible. For the next step, we will focus on the encoder-side complexity reduction. For instance, through the use of a much smaller set of block partition/prediction candidates for some of the references (e.g., LAST2 FRAME and LAST3 FRAME), complexity may be reduced at a sacrifice of the coding gain. We will also investigate the more optimized encoder design specifically applied to the higher resolution videos so that the coding effectiveness on the higher resolution videos may be on par with that on the lower resolution scenarios. Also, it is possible for both the encoder and the decoder to keep track of the update of all the reference frames, and check whether either LAST2 FRAME or LAST3 FRAME belong to the previous GF group. As the current GF group

Appendix

always start with an updated GOLDEN FRAME, it is possible to remove the use of LAST2 FRAME or LAST3 FRAME if they are not in the current GF group, which may greatly help on the encoder speedup whereas incur negligible coding performance degradation.

Explore the future work suggested here in detail with the overall objective of reducing the encoder complexity with negligible degradation of the encoder performance.

Real Media HD (RMHD)

https://www.realnetworks.com/realmediaHD

Real Media HD, successor to its RMVB (Real Media Variable Bitrate) video codec.

Real Networks claims that RMHD codec has achieved a reduction of more than 30% bit rate compared to H.264/AVC, while achieving the same subjective quality. It has been integrated into media players from streaming vendors and OEMs. Real Networks says its royalty fees are straight forward.

Performance comparison of RMHD with H.264, H.265 and VP9 is also described.

RMHD-P1 While the RMHD encoder/decoder details are not known, develop comparison data similar to those shown in charts 1–7. Hopefully, RMHD software can be accessed.

SMPTE

SMPTE ST 2042-1:200x 28/09/2009
Page 9 of 131 pages

Foreword

SMPTE (the Society of Motion Picture and Television Engineers) is an internationally recognized standards developing organization. Headquartered and incorporated in the United States of America, the SMPTE has members in over 80 countries on six continents. The SMPTE's Engineering Documents, including Standards, Recommended Practices, and Engineering Guidelines, are prepared by the SMPTE's Technology Committees. Participation in these Committees is open to all with a bona fide interest in their work. The SMPTE cooperates closely with other standards-developing organizations, including ISO, IEC, and ITU.

SMPTE Engineering Documents are drafted in accordance with the rules given in Part XIII of its Administrative Practices.

http://www.smpte.org   email: eng@smpte.org

About the Society of Motion Picture and Television Engineers® (SMPTE®) For the past 100 years, the people of the Society of Motion Pictures and Television Engineers (SMPTE, pronounced "simp-tee") have sorted out the details of many significant advances in entertainment technology, from the introduction of "talkies" and colour television to HD and UHD (4K, 8K) TV. Since its founding in 1916, the Society has earned an Oscar® and multiple Emmy® Awards for its work in advancing moving-imagery education and engineering across the communications, technology, media, and entertainment industries. The Society has developed thousands of standards, recommended practices, and engineering guidelines, more than 800 of which are currently in force.

SMPTE's global membership today (Jan. 2017) includes 7,000 members, who are motion-imaging executives, engineers, creative and technology professionals, researchers, scientists, educators, and students. A partnership with the Hollywood Professional Association (HPA®) connects SMPTE and its membership with the professional community of businesses and individuals who provide the expertise, support, tools, and infrastructure for the creation and finishing of motion pictures, television programs, commercials, digital media, and other dynamic media content. Information on joining SMPTE is available at www.smpte.org/join.

Also SMPTE WEBCASTS (webinars).

SMPTE adopted VC-1 (named SMPTE 421 M) as the video codec. VC-1 is based on WMV9 developed by Microsoft. VC-1 is similar to H.264.

VC-1

Video coding 1 is based on Windows Media Video (WMV9) developed by Microsoft. See [VC-1,8] cited below for details.

References

VC-1.1. H. Kalva and J.-B. Lee, "The VC-1 video coding standard," IEEE Multimedia, vol. 14, pp. 88–91, Oct.–Dec. 2007.

VC-1.2. J.-B. Lee and H. Kalva, "The VC-1 and H.264 video compression standard for broadband video services," Springer 2008.
VC-1.3. VC-1 Software: http://www.smpte.org/home
VC-1.4. Microsoft website - VC-1 Technical Overview http://www.microsoft.com/windows/windowsmedia/howto/articles/vc1techoverview.aspx#VC1ComparedtoOtherCodecs
VC-1.5. VC-1 Compressed video bitstream format and decoding process *(SMPTE 421M-2006)*, SMPTE Standard, 2006.
VC-1.6. J.-B. Lee and H. Kalva, "An efficient algorithm for VC-1 to H.264 video transcoding in progressive compression," IEEE International Conference on Multimedia and Expo, pp. 53–56, July 2006.
VC-1.7. S. Srinivasan and S. L. Regunathan, "An overview of VC-1". *Proc. SPIE*, vol. 5960, pp. 720–728, 2005.
VC-1.8. S. Srinivasan et al., "Windows Media Video 9: overview and applications" Signal Processing: Image Communication, Vol. 19, pp. 851–875, Oct. 2004.
VC1-P1. See [E393]. Apply the FCDR to VC1 codec both as a post processing operation and as an in-loop (embedded) operation. Implement the projects for VC1 similar to those described in P.5.268 and P.5.269.

WMV-9 (VC-1)

WMV9 of Microsoft

Windows Movie Maker, in Windows Vista or XP can be used, but with limited functionality. It can be used to convert a WMV file to AVI format, for example.

For more precise control of encoding parameters, you can use the Windows Media encoder tool, freely available at http://www.microsoft.com/windows/windowsmedia/forpros/encoder/default.mspx. It is an encoding-only tool.

Source code for such tools is not available.

You can also use commercial video editing tools, such as the ones listed at http://www.microsoft.com/windows/windowsmedia/forpros/service_provider/software/default.aspx#encoding

Source code for VC-1 (WMV9) http://store.smpte.org/VC-1-Test-Material-p/vc-1.htm

http://www.smpte.org/standards/

AOMedia Presents and Demos AV1 at the NAB Streaming Smmit, NYC, Oct. 17–18

Join AOMedia members Bitmovin, Google and Mozilla at the AV1 Happy Hour on Wednesday, Oct. 17, 2018, 5:00–6:00 p.m. at the CMIIP Theater

New York, USA – Oct. 16, 2018 - The Alliance for Open Media (AOMedia) today announced that members will be leading presentations and demos focusing on the AOMedia Video Codec (AV1) from Oct. 17–18, 2018 during the NAB Show's Streaming Summit in New York at the Jacob K. Javits Center. Steaming Media attendees are also invited to the AV1 Happy Hour on Wednesday, Oct. 17, 2018, 5:00–6:00 p.m. in the CMIIP Theater, sponsored by AOMedia, Bitmovin, Mozilla and Google.

Co-produced by streaming industry expert Dan Rayburn, the Streaming Summit at NAB will feature presentations, round-table discussions and fireside chats that will address trends and technologies that are having a direct impact on the continued development, proliferation and profitability of online video.

Track C: AV1, A Win for the Open Web: Demos by Google and Mozilla

When: Wednesday, Oct. 17, 2018, 3:15 p.m.–4:00 p.m.

Where: Location: 3D10

What: Presenters from AOMedia members Google and Mozilla will share a look at AV1, the royalty-free video codec using Chrome and Firefox. AV1 offers a generational improvement in coding efficiency over any web codec and comes backed by nearly 40 companies, including some of the largest in the internet, tech, and video industries. This talk will cover why Google and Mozilla invest in web codecs, the benefits of AV1, where the codec stands going into 2019, and demos of AV1 already available in browsers today as well as on different platforms and for different use cases.

- PRESENTERS:
    - Jai Krishnan, Product Manager, Google
    - Nathan Egge, Video Codec Engineer, Mozilla

406 *Appendix*

AV1 Happy Hour – Sponsored by AOMedia, Bitmovin, Mozilla, and Google

When: Wednesday, Oct. 17, 2018, 5:00–6:00 p.m.

Where: Location CMIIP Theater

What: Attendees are invited to the AV1 Happy Hour. AOMedia members Bitmovin, Mozilla, and Google will deliver a quick general interest update on AV1 and refreshments will be served.

Track B: AV1: How the Codec of the Future is Being Used Today

When: Thursday, Oct. 18, 2018, 10:15 a.m.–11:15 a.m.

Where: Location: 3D09

What: The next-generation video codec AV1 was standardized in the Alliance for Open Media earlier this year, and it is coming to the web in a hurry. Content and encoding providers such as YouTube, Ateme, Harmonic, Netflix, and Bitmovin presented AV1 demos this fall, and both Chrome and Firefox have decode support in browser releases. Join panelists from Google, Mozilla, and Bitmovin to talk about why their companies have invested in AOM, the early gains shown by AV1, and what to expect in 2019 and beyond.

- MODERATOR:
    - John Luther, SVP, Technology, JW Player
- PANELISTS:
    - Nathan Egge, Video Codec Engineer, Mozilla
    - Steve Geiger, Director, Solutions, Bitmovin
    - Jai Krishnan, Product Manager, Google
    - Mike Coward, Director of Engineering, Facebook

Track C: AV1 Benchmarks, 6 Months Post-Launch

Launched in late March 2018 by the Alliance for Open Media (AOMedia), AV1 marks a significant shift in the balance of power in online entertainment. While HEVC and VP9 offer the best compatibility across devices and platforms, the newest standard promises a much higher compression efficiency and thereby also a better user experience. How does this new codec compare to the incumbents in compression and speed? This presentation will showcase the latest performance benchmarks and key encoding techniques that make AV1 an interesting choice for the future of your video delivery.

- PRESENTER
    - Steve Geiger, Director of Solutions, Bitmovin

To set up a press or analyst briefing with an AOMedia spokesperson, please email jhennion@virtualinc.com.

About the Alliance for Open Media

Launched in 2015, the Alliance for Open Media (AOMedia) was formed to define and develop media technologies to address marketplace demand for an open standard for video compression and delivery over the web. Founding Members include Amazon, Apple, Arm, Cisco, Facebook, Google, IBM, Intel, Microsoft, Mozilla, Netflix, and NVIDIA. AOMedia is a Joint Development Foundation Projects, LLC series. The Joint Development Foundation is an independent nonprofit organization that provides the corporate and legal infrastructure to enable groups to establish and operate standards and source code development collaborations.

Alliance for Open Media, 401 Edgewater Place, Suite 600, Wakefield, MA 01880

1. The below link had a lot of information on audio and video codecs. http://wiki.multimedia.cx/index.php?title=Main_Page
2. The link below has a VC-1 reference decoder. It is not the reference decoder but a free one!http://ffmpeg.mplayerhq.hu/download.html
3. Also, Google search provided the reference code for VC 1 - version 6 which might be outdated now and may not be legal to use (I am not sure on this). http://www.richardgoodwin.com/VC1_reference_decoder_release6.zip

VC-2 Standard

The VC-2 standard specifies the compressed stream syntax and reference decoder operations for a video compression system. VC-2 is an intra frame video compression system aimed at professional applications that provides efficient coding at many resolutions including various flavors of CIF, SDTV and HDTV. VC-2 utilizes wavelet transforms that decompose the video signal into frequency bands. The codec is designed to be simple and flexible, yet be able to operate across a wide range of resolutions and application domains.

The system provides the following capabilities:

- **Multi-resolution transforms.** Data is encoded using the wavelet transform, and packed into the bitstream subband by subband. High compression ratios result in a gradual loss of resolution. Lower resolution output pictures can be obtained by extracting only the lower resolution data.
- **Frame and field coding.** Both frames and fields can be individually coded.
- **CBR and VBR operation.** VC-2 permits both constant bit rate and variable bit rate operations. For low delay pictures, the bit rate will be constant for each area (VC-2 slice) in a picture to ensure constant latency.
- **Variable bit depths.** 8, 10, 12, and 16 bit formats and beyond are supported.
- **Multiple colour difference sampling formats.** 444, 422 and 420 video are all supported.
- **Lossless and RGB coding.** A common toolset is used for both lossy and lossless coding. RGB coding is supported either via the YCoCg integer colour transform for maximum compression efficiency, or by directly compressing RGB signals.
- **Wavelet filters.** A range of wavelet filters can be used to trade off performance against complexity. The Daubechies (9,7) filter is supported for compatibility with JPEG2000. A fidelity filter is provided for an improved resolution scalability.
- **Simple stream navigation.** The encoded stream forms a doubly linked list with each picture header indicating an offset to the previous and next picture, to support field-accurate high-speed navigation with no parsing or decoding required.
- **Multiple Profiles.** VC-2 provides multiple profiles to address the specific requirements of particular applications. Different profiles include or omit particular coding tools to best match the requirements of their intended applications. The Main profile provides maximum compression efficiency, variable bit rate coding, and lossless coding using the core syntax. The Simple profile provides a less complex codec, but with lower compression efficiency, by using simple variable length codes for entropy coding rather than the arithmetic coding used by the Main profile. The Low Delay profile uses a modified syntax for applications requiring very low, fixed, latency. This can be as low as a few lines of input or output video. The Low Delay profile is suitable for light compression for the re-use of low bandwidth infrastructure, for example carrying HDTV over SD-SDI links. The High Quality profile similarly

provides light compression with low latency and also supports variable bit rate and lossless coding.

Scope

This standard defines the VC-2 video compression system through the stream syntax, entropy coding, coefficient unpacking process, and picture decoding process. The decoder operations are defined by means of a mixture of pseudo-code and mathematical operations.

VC2-P1. Compare the performance of VC2 lossless coding with other lossless coding techniques described in JPEG/JPEG-LS, JPEG-XR, JPEG-XT, JPEG2000, HEVC, H.264, MPEG-2 and DIRAC-PRO based on standard comparison metrics.

VC2-P2. Performance comparison of SDCT with DCT and DDCT is investigated in [E392]. In VC2 replace wavelets with INTSDCT and implement the performance comparison based on various test sequences including 4K and 8K and different block sizes. See the Tables and Figures in [E392].

ATSC

The Advanced Television Systems Committee, Inc., (ATSC) is an international, non-profit organization developing voluntary standards for digital television. The ATSC member organizations represent the broadcast, broadcast equipment, motion picture, consumer electronics, computer, cable, satellite, and semiconductor industries.

Specifically, the ATSC is working to coordinate the television standards among different communications media focusing on digital television, interactive systems, and broadband multimedia communications. ATSC is also developing digital television implementation strategies and presenting educational seminars on the ATSC standards.

ATSC was formed in 1982 by the member organizations of the Joint Committee on InterSociety Coordination (JCIC): the Electronic Industries Association (EIA), the Institute of Electrical and Electronic Engineers (IEEE), the National Association of Broadcasters (NAB), the National Cable Telecommunications Association (NCTA), and the Society of Motion Picture and Television Engineers (SMPTE). The ATSC members represent the broadcast, broadcast equipment, motion picture, consumer electronics, computer, cable, satellite, and semiconductor industries.

International adopters of the ATSC standard include Canada, Dominican Republic, El Salvador, Guatemala, Honduras, Mexico, and South Korea.

ATSC 3.0 CANDIDATE STANDARDS (Dec. 2016)
Next generation broadcast TV

The A/341 Video standard specifies how to encode video in the ATSC 3.0 system. It uses HEVC (H.265) video compression, the latest MPEG video coding standard, and provides support for 4K Ultra HDTV as well as future capabilities for transmission of information to enable wider colour gamut, higher frame rates and high dynamic range. This is a candidate standard.

Legacy Codec

LC1. J. Chen et al., "Efficient Video Coding Using Legacy Algorithmic Approaches," IEEE Trans. on Multimedia, vol. 14, pp. 111–120, Feb. 2012.

LC-P1. See [LC1] The authors have developed a video codec based on the traditional transform predictive coding aided by motion compensation followed by the QM-coder. While its performance is similar to that of H.264/AVC, the main advantage is its royalty few feature. Implement this legacy codec using various test sequences including ultra HD TV resolution.

DSC by VESA

The Display stream compression (DSC) standard developed by the Video Electronics Standards Association (VESA) addresses low-cost (light weight compression) visually lossless video codec that meets the requirements of display links. The papers by Walls and MacInnis [DSC1, DSC2] provide the background and an overview of DSC problem and its history. Both the algorithmic compression and performance impact of some tools and test results are presented. As a detailed description of DSC is beyond the scope of this book, relevant references and pertinent web sites [DSC3–DSC5] are listed. Also VESA issued a call for technology in order to standardize a significantly more complex codec called advanced DSC (ADSC) [DSC5] that is visually lossless at a lower bit rate than DSC. Peng et al., [OP11] in the overview paper on SCC present not only the details on DSC but also discuss on going work and future outlook on SCC and DSC.

DSC1. F.G. Walls and A.S. MacInnis, "VESA display stream compression for television and cinema applications', IEEE JETCAS, vol. 6, issue 4, pp. 360–370, Dec. 2016.

DSC2. MF. Walls and A. MacInnis, "VESA display stream compression: An overview," in *SID Symp. Dig. Tech. Papers*, Jun. 2014, vol. 4, no. 1, pp. 360–363.

DSC3. Video Electronics Standards Association. (2016). *VESA Display Stream Compression (DSC) Standard v1.2*. [Online]. Available: http://vesa.org

DSC4. (Jan. 2013). *VESA Finalizes Requirements for Display Stream Compression Standard*. [Online]. Available: http://www.vesa.org/news/vesa-finalizes-requirements-for-display-stream-compression-standard/_press.html

DSC5. (Jan. 2015). *VESA Issues Call for Technology: Advanced Display Stream Compression*. [Online]. Available: http://www.vesa.org/news/vesa-issues-call-for-technology-advanced-display-stream-compression/

DSC6. VESA Updates Display Stream Compression Standard to Support New Applications and Richer Display Content, Jan. 2016 [online] Available: http://www.vesa.org/featured-articles/vesa-updates-displaystream-compression-standard-to-support-new-applications-and-richerdisplay-content.

DSC7 N. Jacobson et al., "A new display stream compression standard under development in VESA," Applications of the digital image processing LX, vol. 10396, San Diego, CA, Oct. 2017.

DSC-P1. Go through the review papers on DSC [DSC1, DSC2] and confirm the performance results shown in Tables II–V using the corresponding test images. This project is fairly complex and may require group effort.

DSC-P2. Access the web site on call for technology for ADSC [DSC5] and contribute some proposals that can meet the requirements of ADSC. Follow this up in terms of comparing your proposals with proposals submitted by other groups.

DSC-P3. See [DSC6] and access the web site. Explore how the DSC standard can support new applications and richer display content..

Joint Video Experts Team (JVET)

JVET is exploring video compression beyond HEVC as announced below: (April 2018).

The Joint Video Experts Team – a collaborative team formed by the ITU-T Study Group 16 Video Coding Experts Group and ISO/IEC JTC1 SC29/WG11 (Moving Picture Experts Group, MPEG) – has launched a project to develop a new video coding standard to be known as *Versatile Video Coding (VVC)*.

Video is the main driver of bandwidth use, accounting for over 80 per cent of consumer Internet traffic. The majority of this video is coded using international standards developed in collaboration by ITU-T Study Group 16 and MPEG.

Work on the new VVC standard commenced at a meeting of the Joint Video Experts Team in San Diego, U.S., 10–20 April 2018.

The primary objective of VVC is to provide a significant improvement in compression performance over the existing 'High Efficiency Video Coding' standard (HEVC, published as ITU-T H.265 |ISO/IEC 23008-2). VVC will aid the deployment of higher-quality video services and emerging applications such as 360° omnidirectional immersive multimedia and high-dynamic-range (HDR) video.

The development of the VVC standard is expected to be complete before the close of 2020.

The San Diego meeting evaluated responses to a joint Call for Proposals issued by the two bodies. Responses to the call were received from 32 organizations, with some demonstrating compression efficiency gains over HEVC of 40 per cent or more.

The gain was measured in extensive formal subjective tests conducted by independent test labs. Both 360° omnidirectional video and HDR video were tested as well as conventional dynamic range video. Particular effectiveness was shown on ultra-high definition (UHD) video test material.

The results of this very successful call led to the creation of a first draft of the VVC standard, a test model for simulation experiments, and a technology benchmark set for the VVC project.

The new standard is expected to enable the delivery of UHD services at bit rates that today are used to carry HDTV. Alternatively, using VVC would enable twice as much video content to be stored on a server or sent through a streaming service.https://github.com/Samsung/360tools 360 degree virtual reality video coding.

IEEE DCC 27–30 March 2018 has conducted a special session on "Toward a next-generation international video coding standard', with G. Sullivan and Y. Ye as the session coordinators [E426].

http://mpeg.chiariglione.org/standards/exploration/future-video coding
http://mpeg.chiariglione.org/tags/jvet
JVET JEM software (JEM Joint Video Experts Team, Joint Exploration Model)
https://jvet.hhi.fraunhofer.de
JVET documents http://phenix.int-evry.fr/jvet

JVET1. S.-H. Park and E.S. Jang, "An efficient motion estimation method for QTBT structure in JVET," IEEE DCC, April 2017. (same as [E409]).

JVET2. J. An et al., "Block partitioning structure for next generation video coding," ITU-T SG16 Doc.COM16-C966, Sept. 2015.

JVET3. H. Huang et al., "EE 2.1: Quadtree plus binary tree structure integration with JEM tools," JVET-C0024, Geneva, May 2016.

JVET4. J. Chen, E. Alshina, G.-J. Sullivan, J.-R. Ohm, and J. Boyce, "JVETE1001: Algorithm Description of Joint Exploration Test Model 5," Joint Video Exploration Team (JVET), Geneva, Tech. Rep., Feb. 2017.

(Algorithm Description of Joint Exploration Test Model 1 (JEM 1) document ISO/IEC JTC1/SC29/WG11/N15790, Oct. 2015.)

Ibid: "Algorithm Description of Joint Exploration Test Model 4, document ISO/IEC JTC1/SC29/WG11/JVET-D1001, Oct. 2016.

JVET5. K. Suehring and X. Li, "JVET-B1010: JVET Common test conditions and software reference configurations," Joint Video Exploration Team (JVET), San Diego, Tech. Rep., Feb. 2016.

The best source for a comprehensive overview of the new coding technology developed in the JVET of ITU-T and ISO/IEC is the algorithm description of Joint Exploration Test Model.

The latest available version of that is document JVET-G1001 (http://phenix.intecry.fr/jvet/doc_end_user/documents/7_Torino/wg11/JVET-G1001-v1.zip)

All documents related to this activity can be found at http://phenix.int-evry.fr/jvet/.For getting a summary ,the meeting reports will help (e.g. JVET-G1000).For the tool performance, the reports of Ad Hoc Group 1 (e.g.JVET-G0001) is a good starting point.

JVET6 JVET-G1002.DOCX Draft joint call for proposals on video compression with capability beyond HEVC, (13–21 July 2017).

Details on requirements, test sequences, proposal submission format, evaluation procedures etc. are provided in JVET-G1002.docx.doc

(SDR, HDR and 360 degree video test sequences). Both subjective (DSIS/DSCQS) and objective metrics including weighted PSNR, deltaE100-based, PSNR-L100 etc. will be used in the evaluation process.

Industry/academic/research institutes collaboration is essential to prepare and submit a challenging proposal.

JVET7. J. Chen et al., "Algorithm description of Joint exploration model 7," JVET-G1001, July 2017.

JVET8. Z. Wang et al., "Effective quadtree plus binary tree block partition decision for future video coding," IEEE DCC 2017, pp. 23–32, April 2017.

JVET9. D. Grois, T. Nguyen and D. Marpe, "Performance comparison of AV1, JEM, VP9, and HEVC encoders," SPIE, Applications of the digital image processing LX, vol. 10396, San Diego, CA, Oct. 2017.

JVET10. P. Topiwala, W. Dei and M. Krishnan, "Advanced single–stream HDR coding using JVET JEM with backward compatibility options," SPIE Applications of the digital image processing LX, vol. 10396, San Diego, CA, Oct. 2017.

JVET11. J.-R. Ohm and G.J. Sullivan, "Overview and results of joint video exploration team call for proposals," IEEE PCS 2018, San Francisco, CA, 25–27 June 2018.

JVET12. M. Blaser et al.,

"Geometry-based Partitioning for Predictive Video Coding with Transform Adaptation," IEEE PCS 2018, San Francisco, CA, 25-27 June 2018.
Abstract is repeated here:

Abstract

Rectangular block partitioning as it is used in state of the art video codecs such as HEVC can produce visually displeasing artifacts at low bitrates. This effect is particularly noticeable at moving object boundaries. This contribution presents a comprehensive geometry-based block partitioning framework in a post-HEVC codec for motion compensated prediction, intra-prediction and transform coding as a solution. The method is evaluated on the set of sequences defined by the Joint Call for Proposals on Video Compression with Capabilities beyond HEVC. Our contribution visually improves the quality of object boundaries and performs on par in terms of compression performance with the latest Joint Video Exploration Team (JVET) test model.

JVET13. W. Zhu and A. Segall, "Compound Split Tree for Video Coding," IEEE PCS2018, San Francisco, CA, June 2018.
Abstract is repeated here.

Abstract

During the exploration of video coding technology for potential next generation standards, the Joint Video Exploration Team (JVET) has been studying quad-tree plus binary-tree (QTBT) partition structures within its Joint Exploration Model (JEM). This QTBT partition structure provides more flexibility compared with the quad-tree only partition structure in HEVC. Here, we further consider the QTBT structure and extended it to allow quad-tree partitioning to be performed both before and after a binary-tree partition. We refer to this structure as a compound split tree (CST). To show the efficacy of the approach, we implemented the method into JEM7. The method achieved 1.25%, 2.11% and 1.87% BD-bitrate savings for Y, U and V components on average under the random-access configuration, respectively.

JVET14. M. Karczewicz and E. Alshina, "JVET AHG report: Tool evaluation," Joint Video Exploration Team (JVET), Tech. Rep. JVET-D0001, 2017.

JVET. software repository. Available online: https://jvet.hhi.fraunhofer.de/svn/svn_HMJEMSoftware/branches/HM-13.0-QTBT/

JVET-P1. go through the JVET documents and prepare a detailed proposal in response to JVET-G1002.docx.doc that can meet the goals and objectives of the new compression/coding which is more efficient than HEVC. Make sure it can also address the 360 degree video besides the HDR sequences.

JVET-P2. See JVET-P1. Review audio compression/coding technologies and develop improved audio compression techniques. This includes multi-channel, multi-language and surround sound digital theater systems.

JVET-P3. See JVET-P1 and JVET-P2. Similar to multiplex/demultiplex and lip-synch of current video/audio coding standards, implement this functionality in the new video/audio coding standards.

JVET-P3. See [JVET12]. Implement this codec and compare with the latest JVET test model.

JVET-P4. See [JVET 4 and JVET13]. Zhu and Segall [JVET13] implemented the compound split tree structure (CST) in JEM7 and showed improved compression efficiency for random access configuration. Extend this technique to all intra, low delay P and low delay B configurations.

360 Degree Video Coding

360D1. J. Boyce and Q. Xu, "Spherical rotation orientation indication for HEVC and JEM coding of 360 degree video," SPIE Applications of the digital image processing LX, vol. 10396, San Diego, CA, Oct. 2017.

360D2. P. Hanhart, Y. He and Y. Ye, "Intra prediction using face recognition in 360-degree video coding," SPIE Applications of the digital image processing LX, vol. 10396, San Diego, CA, Oct. 2017.

360D3. S. Gudumasu et al., "segment scheduling method for reducing 360 degree video streaming latency," SPIE Applications of the digital image processing LX, vol. 10396, San Diego, CA, Oct. 2017.

360D4. Z. Xu et al., "Tile-based quality of experience-driven HTTP/2 streaming system for 390 video," IEEE ICME 2018.

RGB <=======> YCbCr Conversion

The mapping from RGB to YCbCr components as recommended by ITU-R is (See Recommendation of itu-r BT.601, http://www.itu.int/rec/r-rec-bt.601," Also [SE40].

$Y = 0.257 \cdot R + 0.504 \cdot G + 0.098 \cdot B + 16$
$Cb = -0.148 \cdot R - 0.291 \cdot G + 0.439 \cdot B + 128$
$Cr = 0.439 \cdot R - 0.368 \cdot G - 0.071 \cdot B + 128$ (8)

and the inverse mapping is

$R = 1.164 \cdot (Y - 16) + 1.596 \cdot (Cr - 128)$
$G = 1.164 \cdot (Y - 16) - 0.813 \cdot (Cr - 128) - 0.392 \cdot (Cb - 128)$.
$B = 1.164 \cdot (Y - 16) + 2.017 \cdot (Cb - 128)$

PSNRAVG

PSNRAVG is a weighted average of luminance (PSNRY) and chrominance (PSNRU and PSNRV) PSNR components. All involved test sequences are in 4:2:0 colour format, for which PSNRAVG is computed as (de facto standard),

PSNRAVG = $(6 \times$ PSNRY + PSNRU + PSNRV$)/ 8$.

Since PSNRAVG also takes the impact of the chrominance components into account, it is supposed to provide more reliable results than the conventional PSNRY metric in the cases when the luminance and chrominance components have dissimilar RD behaviors. See the reference below:

B. Li, G. J. Sullivan, and J. Xu, RDO with Weighted Distortion in HEVC, document JCTVC-G401, ITU-T/ISO/IEC Joint Collaborative Team on Video Coding (JCT-VC), Geneva, Switzerland, Nov. 2011.

Weighted BDBR

Weighted BDBR for 4:2:0 video can be calculated as follows. This is only a de facto measure and not necessarily endorsed by any standards groups. See [E450] for details.

wBDBR=(6 × BDBRY + BDBRU + BDBRV)/8

Colour-Sensitivity-Based Combined PSNR (CSPSNR) See [SE40]

The following is repeated from [SE40]:

"Another method adopted in the HEVC reference software HM derives the combined PSNR from the combined MSE MSEYCbCr. We denote this method as PSNRHM, which is calculated as PSNRHM = 10 log10((2Nb − 1)2/MSEYCbCr), MSEYCbCr = wY MSEY + wCb MSECb + wCr MSECr, (4) wY = NY NY + NCb + NCr , (5) wCb = NCb NY + NCb + NCr , wCr = NCr NY + NCb + NCr , (7) where NY , NCb and NCr are the number of samples of Y, Cb and Cr, respectively. It is based on the assumption that the average MSE is related to the sampling format. For example, for 4:2:0 videos, wY = 2/3 and wCb = wCr = 1/6. For 4:4:4 videos, wY = wCb = wCr = 1/3, which means that the importance of the three components is equal in this case."

L. Zhang et al., "Fsim A feature similarity index for image quality assessment," IEEE Trans. IP, vol. 20, pp. 2378–2386, 2011.

X. Shang et al., "Color-sensitiviy-based combined PSNR for objective video quality assessment." IEEE Trans. CSVT, EA.

Cross-Media Retrieval

Y. Peng, X. Hwang and Y. Zhao, "An overview of cross-media retrieval:

Concepts, methodologies, benchmarks and challenges" IEEE Trans. CSVT, EA

This is an excellent overview paper on cross-media retrieval (query one medium and retrieval in other media) with datasets, state-of-the-art, research topics, etc.

Multiple Description Coding

MDC1 L. Zhao et al., "Multiple description convolutional neural networks for image coding," IEEE Trans. CSVT, EA. Part of the introduction is repeated here.

Introduction

"LARGE amounts of attentions have been paid to various techniques of Internet service and multimedia signal transmission for many years, which not only provide us a convenient manner of communication but also give us many choices for our life style. Meanwhile, the bandwidth of Internet has been accelerated and more stable transmission service is guaranteed by these developments. But there are still some risks of transmission failures, when Internet congestion occurs in the overloaded case or signal packets are conveyed in unpredictable yet unreliable channels [1, 2]. Multiple description coding (MDC) has been studied as a promising technique of source coding to relieve these problems by decomposing signal into multiple redundant subsets, which are transmitted in different channels. Thus, a degraded but acceptable signal reconstruction can be produced after decoding, even though only one description is received at the clients. If more descriptions are available for users, better quality of signal reconstruction can be achieved. MDC has been widely explored in the field of image and video coding [3–21]."

MDC-P1 See [MDC1] The conclusion section is repeated here.

"In this paper, we introduce multiple description image coding framework based on convolutional neural networks. In order to make generated multiple descriptions diverse but have shared structural information, both multiple description distance loss and DSSIM loss are combined together to train multiple description generator network. Besides, two learning algorithms are provided to train our whole framework. More importantly, our method can be applied into many other existing codec, such as BPG, Google's WebP, etc for multiple description image coding." Explore these applications.

Index

2D 11, 25, 61, 170
3D 10, 31, 50, 68

A
AAC 53, 84, 324, 379
ACMMoVid 329
adaptive color
    transform (ACT) 139, 174
  color space conversion 174
  encoder optimization 150
adaptive motion compensation precision 158, 159
adaptive motion vector resolution 139, 142, 150, 168
Advanced Video Coding 19, 182, 184, 379
AHG 28, 391, 393, 415
AI 40, 50, 97, 105
AIF 62
ALF 26, 62
AMVP 42, 129, 254
AOM 21, 55, 197, 397
APIF 62
APSIPA 178, 216, 217
ARM 67, 72, 393, 407
ASIC 109, 119, 136, 310
ASIP 230

ASO 45, 69
ATR 73
ATSC 51, 85, 279, 410
AVC 19, 47, 96, 410
AVS 66, 102, 378, 391

B
BBC 71, 118, 374, 375
BD 6, 10, 19, 61
BD-rate 19, 67, 75, 91
BH 389, 390, 392, 393
BL 103, 116
block vector (BV) 25, 33, 383, 414
BMSB 176, 318, 384
bpp 23, 373
BRISQUE 104
BS 214
BSTM 62
BV 143, 160
BVP 160

C
CABAC 30, 50, 68, 212
CAVLC 33, 266
CBF 161
CCP 50, 114, 119, 241
CE 250, 256, 264, 276
CfE 103
CfP 23, 141, 173, 369
CI 23

CIF 203, 375, 376
COFDM 226, 277
CPU 52, 106, 221, 262
CRA 36
CRI 103
CSVT 30, 182, 212, 232
CTC 291
CU 33, 90, 115, 156
CUDA 71, 124, 282, 371
CWWSSIM 329

D

DASH 45, 55, 273, 377
DATE 245
DCC 54, 222, 250, 255
DCT 26, 61, 381, 409
DCTIF 80, 216
DDCT 29, 59, 63, 377
DF 27, 86
DiOR 140
DIP 329
DIQA 162, 163
DMB 184
DMOS 272, 292
DMVD 77
DPCM 79, 139, 170, 382
DR 375
DSC 51, 78, 293, 368
DSCQS 13, 162, 298, 413
DSIS 13, 66, 113, 413
DSP 53, 78, 220, 351
DST 29, 40, 63, 161
DTM 156
DTT 116, 195, 202, 357
DTV 86, 277, 278
DVB-H 184

E

EBU 10, 82, 289, 375
EC 84, 266

EE 59, 224, 378, 413
EI 217, 283, 349
EL 82, 97, 103, 116
EPFL 346, 361, 367
ETRI 208, 229, 259, 287
EURASIP 182, 192, 258, 349

F

fast intra coding 72, 150, 265, 380
FCC 248
FDAM 253
FDIS 31, 43, 67, 217
FF 320
FIR 26
FMO 45, 69
FPGA 53, 120, 262, 401
fps 4, 7, 130, 371
FSIM 265

G

GPU 106, 124, 282, 371

H

HD 3, 20, 52, 286
HDR 7, 233, 351, 415
HDTV 3, 82, 227, 412
HE-AAC 86, 108, 134, 401
HEIF 320
HEVC 27, 47, 385, 417
HEVStream 130
HHI 65, 71, 318
High efficiency video coding 86, 193, 243, 412
HM 27, 47, 73, 232
HOR 35, 118, 125, 380
HP 45, 93, 117, 337

I

IASTED 60, 209
IBC 141, 167, 169, 395
ICASSP 190, 251, 384, 398
ICCE 55, 196, 309, 362
ICIEA 61, 209, 308
ICIP 29, 206, 309, 391
ICME 177, 224, 317, 416
ICPC 233, 316
ICPR 330
ICTs 201
IEC 19, 139, 207, 249
INTDCT 33, 105, 204, 392
Internet Video Coding 21, 249, 285, 418
intra block copy (IBC) 143, 160, 169, 171
    hash-based 91, 144, 159, 169
    symmetric 42, 99, 169, 200
intra HE 71
intra line copy 171, 179, 386
intra string copy 91, 171, 232, 382
IPTV 45, 234, 280
IQA 162, 299, 303, 386
IS&T 220, 331, 338, 377
ISCAS 62, 176, 209, 259
ISCCSP 183
ISDB-T 184
ITM 92

J

J2K 344, 345
JCI 289
JLS 348
JPEG 11, 54, 345, 357
JPEG-LS 45, 82, 347, 409
JPEG-XR 54, 72, 358, 409

Just Noticeable Difference (JND) 171, 179, 350, 351

L

LAR-LLC 372, 373
LCICT 201, 202
LD 50, 104, 113, 244
LDB 125, 126, 201, 255
LDP 125, 126, 255
LOCO 347
LSC 163
L-SEABI 104

M

MDDCT 62
MDDT 26, 60, 211, 248
MMSP 216, 248, 290
MOMS 68, 69
MOS 81, 165, 286, 360
Motion Compensation 77, 158, 213, 410
Motion Estimation 78, 168, 259, 318
Motion JPEG 59, 327, 341
Motion Vector 123, 133, 150, 254
Mpm 37, 161, 259
MRSE 361
MSP 89, 248, 347, 404
Multi View Coding 186, 273
MVPs 128, 129, 254

N

NAL 65, 120
NAL Unit Header 65, 120
NIQA 162, 163

O

Open Source Software 50, 64, 321, 356

P

P2SM 170
palette mode 142, 145, 160, 168
  coding entries 150
  coding indices 152
palette derivation 145, 146
PCM 36, 64, 317
PCoIP 140
Perceptual Vector
  Quantization 376
Perceptual Video Coding
  93, 196, 259, 330
PPS 147
Prediction Unit 33, 130,
  244, 393

Q

QOE 266, 285, 302, 377

R

Random Access 36, 104,
  206, 416
rate control 101, 166, 189, 263
Rate Distortion 36, 189,
  244, 399
RDOQ 36
RDPCM 79, 153, 154, 155
Real Media HD 54, 402
Reversible Data Hiding 348,
  349, 353, 358
RICT 102, 200
RMVB 324, 402
ROI 52, 84, 237, 318
Rotational Transform 26,
  62, 63, 389
RQT 48, 89
RTC 394
RTP 45

S

sample-based prediction 139
  angular intra
    36, 156, 168, 251
  angular intra with
    edge 157, 168
  weighted 26, 154, 170, 417
SAO 26, 80, 245, 284
SCADA 140
Scalable HEVC 52, 84, 99, 229
Scalable High Efficiency
  Video Coding 252, 262, 268
Scalable Video Coding 40,
  74, 189, 400
SCM 96, 147, 323, 384
Screen Content Coding 51,
  139, 168, 243
screen contents 141,
  169, 173, 383
  coding tools 25, 92, 142, 261
  features 21, 148, 260, 387
screen image quality
  assessment (SIQA) 162,
    168, 177
  objective 76, 163, 285, 301
  subjective 19, 47, 165, 272
SDCT 29, 112, 377, 400
SDSCE 162
segmentation 25, 165, 224, 387
SELC 46
SHV 44, 51, 52
SIMD 53, 90, 215, 231
SIQM 164
SMF 197, 202, 203
SoC 52, 394
SP 45, 86, 199, 351
SPA 184, 219
Spatially Scalable Video Coding
  72, 120, 213

SPEC 166
Special Issues on HEVC 31, 50
SPS 147
SSCQE 162
Stimulus Comparison 162
string matching 142, 170, 178, 383
Structural Similarity 162, 176, 328, 331
Subjective Evaluation 13, 90, 285, 377
Super Resolution 104, 227, 242
SWP 154, 155, 156

T

template matching 77, 154, 169, 251
Test Sequence 10, 142, 288, 416
Test Zone Search 255
TMuC 25, 61, 71, 334
Transform Block 39, 161, 303, 392
transform skip 80, 206, 209, 240
Transform Unit 419
TSF 191
TSM 191, 201, 224
TZSearch 103, 137, 248, 274

V

VCB 122, 123, 143, 279
VCIP 214, 245, 345, 412
VCIR 90, 239, 326, 359
VDI 170
VDP 305, 391
Vector Quantization 406
Vertical 201
VESA 31, 81, 117, 440
Video Coding 40, 296, 303, 447
Video Quality Assessment 55, 276, 320, 361
ViMSSIM 359
VLSI 68, 138, 240, 279
VSB 116, 255, 306, 307

W

WCG 50, 85, 133, 392
Web Video Coding 122
WPP 83, 301
WQVGA 53
WVGA 23, 53, 83, 125

Y

YC_bC_r 53, 132, 321, 447

About the Authors

Humberto Ochoa Domínguez received the Ph.D. degree in electrical engineering from The University of Texas, Arlington in 2004. He is currently working as a professor of computer and electrical engineering in the University of Ciudad Juárez, (UACJ) México. He has published extensively and is a reviewer for referred journals. He is an IEEE member and a consultant for the academy and the industry. Humberto has supervised several masters and doctoral students and served as external examiner for M.S. and Ph.D. students. He is also a reviewer of research proposals from the industry. He has published the books:

H. O. Dominguez and K. R. Rao, "Discrete cosine transform", II Edition, CRC-Press, Taylor and Francis, 2019.

H. O. Dominguez and K. R. Rao, "Versatile video coding", River Publishers, 2019.

K. R. Rao and H. O. Dominguez, "JPEG series", to be published.

K. R. Rao received the Ph.D. degree in electrical engineering from The University of New Mexico, Albuquerque in 1966. He received B.S. E.E from the college of engineering, Guindy, India in 1952. Since 1966, he has been with the University of Texas at Arlington where he is currently a professor of electrical engineering. He, along with two other researchers, introduced the Discrete Cosine Transform (DCT) in 1975 which has since become very popular in digital signal processing. DCT, INTDCT (integer DCT), directional DCT and MDCT (modified DCT) have been adopted in several international video/image/audio coding standards such as JPEG/MPEG/H.26X series and also by SMPTE (VC-1) and by AVS China. He is the co-author of the books "Orthogonal Transforms for Digital Signal Processing" (Springer-Verlag, 1975), Also recorded for the blind in Braille by the Royal National Institute for the blind. "Fast Transforms: Analyses and Applications" (Academic

Press, 1982), "Discrete Cosine Transform-Algorithms, Advantages, Applications" (Academic Press, 1990). He has edited a benchmark volume, "Discrete Transforms and Their Applications" (Van Nostrand Reinhold, 1985). He has co edited a benchmark volume, "Teleconferencing" (Van Nostrand Reinhold, 1985). He is co-author of the books, "Techniques and standards for Image/Video/Audio Coding" (Prentice Hall) 1996 "Packet video communications over ATM networks (Prentice Hall) 2000 and "Multimedia communication systems" (Prentice Hall) 2002. He has co edited a handbook "The transform and data compression handbook," (CRC Press, 2001). Digital video image quality and perceptual coding, (with H.R. Wu) (Taylor and Francis 2006). Introduction to multimedia communications: applications, middleware, networking, (with Z.S. Bojkovic and D.A. Milovanovic), Wiley, (2006). He has also published a book, "Discrete cosine and sine transforms", with V. Britanak and P. Yip (Elsevier 2007). Wireless Multimedia Communications (publisher: Taylor and Francis) Nov. 2008.

"Fast Fourier Transform: algorithms and Applications", with D. Kim and J.J. Hwang (publisher: Springer) 2010. (Also e-book). Also into Chinese by China Machine Press. Also Asian edition (paper back) by Springer India. Also into Korean by A-Jin publishing company.

Some of his books have been translated into Japanese, Chinese, Korean, Russian and Spanish, also published as Asian (paperback) editions (also as e-books).

The book "Video coding standards: AVS China, H.264/MPEG-4 Part10, HEVC, VP6, DIRAC and VC-1", (with D. Kim and J.J. Hwang) is published by Springer, 2014. Translated into Chinese and Spanish.

K.R. Rao, Z.S. Bojkovic and B.M. Bakmaz, "Wireless multimedia communication systems: design, analysis and implementation", CRC Press, 2014. Another book V. Britanak and K.R. Rao, "Cosine-/Sine-Modulated Filter Banks" Springer, Sept. 2017.

K.R. Rao, J.J. Hwang and D.N. Kim, "High efficiency video coding and other emerging standards", River Publishers, August 2017.

H.O. Dominguez and K.R. Rao, "Discrete cosine transform", II Edition, Taylor and Francis, 2019.

V. Britanak and K.R. Rao, "Cosine and sine modulated filter banks", Springer, 2017.

H.O. Dominguez and K.R. Rao, "Versatile video coding", River Publishers, 2019.

K.R. Rao and H.O. Dominguez, "JPEG series", to be published.

He has been an external examiner for graduate students from universities in Australia, Canada, China, Hong Kong, India, Malaysia, Singapore, Thailand, Taiwan and USA. He was a visiting professor in several Universities -3 weeks to 7 and 1/2 months- (Australia, Japan, Korea, Singapore and Thailand). He has conducted workshops/tutorials on video/audio coding/standards worldwide. He has supervised several students at the Masters (115) and Doctoral (32) levels. These former students are well placed in industry and academia. He has published extensively in refereed journals and has been a consultant to industry, research institutes, law firms and academia. He has conducted a 2 week short course 1–12 Jan. 2018 on advanced digital video coding standards in University of Kerala, Trivandrum, Kerala, India under GIAN program approved by MHRD, New Delhi, India.

He was on the NSF and EPA PANELS. He has been a plenary/keynote speaker in several International Conferences (all expenses paid).

He has reviewed 26 book manuscripts for book publishers. His research has attracted thirty four visiting professors from Australia, Canada, China, Japan, Korea, Hong Kong, Hungary, Slovakia, Thailand, USSR, and Yugoslavia for periods ranging from 3 weeks to five years (all fully funded by their Universities or Governments or Foundations)

He is a Fellow of the IEEE (Member # 03911617). He is a member of the Academy of Distinguished Scholars, UTA. He was invited to be a panelist for the 2011 NSF Graduate Research Fellowship Program (GRFP), with service on the following panel:

PANEL NAME: Electrical Engineering Panel

MEETING DATES: Friday Feb 11, 2011 to Sunday Feb 13, 2011

He was invited to be a panelist for the US EPA STAR fellowship program, 14–15 March 2013, Chantilly, VA and in 2016 (virtual panel).